Performance of Elastomeric Components in Contact With Potable Water

Awwa
Research
Foundation

About the Awwa Research Foundation

The Awwa Research Foundation (AwwaRF) is a member-supported, international, nonprofit organization that sponsors research to enable water utilities, public health agencies, and other professionals to provide safe and affordable drinking water to consumers.

The Foundation's mission is to advance the science of water to improve the quality of life. To achieve this mission, the Foundation sponsors studies on all aspects of drinking water, including supply and resources, treatment, monitoring and analysis, distribution, management, and health effects. Funding for research is provided primarily by subscription payments from approximately 1,000 utilities, consulting firms, and manufacturers in North America and abroad. Additional funding comes from collaborative partnerships with other national and international organizations, allowing for resources to be leveraged, expertise to be shared, and broad-based knowledge to be developed and disseminated. Government funding serves as a third source of research dollars.

From its headquarters in Denver, Colorado, the Foundation's staff directs and supports the efforts of more than 800 volunteers who serve on the board of trustees and various committees. These volunteers represent many facets of the water industry, and contribute their expertise to select and monitor research studies that benefit the entire drinking water community.

The results of research are disseminated through a number of channels, including reports, the Web site, conferences, and periodicals.

For subscribers, the Foundation serves as a cooperative program in which water suppliers unite to pool their resources. By applying Foundation research findings, these water suppliers can save substantial costs and stay on the leading edge of drinking water science and technology. Since its inception, AwwaRF has supplied the water community with more than $300 million in applied research.

More information about the Foundation and how to become a subscriber is available on the Web at **www.awwarf.org**.

Performance of Elastomeric Components in Contact With Potable Water

Prepared by:
Thomas D. Rockaway, Gerold A. Willing, Raymond M. Schreck,
and **Kenneth R. Davis**
University of Louisville
2301 South 3rd Street, Louisville, KY 40292

Jointly sponsored by:
Awwa Research Foundation
6666 West Quincy Avenue, Denver, CO 80235-3098

and

U.S. Environmental Protection Agency
Washington, D.C.

Published by:

Awwa Research Foundation

American Water Works Association

IWA Publishing

DISCLAIMER

This study was jointly funded by the Awwa Research Foundation (AwwaRF) and the U.S. Environmental Protection Agency (USEPA) under Cooperative Agreement No. R82967901. AwwaRF and USEPA assume no responsibility for the content of the research study reported in this publication or for the opinions or statements of facts expressed in the report. The mention of trade names for commercial products does not represent or imply the approval or endorsement of AwwaRF or USEPA. This report is presented solely for informational purposes.

CONTENTS

TABLES

FIGURES

FOREWORD

The Awwa Research Foundation (AwwaRF) is a nonprofit corporation that is dedicated to the implementation of a research effort to help utilities respond to regulatory requirements and traditional high-priority concerns of the industry. The research agenda is developed through a process of consultation with subscribers and drinking water professionals. Under the umbrella of a Strategic Research Plan, the Research Advisory Council prioritizes the suggested projects based upon current and future needs, applicability, and past work; the recommendations are forwarded to the Board of Trustees for final selection. The foundation also sponsors research projects through an unsolicited proposal process; the Collaborative Research, Research Applications, and Tailored Collaboration programs; and various joint research efforts with organizations such as the U.S. Environmental Protection Agency, the U.S. Bureau of Reclamation, and the Association of California Water Agencies.

This publication is a result of one of these sponsored studies, and it is hoped that its findings will be applied in communities throughout the world. The following report serves not only as a means of communicating the results of the water industry's centralized research program but also as a tool to enlist the further support of the nonmember utilities and individuals.

Projects are managed closely from their inception to the final report by the foundation's staff and large cadre of volunteers who willingly contribute their time and expertise. The foundation serves a planning and management function and awards contracts to other institutions such as water utilities, universities, and engineering firms. The funding for this research effort comes primarily from the Subscription Program, through which water utilities subscribe to the research program and make an annual payment proportionate to the volume of water they deliver and consultants and manufacturers subscribe based on their annual billings. The program offers a cost-effective and fair method for funding research in the public interest.

A broad spectrum of water supply issues is addressed by the foundation's research agenda: resources, treatment and operations, distribution and storage, water quality and analysis, toxicology, economics, and management. The ultimate purpose of the coordinated effort is to assist water suppliers to provide the highest possible quality of water economically and reliably. The true benefits are realized when the results are implemented at the utility level. The foundation's trustees are pleased to offer this publication as a contribution toward that end.

David E. Rager
Chair, Board of Trustees
Awwa Research Foundation

Robert C. Renner, P.E.
Executive Director
Awwa Research Foundation

ACKNOWLEDGMENTS

The authors of this report thank the Awwa Research Foundation (AwwaRF) for its support of this study, and the AwwaRF project manager Jian Zhang. The authors would also like to acknowledge the input and guidance throughout the course of the project from the members of the AwwaRF Project Advisory Committee:

- Steve Reiber
- Ata Ciechanowski
- Everett Wegerif
- Donovan Larson

A special thanks for the dedicated efforts of our participating utilities for the hours of data tracking, gathering, and review. In particular we would like to thank those involved with the sponsor utilities: Pat Arnette with Greater Cincinnati Water Works; Keith Coombs with Louisville Water Company; and St. Paul Regional Water Services. Without their dedication, support, and input we would not have been able to successfully complete the study.

In addition to the above listed, many other utilities provided materials, information, and comments for the study. We would like to thank South Central Connecticut Regional Water Authority, Portland Water District, Philadelphia Water Department, Greater Vancouver Regional District, City Utilities of Springfield Missouri, and the Public Utilities Bureau of the City of Akron.

We would also like to acknowledge the assistance provided by the University of Louisville research assistants with a particular gratitude to Jahnavi Valleru, Katrina Bredhold, John Spalding, Raja Nagisetty, Randy Schoenbaechler, and Amanda Givan.

And finally, we sincerely appreciate the assistance of Mary Pat Nimon and Larry Bolton with manuscript preparation. Their dedicated and thorough work greatly contributed to this document.

EXECUTIVE SUMMARY

BACKGROUND

For more than 50 years, elastomeric materials have been widely used for components in drinking water distribution systems. Their use has ranged from rubber gaskets, to valves, to hydrants, to fittings—and they have proved to be equally adaptable to a wide variety of piping materials, including cast iron, ductile iron, steel, copper, and PVC. In general, elastomers have performed well in the water distribution system; however, recently an increase of elastomer failures has been noted. Reported problems associated with elastomer failure include:

- Increased leaks and line losses
- Increased maintenance and repair
- Budget planning difficulties
- Customer dissatisfaction

The specific cause of the increased elastomer degradation is not fully understood. While the degradation appears to be coincident with a disinfectant change from chlorine to chloramines, other factors including temperature and pH and are believed to play roles. Additionally, there is a wide variety of elastomer formulations available to the water industry. Each elastomer formulation has a unique reaction when exposed to free chlorine or chloramines; some have been observed to resist degradation, while others degrade rapidly.

Accurately predicting the life expectancy of elastomeric materials is especially important in the water industry because parts are expected to remain in service for very long periods, even for 100-plus years. Unfortunately, little research has been conducted to link elastomer decay rates with actual field conditions to enable better management of existing risks. This research initiative was intended to clarify the factors that contribute to elastomer degradation and provide a quantitative method for predicting the performance and life expectancy of in-service elastomer components.

OBJECTIVES

The primary objectives of this project were:

- To establish a quantitative method for predicting the life expectancy of in-service elastomer components when exposed to potable water
- To identify elastomeric materials with high performance when subjected to chloramines
- To provide the water industry better tools for managing the risks and financial costs associated with elastomer failure

OVERVIEW

Predicting long-term performance, or life-expectancy, for materials is one of the most challenging areas of current research. Often, it is unrealistic to test materials over very long periods of time, thus, accelerated-aging programs are implemented. The goal of this accelerated-aging program was to establish a systematic testing program that reasonably duplicated field conditions

Table ES.1
Common failure modes for typical elastomer components

Component	Failure mode
Resilient seat gate valve	Surface degradation (hardness), swell
Butterfly valve	Swell
Gasket	Swell
Meter seal	Surface degradation, swell, elongation, and tensile strength

in a condensed time period. To accelerate aging, new elastomeric samples were exposed to elevated temperatures and elevated concentrations of chlorine and chloramines. The results of the accelerated-aging process were then compared to actual field samples of similar materials. By overlaying the laboratory-generated degradation curves with real field data, the researchers could build a reliable method for predicting the performance of in-service parts.

The research was accomplished through the completion of four specific tasks. First, the appropriateness of existing elastomer guidelines, standards, and policies from the water industry and related fields were assessed. Second, accelerated-aging tests were conducted on representative un-aged elastomer formulations to define their service degradation curves. (During this testing period, the degradation influence of chlorine, pH, and strain also were assessed.) Third, field samples were matched to the degradation curves to predict degradation factors and remaining service life. Fourth, accelerated-aging tests were performed on the field samples to verify that their ongoing performance would be consistent with the established degradation curves.

Upon completion of the laboratory testing work, life-cycle degradation curves were developed for generic formulations of natural rubber; styrene butadiene rubber (SBR); nitrile (NBR); neoprene; ethylene propylene diene monomer, sulfur-cured (EPDM-S); and ethylene propylene diene monomer, peroxide-cured (EPDM-P). For each material, a unique degradation curve was identified for surface hardness, elongation, and swell (weight and volume change). For this review, the research team defined critical failure of the laboratory samples as 50 percent decrease degradation for hardness, elongation, and a 50 percent increase in volume for swell. As the degradation curves and the resulting preferential order of the elastomer selection were dependent on failure mode, it is important to consider the intended application of the elastomer material. Common failure modes for typical elastomer components within the water industry are shown in Table ES.1.

RESULTS

The master degradation curves for the various elastomers provide insight into the expected lifespan of rubber parts within a simplified chloramine system. However, how the elastomers perform in a real-world application is dependent on a variety of factors. Most importantly, the function of the part must be assessed in relation to the critical elastomer failure mode. For example, in some conditions, it may be entirely suitable to form a gasket out of a highly swelling material and have adequate long-term performance. However, it is unlikely that the same material would perform as well on a gate valve where swelling would adversely affect the closure operation.

Table ES.2
The 50 percent swell degradation of common elastomers

Elastomer name	Number of years to reach 50% of weight change
EPDM-P	>30
EPDM-S	21.9
Nitrile	3.9
SBR	2.3
Natural rubber	1.9
Neoprene	0.4

The degradation curves from the simplified chloramine systems appeared to over-predict the elastomer degradation rates, and some common materials showed 50 percent failure occurring in less than five years. Significant work was undertaken to reconcile the laboratory degradation curves with the field degradation curves. The limited number of field samples and complex exposure conditions complicated the analysis. While the exact time to failure may be uncertain, the relative performance of the elastomers was confirmed.

The tables that follow should be used as a general guide to selecting elastomers with respect to the particular parameters tested. Formulators can modify elastomer compounds to meet certain performance criteria, and specific materials may perform significantly differently than the generic formulations used in this study.

Elastomer Swell Performance

Table ES.2 compares the 50 percent swell degradation values for the elastomers evaluated in this study when exposed to a 1 ppm chloramine solution. It is noted that EPDM-P does not have a discrete value associated with the time to reach 50 percent weight gain. The EPDM-P material was observed to be very resistant to the accelerated degradation and thus long-term performance could not be estimated with the test conditions.

The change in mass is an important parameter to consider when selecting elastomers for a specific application, as it correlates directly to swell and breakdown of the internal elastomer structure. In applications where long-term elastomer size is critical because of clearance or other issues, it is important to select an elastomer that will have limited swell. However, in water utility applications where swell is not a critical factor, such as gaskets, high-swell elastomers may be suitable, provided they meet other critical performance requirements.

Elastomer Strain Performance

Table ES.3 compares the 50 percent strain degradation values for the elastomers evaluated in this study when exposed to a 1 ppm chloramine solution. EPDM-P does not have a discrete value associated with the time to reach 50 percent strain. The EPDM-P material was observed to be very resistant to the accelerated degradation and thus long-term performance could not be estimated with the test conditions.

Table ES.3
The 50 percent strain degradation of common elastomers

Elastomer name	Number of years to reach 50% of initial breaking strain
EPDM-P	>30
Natural rubber	17.4
Nitrile	13.3
SBR	3.4
Neoprene	1.8
EPDM-S	1.1

The change in breaking strain often is considered in elastomer selection because it correlates to the integrity of the elastomer's molecular backbone structure. In applications where long-term elastomer elongation or strength is critical, such as diaphragms, it is important to select an elastomer that will have limited loss of strain over time. However, in water utility applications where strength or strain is not a critical factor, elastomers that exhibit a loss of strain with time may be suitable, provided they meet other critical performance requirements.

Elastomer Hardness Performance

Table ES.4 compares the 50 percent surface hardness degradation values for the elastomers evaluated in this study when exposed to a 1 ppm chloramine solution. Accelerated-age predictions could only be completed for natural rubber and neoprene elastomers because of the limited hardness degradation observed within the other samples. Without reasonably consistent and predictable hardness values, the time-accelerated degradation could not be calculated. This same variability also affected the natural rubber and neoprene hardness values. Therefore, the time required to reach 50 percent of the initial hardness value should be used as a relative ranking only.

Decreasing the surface-area-to-mass ratio of a part has the benefit of increasing the hardness of that component. But certain elastomers, like SBR, tend to undergo large changes in hardness over time due to the mechanisms of carbon black loss.

In applications where long-term elastomer surface hardness is critical, it is important to select an elastomer that will have limited loss of hardness over time. An application of this in the water utility industry would be in cases where when considering wear resistance of an elastomer part. However, in water utility applications where hardness is not a critical factor, elastomers that exhibit a loss of hardness with time may be suitable, provided they meet other critical performance requirements.

CONCLUSIONS

Based on their research and observations of elastomeric performance, the project team arrived at the following conclusions:

Table ES.4
The 50 percent hardness degradation of common elastomers

Elastomer name	Number of years to reach 50% of initial hardness value
Natural rubber	8,260 years
Neoprene	273 years
EPDM-P	Indeterminate
Nitrile	Indeterminate
SBR	Indeterminate
EPDM-S	Indeterminate

- Overall, the EPDM-P materials performed significantly better than any of the test elastomers exposed to chlorine and chloramine disinfectants. The EPDM-P material exhibited very little degradation in any of the measured parameters throughout the testing. The exception was at the most extreme testing condition (70°C temperature and 60 ppm average concentration), in which the EPDM-P materials degraded rapidly. This rapid degradation likely was due to higher order failure modes occurring within the sample and not anticipated to occur within normal water distribution systems.
- The presence of loop strain actually *reduced swelling* (less weight and volume change) and *improved tensile performance* for all of the elastomers tested. This behavior, combined with the small surface-to-mass ratio and limited exposure area, is the likely reason that pipe gaskets made of any elastomeric material have been observed to have adequate long-term performance.
- In general, 30-day testing of un-aged samples under the specified test conditions showed that at low pH, where dichloramine species is preferentially formed, performance degradation was higher than in the corresponding monochloramine condition. (It should be noted that the differences were not extreme, generally less than 10 percentage points, and may be within typical elastomer sampling variability.) Except for EPDM-S, all elastomer materials tended to exhibit higher swelling, loss of strength, and greater softening in the dichloramine environment. SBR and natural rubber tended to be more sensitive to pH than the other elastomers tested.
- The surface-area-to-mass ratio was observed to have little influence on the bulk mechanical property degradation of a given elastomer. The hardness values, which are highly surface specific, tended to follow the same trends regardless of surface-area-to-mass ratio. There were a few exceptions, namely, SBR and natural rubber at high temperature and concentration. These exceptions could be due to a change in the surface microstructure as the new samples were prepared differently from the original 0.080 inch samples, or they could be due to some other aging mechanism. Further studies could aid in determining the exact cause of these departures. While the surface-to-mass ratio does not influence the rate of swelling, it is important to note that components that have a higher mass will swell to greater extents.
- Elastomer materials exposed to monochloramine under the test conditions specified were observed to have a much greater swelling and loss in hardness than those

exposed to chlorine at the same conditions. Tensile performance loss also was much worse in monochloramine than chlorine. The difference in performance between the two environments is large (10 to 50 percent) for all of the elastomers with the exception natural rubber, which showed a change in performance in the range of 1 to 10 percent.

- The materials exposed to dichloramine also degraded substantially more than those in chlorine under the same conditions.
- The Atomic Force Microscope testing indicated that loss of carbon black from within the elastomer matrix appeared to be closely related to the degradation of the measured performance parameters.

This research confirmed that elastomeric components must be considered an "engineering material." As such, the selection of the appropriate elastomeric material for a specific application must take into consideration the critical failure mode for the component. An elastomeric material choice should be based on the anticipated product application and associated failure condition.

CHAPTER 1
INTRODUCTION

BACKGROUND

For more than 50 years, elastomeric materials have been used ubiquitously in drinking water distribution systems. Elastomeric materials have successfully been utilized in a variety of applications ranging from rubber gaskets, to valves, to hydrants, to fittings—and have been equally suitable to a wide variety of piping materials, including cast iron, ductile iron, steel, copper, and PVC. In general, elastomers have performed well in the water distribution system. However, as an increasing number of water utilities have changed their disinfectant from chlorine to chloramines, some elastomers have been observed to degrade more quickly. Elastomers that degrade more quickly than expected create service problems, make it difficult for utilities to cost-effectively plan preventive maintenance programs, and negatively affect customer relations.

Since the early 1900s, water systems throughout the United States have included disinfection in their purification process. Originally, free chlorine was the predominant disinfectant used because it was "effective, economical, and commonly available" (Sarai 2005). However, chlorine residuals are believed to form trihalomethanes (THMs)—chemicals formed when organic matter in water reacts with chlorine (Sarai 2005). THMs have been identified by the U.S. Environmental Protect Agency (USEPA) as carcinogens with a current limit of 80 ppb (USEPA 2006). However, THM formation can be reduced by adding ammonia to the disinfectant process; the ammonia combines with free chlorine to produce chloramines. Under most conditions, the chloramines form fewer disinfection byproducts but still provide effective disinfection (AWWA 2006).

One of the limitations of the disinfection process in general is that the free chlorine is believed to attack the elastomer matrix. While both free chlorine and chloramines contribute to elastomer degradation, chloramines appear to attack elastomer components much more aggressively than free chlorine (Reiber 1993a, b). In some situations, the chloramine residual has had significant negative impacts on the water distribution infrastructure and plumbing systems. Figure 1.1 shows the increasing effects of chloramine-induced degradation on a rubber component during testing conducted by the appliance industry.

In some southern cities, the chloramine ion is believed to be responsible for reducing the usable life of rubber components from five or six years to six months. In several Midwestern cities, the increased degradation rate was observed four and ten years after switching to the chloramine system (Reiber 1993a, b). For the public water utility, chloramine-induced elastomer degradation may cause:

- Increased leaks and line losses
- Increased maintenance and repair
- More expensive replacement parts
- Budget planning difficulties
- Customer dissatisfaction

The rate that elastomers degrade when exposed to potable water is dependent on a number of poorly understood factors. In addition to chlorine/chloramines concentration, temperature and pH and are believed to play roles, but those have not been adequately defined. Additionally, there is a wide variety of elastomer formulations available to the water industry, ranging from nitriles (NBR), to neoprenes, to EPDMs. Each elastomer formulation has a unique reaction when exposed to free chlorine or chloramines; some have been observed to resist degradation, while others degrade rapidly.

This research initiative was intended to clarify the factors that contribute to elastomer degradation and provide a quantitative method for predicting the performance and life expectancy of in-service elastomer components.

PROGRAM OVERVIEW

Very little research has been conducted to link elastomer decay rates with actual field conditions. Most of the related research from within the water distribution community and the elastomer industry has been devoted to determining if existing products are susceptible to degradation and to developing new products that resist degradation. While this information is undoubtedly valuable for future construction, it is of limited usefulness when applied to the numerous elastomer components already in use in virtually every water distribution system. Knowing degradation rates of these in-service elastomers would be immediately practical and enable water utilities professionals to better manage existing risks. This project is intended to assist water industry professionals by providing them with a quantitative method for predicting the life expectancy and dependability of elastomer components now in service. In addition, the research findings will make it easier to select elastomers for future projects.

Elastomer materials installed in water distribution systems are expected to have design lives similar to their surrounding materials. For diaphragms a reasonable design life may be several months to years, but for gaskets and valves, the design life may be 50 or 100 years. It would be impractical to obtain examples of all parts now in service, and unrealistic to test materials over a period of many years. Therefore, researchers selected un-aged samples of elastomer formulations that were representative of components currently in use. They established a systematic testing program that would duplicate field conditions in a condensed (or accelerated) time period. Using increased temperatures and increased concentrations of chlorine and chloramines, these accelerated-aging tests were conducted in the lab on un-aged samples. The results were then compared to actual field samples. By overlaying the degradation curves, the researchers could build a reliable method for predicting the performance of in-service parts.

The research was accomplished by first assessing the appropriateness of existing elastomer guidelines and performing technical reviews of standards and policies from related industries. Second, accelerated-aging tests were conducted on representative un-aged elastomer formulations to define their service degradation curves. Third, field samples were matched to the degradation curves to predict degradation factors and remaining service life. Fourth, accelerated-aging tests were performed on the field samples to verify that their ongoing performance would be consistent with the established degradation curves.

PROGRAM OBJECTIVE

Although elastomers have been used successfully in water distribution systems for more than 50 years, the increasing use of chloramines appears to have increased the rate at which elastomer components degrade. The primary objective of this project was to derive a quantitative method for predicting the life expectancy of existing in-service elastomer components when exposed to potable water. The goal was to give water industry professionals better tools for managing the risks and costs associated with elastomer degradation. When long-term performance can be predicted, professionals can better forecast financial expenditures associated with repair or replacement. Predictive and preventive maintenance can be enhanced, resulting in improved management and customer service.

Elastomeric materials can be chosen specifically for certain uses and conditions, and this research will give water industry professionals information to assist in those selections. Additionally, the research identified elastomeric materials that performed well when exposed to chloramines. Based on these results, the water industry may choose to modify industry standards for certain types of elastomer applications.

CHLORAMINE USE IN THE WATER INDUSTRY

Evolution of Chloramine Use

Over the past 100 years, free chlorine has been the primary disinfectant used in water distribution systems. It has proven effective against bacteria and waterborne microbes, with only minor impact on water taste and odor. It is relatively inexpensive, readily available, and easily applied to water. However, recent work by the USEPA has indicated that prolonged exposure to trihalomethanes (THMs) from the chlorination process may lead to adverse health effects. To reduce the risk associated with THMs, many water distribution systems have switched from free chlorine to chloramines.

Chloramines are produced when chlorine reacts with ammonia to produce a series of chlorinated ammonia compounds. The type of chloramine formed depends on the pH and chlorine-to-ammonia ratio. At a ratio of chlorine-to-ammonia of between 4 and 5 and at a pH 8, monochloramine is the predominant chloramine. At a pH of between 5 and 8, both mono- and dichloramines form; at pH of between 4.4 and 5, dichloramine is predominate; below pH 4.4, trichloramine is the predominate species. Monochloramine (NH_2Cl) changes to dichloramine ($NHCl_2$) as the chlorine-to-ammonia ratio progresses, then to trichloramine ($NHCl_3$), and eventually to nitrogen gas or a variety of nitrogen-containing chlorine-free products. A simplified formation of the chloramines can be depicted as a step-by-step process (Sarai 2005):

$$NH_3(aq) + HOCl \rightarrow NH_2Cl + H_2O \quad \text{(Monochloramine)}$$
$$NH_2Cl + HOCl \rightarrow NHCl_2 + H_2O \quad \text{(Dichloramine)}$$
$$NHCl_2 + HOCl \rightarrow NCl_3 + H_2O \quad \text{(Trichloramine)}$$

The water industry prefers monochloramine for disinfectant purposes because it has a longer residual life and has less impact on taste and odor than dichloramine or trichloramine. However, research sponsored by AwwaRF indicates that all three chloramine compounds may accelerate elastomer degradation rates, and monochloramine appears to be the most aggressive agent (Reiber 1993a).

Prevalence of Chloramine Disinfection and Its Effects

Through the years, the prevalence of chloramine use in water treatment has varied. In 1936, 16 percent of U.S. water treatment facilities used chloramines. The number declined in

World War II because of a shortage of ammonia, which was needed for fertilizer, explosives and other uses. The decline continued through the 1960s to a low of 2.6 percent (San Francisco Public Utilities Commission 2006).

In 1974 the U.S. Congress passed the Safe Drinking Water Act (SDWA) to regulate public drinking water and to authorize the USEPA to set water standards. In 1979 the USEPA identified that disinfectant byproducts (DBPs) from the chlorination process may lead to adverse health effects and set a 100 μg/L maximum containment level for THMs. Subsequently, many water distribution systems began switching their disinfectant from free chlorine to chloramines in order to comply.

By 1990, approximately 20 percent of treatment facilities were using chloramines. In 2002, USEPA data from 353 treatment plants showed more than 46 percent were using chloramines. The USEPA estimates that the percentage of water treatment facilities using chloramines will likely increase to 60 percent as more stringent regulations gradually go into effect (Kirmeyer et al. 2003).

Chloramines and Elastomers

As chloramination has become more widespread, several utilities have reported problems with substantially higher rates of elastomer failures. Specific water quality conditions appear to enhance the oxidative effect of chloramines. The following is a list of failure incidents in chloraminated systems (Reiber 1993a and Simmons and Evanson 1988):

- Fort Lauderdale, Fla.: Rubber parts in toilet tanks were breaking down after only six months.
- Topeka, Kan.: City inspectors found that gaskets and diaphragms on backflow-prevention assemblies were beginning to deteriorate.
- Ramona, Calif.: O-rings used in chicken watering assemblies were replaced after six months due to severe swelling.
- Minneapolis, Minn.: Aerator washers swelled shortly after chloramine changeover.
- Ingleside, Tex.: Failure of seating materials on large valves in backwash assemblies was reported.

It should be noted that not all utilities have reported an increased elastomer degradation rate due to chloramination. In 2006, Manchester Water Works in New Hampshire was considering changing their secondary disinfectant from chlorine to chloramine in order to meet the new regulations. They conducted an informal survey of their peer utilities to assess the potential effect of chloramines on similar water systems. Eight utilities responded that they had experienced few or no problems in systems up to 25 years old (Bowen 2006).

In recent years, a significant amount of work has been undertaken to understand elastomer performance when exposed to chloramine. In 1993, AwwaRF commissioned a study on the *Chloramine Effects on Distribution System Materials* (Reiber 1993a). The study concluded:

- Chloraminated waters have a significantly more aggressive effect on elastomers than waters with equivalent concentrations of free chlorine.
- Elastomer formulations vary widely in their resistance to chlorine.
- The swelling, or water absorption, test is a sensitive and analytically simple methodology for assessing the degree of elastomer degradation.
- Temperature plays a critical role in determining the rate of chloramine attack.

Table 1.1
Bond disassociation energy of common bonds

Bond type	Bond disassociation energy (kJ mole-1)
-C-C-	351
C-S-C-	285
-C-S-S-C-	267
-C-S_X-C-	<267

In addition, the AWWA Committee on Gasket Materials has identified that many types of elastomer compounds, especially those without antidegradants, are at risk in chloraminated water distribution systems (AWWA Committee on Gasket Materials 1998).

Upon conclusion of the work, however, it was not possible to develop a standard for chloramine resistance that would apply to all water system applications. Due to the extensive variety of materials available and exposure conditions, a pass-fail test to assess an elastomer material suitability for a particular application could not be defined (AWWA Committee on Gasket Materials 1998).

Chemical Attack Mechanism

The exact chemical attack mechanism inducing elastomer degradation during chlorine or chloramines exposure is not clearly understood. Much of the uncertainty is due to the complex structure of the elastomer itself and its unique properties. Both synthetic and natural rubber elastomers consist of long, intertwined polymer chains that are occasionally bonded or "cross-linked." The cross-links are small polymer chains that occur at regular intervals on the long chains, linking one long chain to another. These cross-links make the long chains of molecules act more as a group than as individuals. This macromolecular structure is stretchable under low stress and has a high degree of shape memory (International Institute of Synthetic Rubber Producers 2006a).

The degradation of elastomers is believed to be caused primarily by chemical cross-link breakages. It is believed that chloramines attack the cross-linking bonds, and the rate of degradation is related to the type of chemical bond. Cross-link bonds that break easily have a low "bond disassociation energy," and cross-link bonds that are difficult to break have a high "bond disassociation energy." The chemical attack is likely to be more severe in elastomers with cross-links that have lower bond disassociation energy. As shown in Table 1.1, the type of bonding can be an indicator of elastomer resistance to chlorine and chloramines resistance.

The type of cross-link bonds that are formed depends upon the base elastomer polymer and on the curing agent. Natural rubber and most synthetic elastomer materials are sulfur-cured. However, some specialized elastomers (some formulated for chloramines resistance) may be cured with peroxide. The selection of curing agent has a significant effect on the formation of cross-link bonds.

In all rubbers, cross-links are formed when the excess curing sulfur attaches to the double-bonded carbon in adjacent polymer chains to form a bridge. The bridge, or cross-link, can be short and of the form C-S-C, or relatively long and of the form C-S_X-C, where "x" is the number of sulfur nodes present in the cross-link. The longer cross-link chains are more susceptible to change as the relatively weak sulfur bonds may break and move to new bonding sites.

Another commonly used elastomer curing agent is peroxide. While significantly more expensive than sulfur, the peroxide curing agent forces a C-C cross-link bridge between adjacent polymer chains. As the C-C bond has a much higher bond disassociation energy, the peroxide cured elastomers are less susceptible to change.

The natural rubber, SBR, nitrile (NBR) and neoprene backbone polymer chains all contain double-bonded carbon atoms that are available to form cross-links during curing. However, not all of the available sites form cross-linkages, and ultimately only a small number of cross-linkages are completed. Thus, the chlorine and chloramine ions attack both the weak cross-link bonds and the available carbon sites (open sulfur bond locations) on the backbone polymer chain. In addition to removing the cross-linkages, the backbone chain length is reduced, which further reduces polymer entanglement to the point that parts of the network break away.

Due to its chemical composition, EPDM has proved to be a very durable elastomer material. Unlike many elastomers, the double-bonded carbon atoms within EPDM elastomers are not contained within the polymer "backbone" chain. Rather, the double-bonded carbon atoms hang off the polymer chain as a pendant. Thus, the sites available for cross-linkage are offset from the backbone, and chemical or physical attack of the cross-linkage chain does not affect the base polymer backbone or the molecular chain length. Therefore, the material properties derived from chain entanglement will not be altered by chemical or physical attack (Sartomer Company 2002).

The pendent double-bond carbon sites within EPDM that are available for cross-linkage can be formed using either sulfur or peroxide. When using sulfur cross-links in EPDM, large amounts of oil and reinforcement fillers are typically used. While these additives lower the cost of the finished product, the amount of rubber in the network is significantly reduced. Since EPDM contains relatively few cross-link sites compared to other elastomers, the material properties are primarily derived from the few cross-linkages and molecular entanglement. If cross-links are broken, then the material properties are sustained by molecular entanglement. However, long-term material performance may suffer as entanglement is subject to thermodynamic creep and relaxation.

During peroxide curing of EPDM, peroxide is lost because of its immediate interactions with the acid surface of some fillers and with the small-molecule double bonds in some oils. To overcome this reaction, huge quantities of very costly peroxide must be used in order to have sufficient peroxide available to form the cross-link chains. Therefore, there is typically more rubber included in EPDM-peroxide-cured compounds to increase the available cross-linkage sites and to form more carbon-carbon bonds (high disassociation energy). However, once the C-C site breaks, material properties are dependent on polymer entanglement and thus subject to thermodynamic creep and relaxation.

In addition to selecting the base polymer material and curing agents, elastomer compounders tailor products to meet specific needs by adding vulcanizing agents, antidegradants, accelerators, retarders, pigments, plasticizers, processing aids, and fillers—all of which affect the cross-links. However, the exact formulations are proprietary and closely guarded.

Degradation also is related to elastomer permeability and fluid absorption. Any fluids in contact with the elastomer will permeate the material, and in time, the elastomer swells. The presence of absorbed fluid can weaken the elastomer as the fluid molecules push apart the polymer molecules (plasticization). Additionally, the contacting fluid may leach out soluble elements from the elastomer structure. Or, the fluid may be chemically aggressive, first attacking the elastomer surface, and after absorption, attacking from the inside, causing further deterioration in properties and performance of the component (Loadman 2005, Nagdi 1993).

ELASTOMER USE IN THE WATER INDUSTRY

Development of Elastomers

Long before the arrival of European explorers, rubber was known to the indigenous peoples of the Americas. Natural rubber had been obtained by coagulating the latex produced of certain plants, particularly the Brazilian rubber tree (*Hevea Brasiliensis*). Most frequently, the rubber material was used to coat fabrics, dip feet (to make shoes), or to form balls for games. Rubber was not known to the Europeans in the early 1500s when explorers reported seeing Mexican tribes playing with elastic balls. It wasn't until 1770 that rubber was used in its first commercial product; pencil erasers (Loadman 2005).

The rubber material of the early 1800s was much different from the product we think of today. Natural rubber tended to become brittle in the cold and gummy if left in the sun. In 1842, however, Charles Goodyear found the secret of vulcanization—a process that made the material more workable and usable. In the mid-19th century, uses for vulcanized rubber included seals for sewers, conveyor and transmission belting, flooring, inflatable boats, toys, hollow balls for golf and tennis, and many items associated with clothing—including the famous MacIntosh waterproof coat.

Rubber—and later, synthetic rubber—developed in tandem with the automobile industry. In 1875, the Michelin brothers were the first to use vulcanized rubber pneumatic tires on a car they entered in an historic race from Paris to Bordeaux and back. Many well-known companies in the automobile tire industry were founded near the turn of the century, including:

- B.F. Goodrich, 1872
- Michelin & Co., 1889
- United States Rubber Co., 1892
- Goodyear Tire & Rubber Co., 1898
- Firestone Tire & Rubber Co., 1899

The escalating demands for rubber worldwide, fueled by the Industrial Revolution, created an economic boon in northern Brazil where the tree was plentiful. However, in 1876, the British smuggled rubber-tree seeds out of the country and, through grafting, developed more resistant varieties. These grafts were sent to their colonies in Asia where massive and efficient rubber plantations were established—significantly increasing efficiencies and all but eliminating the Brazilian competition. The automobile industry and others became reliant on the low-cost Asian sources of natural rubber. Henry Ford attempted to re-establish a rubber plantation in Brazil that incorporated the Asian-style efficiency, but it failed (International Institute of Synthetic Rubber Producers 2006b).

During World War II, three months after the attack on Pearl Harbor, the Japanese invaded Malaysia and the Dutch East Indies and took over natural rubber production, gaining control of 95 percent of the world rubber supplies. Rubber was critical to U.S. success as each Sherman tank contained a half ton of rubber and each warship contained 20,000 rubber parts. The crisis prompted the largest recycling campaign ever recorded, as U.S. citizens contributed rubber items large and small to the war effort (International Institute of Synthetic Rubber Producers 2006b).

United States' chemists and engineers were called upon to develop a synthetic rubber that would have elastomeric qualities similar to natural rubber. The nation's war effort depended on its ability to manufacture over 800,000 tons annually of products from a material that had barely been developed. The research built upon some early advancements and produced a variety of new elastomeric materials ranging from neoprene, to butyl, to EPDM (Plastics Historical Society 2006).

Historical Elastomer Use in the Water Industry

The water industry has benefited from many of the advancements and achievements of elastomer research supported by the automotive industry. Once the water-sealing capabilities of vulcanized rubber were identified, rubber seals and gaskets slowly replaced lead and oakum in pipe joints and other applications. While many new materials were available, the water industry primarily focused on the "water sealing" properties of the elastomers and did not typically differentiate between materials. As such, vulcanized natural rubber was used in water system construction into the 1950s and 1960s. After that time, however, United States drinking water regulations forced a changed from natural rubber to synthetic elastomer products because natural rubber appeared to promote the growth of bacteria (Leeflang, 1968). After this, the United States gravitated toward SBR as the standard gasket material. In the mid-1990s, Europe switched to synthetic EPDM as their preferred gasket material for drinking water applications, likely to provide increased resistance to the ozone disinfection process. However, in other parts of the world, including Mexico, natural rubber is commonly used, although it is rapidly being replaced by SBR (Dittel 2006).

The following is a timeline that combines important benchmarks and historic snapshots to provide a sense of how synthetic rubber was developed and implemented in the water industry. The timeline is compiled primarily from information provided by the Plastics Historical Society (2006), the International Rubber Research and Development Board (2007), Loadman (2005), and other sources as noted:

1920s: German company I.G. Farbenindustrie develops a synthetic rubber that has since become known as styrene butadiene rubber (SBR).

1929: First commercial synthetic rubber, polysulphide rubber (Thiokol) introduced.

1931: Polychloroprene rubber was marketed by DuPont (Neoprene).

1937: Roll-on joint developed for the water industry. Joint construction included a rubber gasket, restriction ring, braided jute, and bituminous compound.

1939: Nitrile (NBR) rubber was marketed in the United States, providing a source of oil-resistant rubber.

1945: During war time, a large synthetic rubber industry is established in the United States, with an annual output of nearly 830,000 tons of elastomers.

1953 and 1954: Karl Ziegler in Switzerland and Giulio Natta in Italy, develop a process to produce high molecular weight, high crystallinity synthetic polyisoprene and polyethylene. This same technique later enables the development of solution polymerized SBR, polybutadiene and ethylene propylene rubbers (EPDM).

1953: ASA/AWWA C111/A21.11 is a new standard issued for rubber gasket joints (AWWA 1953). The standard describes only natural rubber.

1954: A Cast Iron Pipe Research Association (CIPRA) standard for "roll-on" gaskets calls for rubber or synthetic rubber.

1956: Push-on rubber joint is developed. This joint consists of a rubber gasket inserted into a pre-formed groove on a bell-and-spigot connection.

1964: The ASA/AWWA C111 standard now recognizes both natural and synthetic rubbers (AWWA 1964).

1983: U.S. Pipe patents resilient seat gate valve using EPDM elastomer (Dunn 2007).

1985: Revisions to ANSI/AWWA C111/A21.11 includes the deletion of natural rubber as a gasket material (AWWA 2000).

1990: Revisions to ANSI/AWWA C111/A21.11 identify SBR as the standard material for gaskets and indicates the availability of other elastomers for special service applications (AWWA 2000). A section on permeation is added to the standard. The revision calls for the name of the country in which the gasket was manufactured to be molded into or permanently affixed to each gasket.

1995: A statement addressing physical requirements of elastomers other than SBR is added to the standards.

Late 1990s: Responding to the use of chloramines, Neptune Meter Co. switches from EPDM to Santoprene (ExxonMobile, 2007) for their seal rings (Noel 2006).

Current Elastomer Materials Available to the Water Industry

Elastomers and thermoplastics are revolutionizing the manufacturing industry; they appear as gaskets, coatings, membranes, hinges, structural materials, and shock-absorbing rigid bodies. They have become indispensable to the manufacturing industry and, in particular, to the portion of the industry producing valves, pipes, fittings, and the host of appurtenances required by the water works field.

There are more than 25 base polymers in widespread commercial use, and from these, thousands of elastomeric formulations are possible. The elastomer types commonly used in the water industry are much more limited, generally breaking down into categories reflecting eight base polymers:

Natural Rubber—Natural Polyisoprene

Natural rubber was the first elastomeric product adapted for use in the water distribution industry. It was relatively easy to form and was effective in creating watertight fittings. However, natural rubber is not commonly used in today's water distribution systems because of its susceptibility to severe ozone-and disinfectant-induced degradation. Natural rubber is a good gasket material due to excellent resilience, tear strength and wear resistance (Allstate Gasket 2007). Suitable applications include heavy-duty tires, vibration dampers, gaskets, seals and rolls (Timco Rubber 2007, Trelleborg 2007, Nagdi, 1993).

Styrene Butadiene Rubber (SBR, Buna-S)

Styrene butadiene rubber (SBR) was developed during World War II as a synthetic alternative to replace natural rubber. SBR is a low-cost material that has good mechanical properties and ozone resistance, although not as outstanding as those of natural rubber. It can be compounded to provide very fine abrasion, wear, and tensile qualities. However, it is susceptible to oxidant attack and has poor resistance to organic solvents. In the water industry, SBR is commonly used in pipe gaskets and other water applications where only a portion of the part is exposed to potable water. It is widely used as a gasket material (Allstate Gasket 2007, Nagdi, 1993).

Nitrile (NBR) (NBR, Buna-N)—Acrylonitrile (NBR)-Butadiene

Nitrile (NBR) rubber is an oil-resistant elastomer that was formulated shortly after World War II. In addition to its resistance to petroleum oils and aromatic hydrocarbons, NBR is highly resistant to mineral oils, vegetable oils, and many acids. For this reason, it has wide application in machinery and instrumentation. It also has good elongation properties as well as adequate resilience, tensile, and compression set. While typically more expensive than SBR, municipal utilities frequently use nitrile (NBR) rubber where a combination of oily and water residues is anticipated. Typical application for Nitrile (NBR) rubber includes seals, O-rings, gaskets, diaphragms, tank linings, boots, valve seats, pump impellors and check valves (Lutz Sales Co 2007 and Reiber 1993a).

Neoprene (Chloroprene Rubber)—Polychloroprene

Neoprene is one of the best all-purpose elastomers for unique conditions where resistance to sunlight, oxidation, heat and many petroleum derivatives are of prime importance. It has a high degree of flame resistance because of its chlorine content it provides excellent resistance to many oils, chemicals, water acids, and alkalis. It is generally attacked by strong oxidizing acids, esters, and ketones as well as chlorinated, aromatic and nitro hydrocarbons. It is used in a wide variety of environments, such as in wetsuits, electrical insulation, and car fan belts. Its chemical inertness makes it well suited for industrial applications such as gaskets, hoses, and corrosion-resistant coatings. It can also be used as a base for adhesives (Accurate Products 2007, Nagdi, 1993).

EPDM—Ethylene Propylene Diene Monomer

EPDM is a relatively new elastomer product formulated by ExxonMobile Chemical in 1988 (ExxonMobil Chemical 2007). EPDM rubber is ideal for outdoor applications because of its excellent resistance to ozone, oxidants, and severe weather conditions. Its primary limitations are its susceptibility to hydrocarbons and its cost. Within the water industry, EPDM is widely used for O-rings, valve seats, flat gaskets, and pond liners. More recently, peroxide-cured EPDM is being used where chloramines resistance is important.

Butyl Rubber (IIR)—Isobutylene-Isoprene

Butyl rubber has been available to industry for more than 60 years. The primary attributes of this material are excellent impermeability/air retention and good flex properties. It displays outstanding resistance to many chemicals, including oxygenated solvents and ozone; it is also resistant to water, steam, sunlight, and heat. Within the water industry, butyl rubber is primarily used for diaphragms, construction sealants, and hoses.

Fluorocarbon (FKM)—Fluorinated Hydrocarbon

Due to its excellent chemical resistance and outstanding upper temperature limit, fluorocarbon rubber has become one of the most widely used seal materials in the world. The material is resistant to the majority of greases, fuels, oils, and chemicals. The high cost of fluorocarbon rubber is usually justified by its all-around performance. Fluorocarbon is used for O-rings, gaskets, seals, and chemical feed pumps.

Sialstic (SI)—Polysiloxane

Silicones have many unique properties that most organic elastomers are unable to copy. One excellent example is the wide operating temperature range, from the very low temperature of –100°C to the very high temperature of 316°C. Silicone rubber is also oil resistant and resists abrasion well. These qualities make silicones excellent for many automotive, industrial, and high-temperature applications. However, due to their expense, they have seen limited application to the water industry.

ELASTOMER STANDARDS AND TESTING METHODS IN OTHER INDUSTRIES

During the development of natural and synthetic elastomer materials, it became necessary to develop a systematic method of rating material performance. To meet this need, an entire testing discipline evolved that was specifically designed to quantitatively assess elastomeric performance. Standard test procedures were established to assess both physical and chemical performance. At present, ASTM has over 150 standardized test methods for elastomer testing. In addition, the automobile, appliance, and other industries also have their own unique set of testing protocols designed to test materials for their specific applications. In all, hundreds of standard test methods are available to assess elastomer performance. The standard test methods available to the elastomer industry are generally either chemical tests designed to identify elastomeric composition or physical tests designed to assess material performance.

Many elastomer compounders are acutely aware of the water industry's desire to identify an elastomeric material resistant to chloramines degradation. Several compounders have developed their own testing protocol to demonstrate that their products are, not only resistant to the degradation caused by chloramine attack, but superior to other products. One such study, *Resistance to Aqueous Solution* performed by ExxonMobil Chemical, used temperature and concentration to accelerate aging. Three grades of Santoprene were compared to EPDM, NBR, and silicone thermoset rubbers for hardness, tensile strength, ultimate elongation, 100 percent modulus, and weight before and after accelerated-aging tests. The accelerated–aging test specifications were reported to involve immersion of these seven elastomers in an aqueous solution with a chloramine concentration of 1,000 ppm at an elevated temperature of 70°C and two pH values (5.5, slightly acidic, and 8.5, slightly alkaline). These conditions were maintained over 180 days with testing occurring at 14, 30, 60, 90, 120, and 180 days. While many testing details are unavailable, it was reported that Santoprene has a resistance to chloramine attack (ExxonMobile Chemical 2007).

Emerging Materials

Advancements in processes and compounds continue to result in new elastomeric materials with specific uses. While these new products are immediately available, the water industry is routinely cautious with new products in contact with water systems. Much of the caution is associated with the very long expected life of the in-use products and public health concerns. However, there are several new elastomer materials that have entered the market that appear adaptable to the water industry. These materials must meet long-term performance tests and appropriate water quality assessments prior to their incorporation.

Many companies are modifying the manufacturing process of existing products to improve the performance of elastomers for water utility applications. For example, slightly increasing the ratios of sulfur and accelerator will result in a compound with greater cross-link density. A secondary accelerator added to the formulation will improve oxidation resistance. Adding microcrystalline paraffin wax will reduce oxidation and ozonation as it migrates to the surface of the gasket after cure to act as a barrier. And finally, a strong chemical antioxidant/antiozonant may be added to the formulation to meet the gasket-aging requirements set by the AWWA C111 standard (AWWA 2000). Thus, a variety of additives or modifications to the curing process can significantly affect the behavior of the finished elastomer product.

Several new EPDM-based products are available that have been formulated specifically for chloramine resistance in aggressive water utility applications. These new products are designed to work effectively in high-flow conditions with either hot or cold water. Typical applications for these new materials include faucets, showerheads, aerators, water-filtration components, irrigation systems, ice makers, water meters, valves, fittings, and many other uses (Parker Hannifin Corporation 2006).

One of the newest materials available to the water industry is *Santoprene*, an advanced, cross-linked thermoplastic elastomer produced by Exxon Mobil Chemicals. It is processed as a plastic, yet performs with the flexibility and durability of rubber (ExxonMobile Chemical 2004).

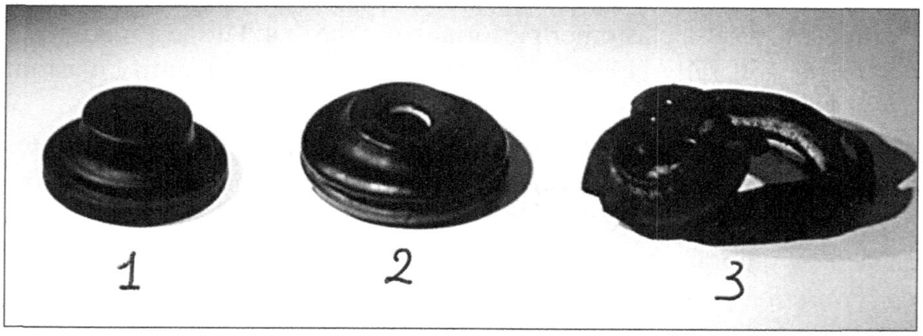

Figure 1.1 Example of chloramines-induced elastomer degradation of a rubber component from the appliance industry (Courtesy of Kenneth R. Davis)

CHAPTER 2
LABORATORY TESTING PROGRAM

OVERVIEW

The purpose of the laboratory test program was to assess the chloramine-induced degradation characteristics of six elastomer materials typically used in the water industry. To assess degradation within a reasonable timeframe, exposure time was accelerated using elevated temperatures and chloramine concentrations. To account for varying water conditions, degradation characteristics were assessed for differing water temperatures, chloramine concentration, pH, and strain. The laboratory generated degradation curves were then compared to degradation curves generated from field return samples. Thus, the laboratory test program involved the following key elements:

- Accelerated-aging tests of selected un-aged elastomer samples using temperature and concentration of chloramines and chlorine as accelerants
- Testing and accelerated-aging tests of samples returned from field service
- Accelerated-aging tests of un-aged samples under various pH conditions
- Accelerated-aging tests of un-aged samples under constant strain conditions

Elastomer degradation was assessed by quantifying, through chemical and physical testing, the performance of six typical elastomer materials. Un-aged sample materials were artificially aged by exposing them to elevated concentration and temperature levels. Laboratory studies were performed to assess degradation rates for differing pH, disinfectant (chlorine or chloramines), and strain conditions. During the accelerated-aging process, degradation indicators (weight, volume, tensile strength, elongation, and change in surface condition) were recorded. These physical tests were specifically selected to quantitatively measure material performance or other identifying characteristics that would affect field service. The changes in material performance were used to assemble a degradation curve for similar materials.

The laboratory-derived degradation curves were compared to the degradation characteristics of field samples that had been in water distribution systems for between one and 50 years. This comparison correlated the laboratory study with real-world exposure patterns. The results allowed researchers to predict degradation characteristics for a period of 30 to 80 years.

To develop a complete ranking table for a specific physical parameter, the degradation trends of all five elastomers had to be known. These trends were discovered for the parameters of breaking strain and percent mass change. Breaking strain and percent mass change are parameters indicative of elastomer performance in varied applications. For these two parameters, the degradation curves resulting from the accelerated-aging tests could be used to predict the degradation rates and remaining service life of the in-service elastomer components.

LABORATORY ELASTOMER SELECTION AND FORMULATION

Currently, a wide variety of elastomers is available for use in the water industry. When all base compounds, materials, and admixtures are considered, the number of elastomeric options available is almost limitless. After assessing typical materials and interviewing water utilities and elastomer compounders, six typical backbone polymers were selected for evaluation (Table 2.1).

Table 2.1
Base elastomer materials included in the elastomer testing program

Base polymer	Description
Natural rubber	Used in early water distribution systems
Styrene butadiene rubber (SBR)	Found commonly in distribution system gaskets
Nitrile (NBR) rubber	Commonly used today
Chloroprene rubber (neoprene)	Found occasionally in use today
Ethylene propylene diene monomer (EPDM-S = sulfur cured)	Widely used today
Ethylene propylene diene monomer (EPDM-P = peroxide cured)	Widely used today

Natural rubber, SBR, nitrile (NBR), neoprene and EPDM-S are basic elastomeric compound formulations, and it was anticipated they would be representative of elastomer parts commonly used by the industry. Peroxide-cured EPDM was added to the degradation tests as it is a relatively new product that is commonly recommended for chloramine environments. Specialty elastomers such as silicone and fluorocarbons were not included, primarily due to low usage in commercial potable water systems.

The elastomer compounds selected for degradation testing were specifically formulated without the proprietary anti-degradation additives commonly used in most commercial compounds. While additives can influence material performance, it was determined that additive formulations are not easily obtained because they are closely protected trade secrets and used to differentiate one produce from another. However, using base formulations was considered a reasonable simplification because:

- Elastomer formulations vary throughout production life due to the price and availability of components. For example, it is unlikely that a nitrile (NBR) rubber gasket manufactured by Minnesota Rubber Company in 1985 will be materially identical to a new nitrile (NBR) rubber sample molded today. Thus, one specific elastomer formulation with additives would not appropriately represent all elastomeric products.
- Elastomer materials generally follow "family trends" regardless of the additives. Thus, two SBR formulations with differing additives will behave more similarly to a base SBR material than to another polymer backbone structure.
- Elastomer compounds are generally formulated to meet base performance requirements. The additives are included in the recipes to improve performance or to reduce costs. Thus, formulating the test samples without additives should conservatively predict material performance. Actual performance of the field elastomers can be expected to be proportionately better because of the anti-degradation additives.

Tables 2.2 through 2.7 present the elastomeric formulas used to compound the elastomeric materials. To build on previous AwwaRF work, the elastomers compounded for this study were very similar to the formulations used in *Chloramine Effects on Distribution System Materials* (Reiber 1993a). Ashtabula Rubber Company formulated these samples within their test laboratory.

Table 2.2
Natural rubber sample formulation

Ingredient	PHR	Function
SMR CV 60	100	Standard Malaysia rubber (latex)
SRF N762	70	Carbon black
Sun 4240 (or equiv.)	2	Extending oil
Sulfur	2.5	Primary vulcanizing agent
Zinc oxide	5	Curing system additive (activator)
Santocure	0.7	Accelerant—thiazole based

Table 2.3
SBR sample formulation

Ingredient	PHR	Function
SBR 1502	100	Styrene butadiene rubber base polymer
SRF N767	90	Carbon black
Sun 4240	10	Extending oil
Sulfur	1	Primary vulcanizing agent
Zinc oxide	5	Curing system additive (activator)
Stearic acid	1	Curing system additive
Santocure	1	Accelerant

Table 2.4
Nitrile (NBR) sample formulation

Ingredient	PHR	Function
Paracril BJLT	100	Acrylonitrile (NBR)/butadiene (NBR) resin—Uniroyal
SRF N762	100	Carbon black
Plasticizer SC	8	Harwick plasticizer
Sulfur	0.3	Primary vulcanizing agent
Zinc oxide/stearic acid	5.1	Curing system components
Santocure	1.0	Accelerant
M Tuads	1.0	Accelerant

Table 2.5
Neoprene sample formulation

Ingredient	PHR	Function
Neoprene W	100	DuPont—nonsulfur modified chloroprene-CR
SRF N762	60	Carbon black
Sun 4240	10	Extending oil
Sulfur	.25	Primary vulcanizing agent
Zinc oxide	5	Curing system additive (activator)
Stearic acid	0.5	Curing system additive
DOTG/Monex	2.5	Accelerants/processing aid

Table 2.6
EPDM-S sample formulation

Ingredient	PHR	Function
Nordel 1070	100	Ethylene-propylene diene copolymer
FEF N550	100	Carbon black
Sun 2280 (or equiv.)	110	Extending oil
Sulfur	2	Primary vulcanizing agent
Zinc oxide	5	Curing system additive (activator)
Stearic acid	2	Curing system additive
Altax/Butyl Z/M Tuads	4.8	Accelerants

Table 2.7
EPDM-peroxide sample formulation

Ingredient	PHR	Function
DSM Keltan 27	100	Ethylene-propylene diene copolymer
SRF N762	75	Carbon black
Zinc oxide	5	Curing agent
Sun 2280	20	Extending oil
Dicup 4OKE	10	Accelerant

Table 2.8
Initial properties of un-aged lab samples

Compound type	Tensile strength (psi)	Elongation (%)	Hardness (Shore A)	Density (gr/cc)
Natural rubber	3036	396	65	1.1765
EPDM-S	1663	200	72	1.1869
Nitrile (NBR)	2214	394	70	1.2160
SBR	2404	425	65	1.2227
Neoprene	2770	375	70	1.3880
EPDM-P	2697	245	63	1.1184

Properties of Laboratory Samples

Prior to testing, preliminary physical property tests were conducted on each of the elastomer samples. The purpose of the testing was to characterize the basic elastomer parameters in the "as-received" condition. Table 2.8 shows the initial properties of the laboratory samples.

ACCELERATED AGING USING TEMPERATURE AND CONCENTRATION

Overview

For material performance testing, the rate of elastomer degradation is often accelerated by exposing the elastomer samples to elevated temperatures and concentrations. For this study, samples were cut into dog bone coupons and were submerged in chloramine solutions with varying concentrations (Figure 2.1). The temperatures were elevated using a water bath. At systematic intervals, the samples were temporarily removed from the bath and subjected to performance tests.

ASTM D-6284 (2002) provides a standard test method where 70°C temperature and 50 ppm initial concentrations are used to accelerate the aging process. Elastomer performance is measured on a periodic basis by performing physical tests on the degrading specimens. The degradation patterns observed in the accelerated testing can then be applied to real time applications.

The test procedures used in this study expanded the standard ASTM testing protocol with additional test conditions (temperature, chlorine and chloramines) as defined in the matrix (Figure 2.2). By including a variety of test conditions, independent concentration and temperature effects could be isolated. It is noted that ASTM D-6284 specifies an initial 50 ppm concentration at 70°C and the test protocol used in this study defines the "ASTM" procedures as an average 30 ppm concentration at 70°C. As discussed subsequently, the two conditions are essentially the same due to the loss of chlorine ion with time.

Temperature

The 23°C, 45°C, and 70°C temperatures used to accelerate elastomer degradation were selected to provide a wide range of acceleration factors. The 70°C temperature was selected as the "high" temperature used in the temperature-degradation testing, as it correlated with accelerated elastomer degradation specified in ASTM protocols. The 23°C test condition provided a consistent baseline and was a reasonable "control" for comparison purposes. And 45°C was selected as reasonable intermediate test temperature.

Initial testing plans considered using 83°C as the higher temperature to achieve a more rapid degradation and to coincide with extreme hot water temperatures of household plumbing systems. However, at 83°C, degradation progressed too rapidly and almost certainly induced failure modes not representative of typical water distribution systems. Additionally, the chlorine and chloramines solutions were very unstable at this temperature and exhibited dramatic decay of chlorine concentration.

Concentration

Similar to the three temperatures, 1 ppm, 30 ppm and 60 ppm chloramines concentrations were selected to accelerate elastomer degradation. A 1 ppm concentration was used for the baseline value because it was considered representative of typical water distribution systems. For the high end, a 30 ppm solution was included because it was consistent with ASTM test protocol at 70°C. As discussed in later sections, the initial starting concentration was consistent with ASTM protocol and the 30 ppm represented the weighted concentration average during a 24-hour period. The 60 ppm concentration was selected as a reasonable increment to provide a high concentration value for the data set.

Supplemental acceleration tests at different temperatures and concentrations were not proposed for the accelerated chlorine testing due to the extensive work already completed on chlorine-immersed elastomer components and the difficulty of maintaining high chlorine concentration levels. As such, only the ASTM test procedure (30 ppm and 70°C) was used to assess chlorine-induced degradation in this study.

Solution Replacement and Test Intervals

During the tests, the solutions were replaced with fresh solution at 24-hour intervals for all test conditions. Testing intervals were set as follows: 3 days, 6 days, 12 days, 20 days, and 30 days. At these times, test coupons were removed and tested for the parameters discussed below. Note: For the tests requiring destructive testing of the samples, the sample material was returned to the bath to maintain a consistent chlorine-to-elastomer ratio. All tests were performed in triplicate.

Testing of Un-aged Samples

During the accelerated-aging process, the samples were temporarily removed from the testing solutions for performance testing. These performance tests were then plotted as a function of time to develop degradation curves for each elastomer sample. It was anticipated that these degradation curves defined in the laboratory could be matched to the observed field conditions to estimate future field performance.

The performance tests performed on the samples during aging were selected to quantitatively assess the amount of elastomer degradation. The following performance tests were performed on the elastomer samples. Detailed test procedures for these methods are available from ASTM.

- Change in Shore A hardness
- Change in mass
- Change in volume
- Surface condition
- Tensile strength
- Elongation
- Turbidity of test solution, measured after each day's testing.

ACCELERATED-AGING TESTS OF FIELD RETURN SAMPLES

The laboratory accelerated-aging testing of the un-aged samples provided an estimate of real-world performance. However, it was important to verify these trends based on actual field data. In support of this analysis, utilities were requested to retrieve elastomer samples that had been exposed to their water system and ship them to the laboratory for testing. It was anticipated that samples that had been in use for longer periods of time should also have greater degradation. In this fashion, the laboratory generated elastomer degradation curves could be calibrated based on actual field data.

Retrieval and Hydration

Utilities that participated in the field retrieval program provided elastomeric parts that had direct exposure to potable water within their distribution systems. The following nine utilities participated in the field sample program:

- Akron, Ohio
- Springfield, Mo.
- Cincinnati, Ohio
- Vancouver, British Columbia
- Louisville, Ky.
- Philadelphia, Pa.
- Portland, Maine
- St. Paul, Minn.
- South Central Connecticut Regional Water Authority

These participants provided aged field samples that had been exposed to a variety of environmental factors. Critical historical data—including location, service application, pH, chlorine/chloramine concentration, flow rate, and temperature—were recorded for all the samples as they were removed from the field. In this way, the program's analyses could identify trends or "normalize" data to consistent environmental conditions.

Ultimately, the utility participants provided 226 elastomer samples installed in the 1920s through 2005. (Some of the samples arrived after testing was complete and are not reflected in the charts below.) Samples were primarily retrieved from meter vaults, pipe joints, valves, and hydrants, and generally ranged in age from 5 months to more than 25 years of exposure (Figures 2.3 and 2.4).

As the samples were received, only limited information was available about the formulation of the elastomer materials. Since the objective was to assess degradation within each base elastomer type, it was critical to identify the base elastomer formulation of each field sample. Portions of each sample were subjected to infrared spectrometry testing (FTIR). The resulting FTIR spectrum was compared to library spectrum of common elastomer materials. The unique spectra for EPDM, natural rubber, SBR, and nitrile (NBR) materials were easily identified. Distinguishing between EPDM-sulfur and EPDM-peroxide materials was more difficult because of their similar spectra. Supplemental laboratory work was completed to distinguish the differences between EPDM-sulfur and EPDM-peroxide by using the electron dispersion spectrometry (EDAX-EDS) and looking for elemental sulfur introduced in the curing process. By using the combined FTIR and EDAX-EDS test results, the field samples were identified as indicated in Figure 2.5. Interestingly, no EPDM-P samples were identified within our field returns. As EPDM-P is a relatively new product designed for specialty applications, it would not be commonly replaced by utilities.

While efforts were made to preserve the moisture content of the samples, some moisture change during transport was inevitable. Preliminary testing of the field samples indicated that there was a significant change of measurable parameters (weight, volume, hardness) because the field samples had dried during shipment to the laboratory. It was determined that if samples were tested in the "as received" condition, they would not be representative of in-service characteristics due to sample drying.

In order for the field data to more closely represent in-service conditions, the field samples were rehydrated in de-ionized water in an effort to return them to their normal in-service condition. Weight and volume changes were monitored during preliminary rehydration tests to determine typical patterns and the time necessary to achieve full sample rehydration. Testing indicated that rehydration is a continuing process and that some samples might not achieve equilibrium conditions after 180 days (Figure 2.6). However, it was observed that water absorption slowed dramatically and leveled off after 30 days of water absorption for all elastomer materials, as is evident in Figure 2.6. Therefore, a standard 30-day rehydration period was incorporated into subsequent testing. The various swelling behaviors previous to the 30-day leveling off point could have been due to the age of the samples, percentage of degradation at the initial stage or the material type.

Test Methods

Once rehydration was complete, the mass, volume, and hardness values of the field samples were recorded (Appendix B). (Note: Tensile and elongation data could not be obtained from the field samples because the tests required dumbbell coupons that could not be accurately cut from the gaskets, meter seals, valves, and other irregularly-shaped components.)

After the 30-day rehydration, the data obtained from similar field samples were combined to form a degradation curve for the field materials only. The test procedures are defined in the matrix shown in Figure 2.7.

Accelerated Laboratory Testing: pH Effects

Based on tacit evidence as assessed through conversations with various utilities, pH is believed to have an influence on chloramine-induced elastomer degradation. To assess the effects of pH, a test program was developed based on the most commonly occurring water system conditions. Water quality data was reviewed to assess the ranges of pH within typical water distribution

Table 2.9
Distribution of raw water quality by plant-month*

Parameter	Units	Minimum	Percentiles 10th	25th	50th	75th	90th	Maximum	Mean
Temperature	Degrees Celsius	0.6	7	11	16	22	26	43	16.3
pH	Standard units	5.2	6.7	7.2	7.7	8.1	8.3	14†	7.6
Alkalinity	mg/L as $CaCO_3$	1	12	24	80	121	168	459	82
TOC	Mg/L	0.70	1.4	2.0	2.7	3.8	5.3	25	3.1
UV absorption at 254 nm	1/cm	0.009	0.027	0.042	0.072	0.116	0.177	1.10	0.096
Bromide	Mg/L	0.020	0.020	0.020	0.027	0.054	0.119	28.0†	0.069
Hardness	mg/L as $CaCO_3$	1	8	20	62	113	168	486	75
Calcium	mg/L as $CaCO_3$	3	15	33	105	168	250	643	116
Ammonia	mg/L as N	0.0	0.0	0.0	0.1	0.1	0.3	2.3	0.13

Source: McGuire, McLain, and Obolensky 2002

* Total number of plant-months, N=2158

† Though the AUX1 database was subjected to a validation process, questionable outlier data, such as those indicated, remained in the database and were transferred to AUX8: this is the primary reason that records containing the highest absolute residuals were eliminated from the calibration process, as described under the subheading "Approach."

Table 2.10
Starting pH values

pH (starting value)	Monochloramine (percent)	Dichloramine (percent)
8.3	90	10
6.0	35	65

systems across the United States. As is shown in Table 2.9, 75 percent of the water utilities surveyed in 2002 have a pH greater than 7.2 (McGuire, McLain, and Obolensky 2002).

As discussed previously, when ammonia is added to chlorine, mono-, di- and trichloramine compounds form preferentially based on the pH and reactant ratio (Cl_2:NH_3). Figure 2.8 describes the distribution of the species as a function of pH, assuming a molar ratio (Cl_2:NH_3) of 2.0. A pH less than 6.0 would be desirable to maximize dichloramine distribution. Trichloramine is not included in the figure because only relatively small amounts of the compound are formed in typical water chemistries.

Because pH controls the formation of mono- and dichloramines, the pH effects study can more appropriately be described as assessing the effect of monochloramine-induced elastomer degradation and dichloramine-induced degradation. Thus, for the purpose of evaluating the pH effect on the six elastomers, a starting pH of 8.3 (preferential for monochloramine) was chosen. This provided an ultimate distribution of species as shown in Table 2.10. For dichloramine testing, a starting pH of 6.0 (preferential for dichloramine) was used.

When performing accelerated-aging tests, temperature is often used as an accelerant. However, when assessing the pH effects on elastomer degradation, elevated temperatures could not be included because the lower pH test solution was not stable over a 24-hour period. Figure 2.9 shows three initial 60 ppm chloramine solutions at a pH of 6.0. The solutions were maintained at 23°C, 45°C and 70°C for 24 hours, and pH values were recorded. Because a 6.0 pH could not be consistently maintained at higher temperatures, a 23°C and 60 ppm concentration was selected as the most suitable combination to assess pH effects on elastomer degradation.

Buffers

To achieve the high pH condition (monochloramine predominant), a commercially available phosphate buffer system was used as specified in the ASTM D-6284 test method. For the low pH condition (dichloramine predominant), a custom buffer solution was developed at 0.5 moles through the Henderson-Hasselbalch relationship, using the same phosphate acid-conjugate base system. A qualitative comparison with previous AwwaRF results (Reiber 1993a) showed general trend agreement.

Constant Strain Effects

Because elastomers have many applications in the water industry, the accelerated-aging test program assessed the effect of constant strain on long-term elastomer performance. The test program goal was to determine if elastomers under strain would have a different rate of degradation when exposed to elevated temperature and chlorine/chloramines concentrations.

The loop configuration was selected as the most suitable method to assess strain effects because a variety of strain conditions could be created within a single sample. When the loop is formed, the inside of the sample is in compression and the outside of the sample is in tension. In addition, the amount of strain on the sample varies from 0 to 25 percent, depending on the position around the strain loop. Varying strain conditions created by the loop provide a method to collect specific hardness data for low, medium, and high strain areas. The sample configuration during the strain testing is shown in Figure 2.10.

A test condition of 45°C temperature and 30 ppm average concentration was selected for the strain testing. Based on earlier (zero strain) testing of un-aged samples, it was concluded that more extreme temperature and concentration (such as 70°C, 30–60 ppm) would likely cause some elastomers to break during the 30-day test. Strain data was obtained from physical locations corresponding to "zero" strain, "medium" strain (from locations corresponding to 12 percent strain), and "high" strain (from locations corresponding to 25 percent strain). It must be noted, however, that as the elastomers swelled during testing, strain configuration changed, and the actual strain at these physical locations shifted. Thus, the zero, medium, and high strain designations represent *approximate* estimates of strain. The test method details are included in the appendix and Figure 2.11 shows the stainless steel test fixtures used in the submersion testing.

LABORATORY TESTING—SPECIAL CONSIDERATIONS

The research project evolved as various issues and concerns came to light. Many of these issues became determining factors in the final laboratory program. The following issues were significant factors in the development and completion of the final testing protocol.

Table 2.11
Alternative formulas for monochloramine

Method	$[Cl_2/NH_3]_{molar}$	$[Cl_2/NH_3]_{wt}$
ASTM	0.5	2
ALT	1	4

Test Solution Concentration and Molar Ratios

Forming the chloramine compounds in the laboratory is technically challenging. While there are several recipes available, it is important to select a formulation method that is consistent with what is standard in the water industry.

To form monochloramine (NH_2Cl) it is simply a matter of combining ammonia (NH_3) with hypochlorous acid (HOCl). However, as discussed previously, the products of this reaction vary with pH, contact time and the initial Cl_2 to NH_3 ratio. As described by Snoeyink and Jenkins (1980), the initial ratio of Cl_2/NH_3 can have a significant effect on the form of the aqueous chlorine produced (NH_2Cl or free chlorine) and the amount of total residual chlorine formed. As shown in Table 2.11, a molar ratio of Cl_2/NH_3 greater than 1.5 will preferentially form free chlorine as opposed to monochloramine. An initial molar ratio of 1.0 will yield the maximum residual chlorine (in the form of NH_2Cl). Initial Cl_2/NH_3 molar ratios less than 1.0 will have excess NH_3 and other forms of chlorine present. Most water utilities attempt to reduce the amount of free ammonia within their distribution systems due to potential nitrification (AWWA 2006) and because the effects of NH_3, or free nitrogen, on elastomer components are poorly understood.

Two separate monochloramine formulation recipes were considered for the test program: the ASTM D6284 method ("ASTM") and an alternative method ("ALT") based on input from participating utilities. The ASTM and ALT methods both yield monochloramine in the form of NH_2Cl; however, they have different Cl_2 to NH_3 molar ratios as identified in Table 2.11.

The initial molar ratios are shown plotted in Figure 2.12. Both formulas will preferentially form NH_2Cl as desired. However, the ASTM method has significantly more residual NH_3.

Since excess ammonia is not encouraged in distribution system operation, the "ALT" method may be more typical of a true distribution system. However, to maintain consistency with the ASTM reference procedure, the laboratory testing program used a 2:1 weight ratio.

Test Solution Concentration and "Average Value"

Maintaining a consistent chlorine or chloramines concentration throughout the testing program was essential. However, the degradation studies completed by the appliance industry and work performed by Reiber (1993a) have concluded that a substantial amount of free chlorine (in both chlorine and chloramine solutions) is fairly quickly driven out of solution, thus lowering the effective concentration.

Preliminary testing indicated that characterizing test solutions by initial concentration value did not provide sufficient detail to fully describe the chlorine or chloramines concentrations throughout the testing interval. Therefore, a concentration "weighted average" was determined to be the most effective terminology to express exposure concentration. The concentration decay rate is dependent on temperature, as is shown in Figure 2.13 for a monochloramine solution at 23°C, 45°C and 70°C temperatures. Thus, to maintain a 60 ppm weighted-average monochloramine

Table 2.12
Summary of starting/average concentrations

Temperature (°C)	Average concentration	Starting concentration (RT)	7.0 buffer cc/9 liters	NaOCl cc/9 liters	NH_4OH cc/9 liters	7.0 buffer cc/liter
"High" concentration		+/–5 ppm				
70	60 ppm	120 ppm	36	19.8	34.2	2–8 cc/L
45	60 ppm	70 ppm	36	12.6	19.8	2–8 cc/L
23	60 ppm	60 ppm	36	9.9	15.8	2–8 cc/L
"Mid" concentration						
70	30 ppm	60 ppm	36	9.9	17.1	2–8 cc/L
45	30 ppm	35 ppm	36	6.3	9.9	2–8 cc/L
23	30 ppm	30 ppm	36	5.0	7.9	2–8 cc/L
"Low" concentration						
70	1 ppm	1.50 ppm	36	0.25	0.43	2–8 cc/L
45	1 ppm	1.25 ppm	36	0.23	0.35	2–8 cc/L
23	1 ppm	1.20 ppm	36	0.20	0.32	2–8 cc/L

concentration through the 24-hour period, a unique starting concentration was required for each temperature. Due to the instability of the chlorine at high temperatures, a higher initial concentration was required to maintain the average.

The chloramine test solutions were replaced at 24-hour intervals, and the actual concentration varied during that period. The "saw tooth" graph below, Figure 2.14, shows the concentrations of chloramines during the 30-day testing period. A summary of the starting/average concentrations used for the various testing scenarios is shown in Table 2.12.

Crossover Effect

During accelerated testing of un-aged sample coupons at the 30 ppm and 60 ppm concentration levels, an unexpected trend in the data was encountered. For the parameters associated with swelling and softening (change in mass and change in volume), the 45°C results crossed the 70°C results at approximately the midpoint of the aging period. Counter to expectations, this indicated higher swelling values and greater loss of hardness with time for the elastomer exposed to 45°C than to 70°C. Figure 2.15 is a representative graph of the observed phenomena.

This effect was prominent in natural rubber, SBR, and neoprene. Nitrile (NBR) rubber exhibited the phenomenon only during 30 ppm tests but not during 60 ppm. Neither of the EPDM rubbers showed this effect during any of the testing. It was also observed that the tensile properties for all materials did not exhibit this behavior. As summarized in Table 2.13, only changes in mass, volume and, to a much lesser extent, hardness displayed this "crossover" behavior. This behavior was not detected at the 1 ppm concentration condition.

A retest was conducted at a chloramine concentration of 60 ppm for the four materials that exhibited the phenomenon—EPDM compounds were excluded. This retest confirmed the previous test results. A possible explanation for this behavior could be that both the volume change and mass change are determined from mass measurements. Mass measurements were

Table 2.13
Crossover effects

Elastomer	Change in mass	Change in volume	Hardness	Tensile strength	Elongation
Natural rubber	√√	√√	√√	X	X
SBR	√√	√√	X	X	X
Nitrile (NBR)	X	X	√	X	X
Neoprene	√√	√√	√√	X	X
EPDM-S	X	X	X	X	X
EPDM-P	X	X	X	X	X

√—exhibited crossover effect at 30 concentration only
√√—exhibited crossover effect at 30 and 60 ppm concentration
X—did not exhibit crossover under any conditions

done immediately as samples were removed from the test solution—no drying except to remove liquid water from the surface of the test specimen. The mass determination was therefore the mass of the sample and the mass of the solution absorbed. In the cases where crossover occurred, at elevated concentration and temperature, it is likely that the elastomer was undergoing degradation that involved loss of base material, while it continued to absorb test solution. Material loss will occur more rapidly at higher temperatures and concentrations. Therefore, although the mass of solution absorbed could have been increasing, the *total mass* measured could have been "flattening" or actually *decreasing* as shown in Figure 2.16.

It is the conclusion of the research team that this phenomenon is likely the result of testing at conditions *too severe* for accelerated aging of the elastomer type. Consequently, the data has not been used in the accelerated life projections.

Figure 2.1 Dog bone coupons in the laboratory are prepared to be submerged in the chloramine solution. Temperature will be raised using a water bath.

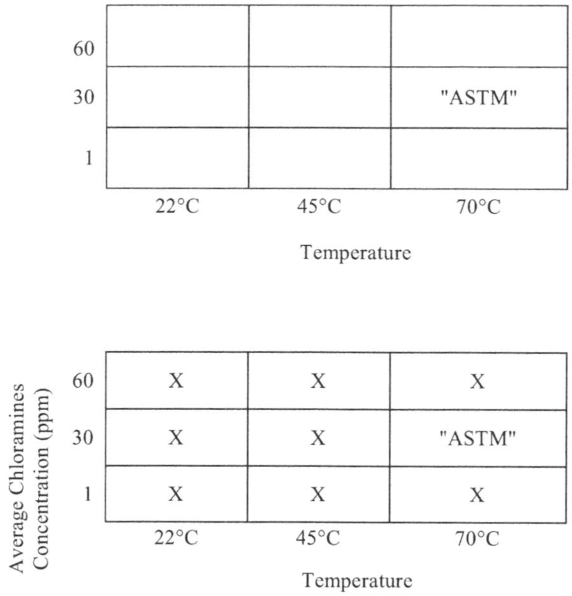

Figure 2.2 Accelerated-aging test schedule for un-aged samples

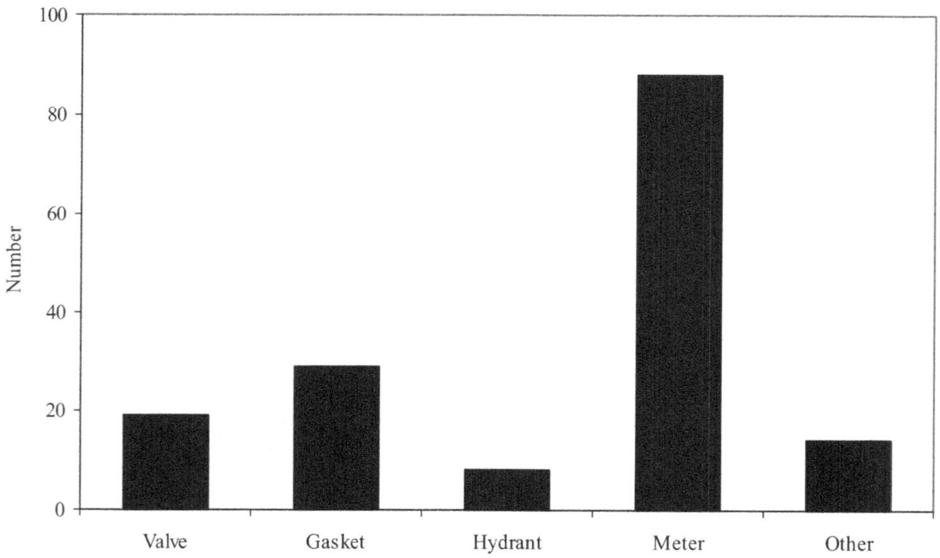

Figure 2.3 Types of samples returned from the field

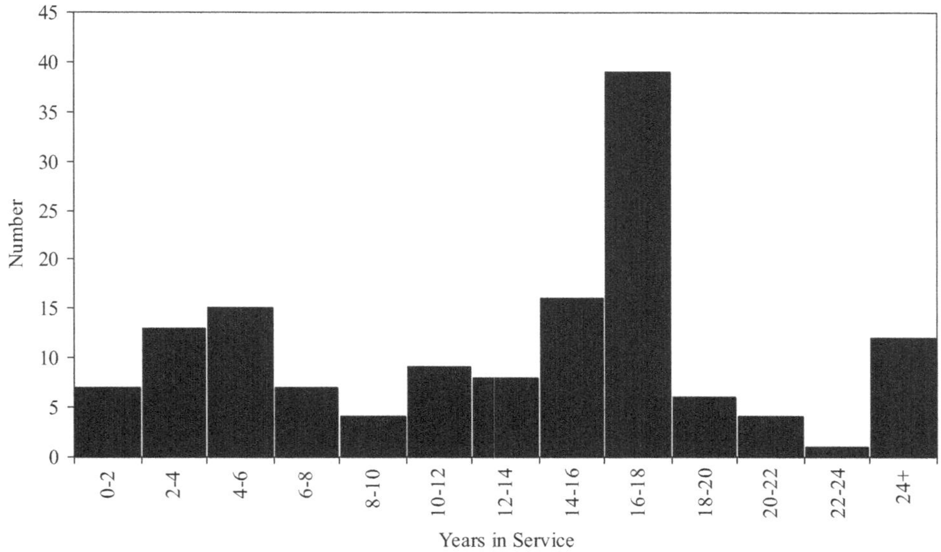

Figure 2.4 Number and ages of field samples

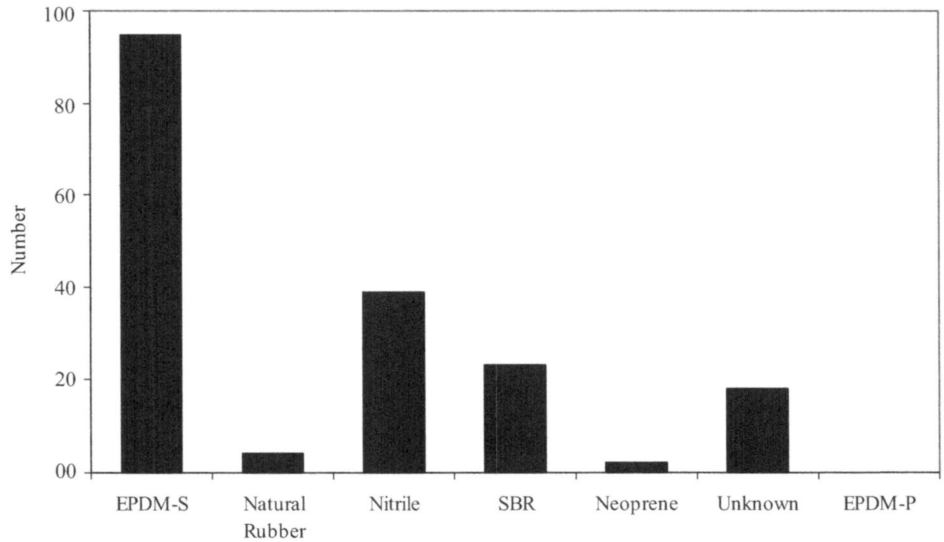

Figure 2.5 Field sample formulations identified

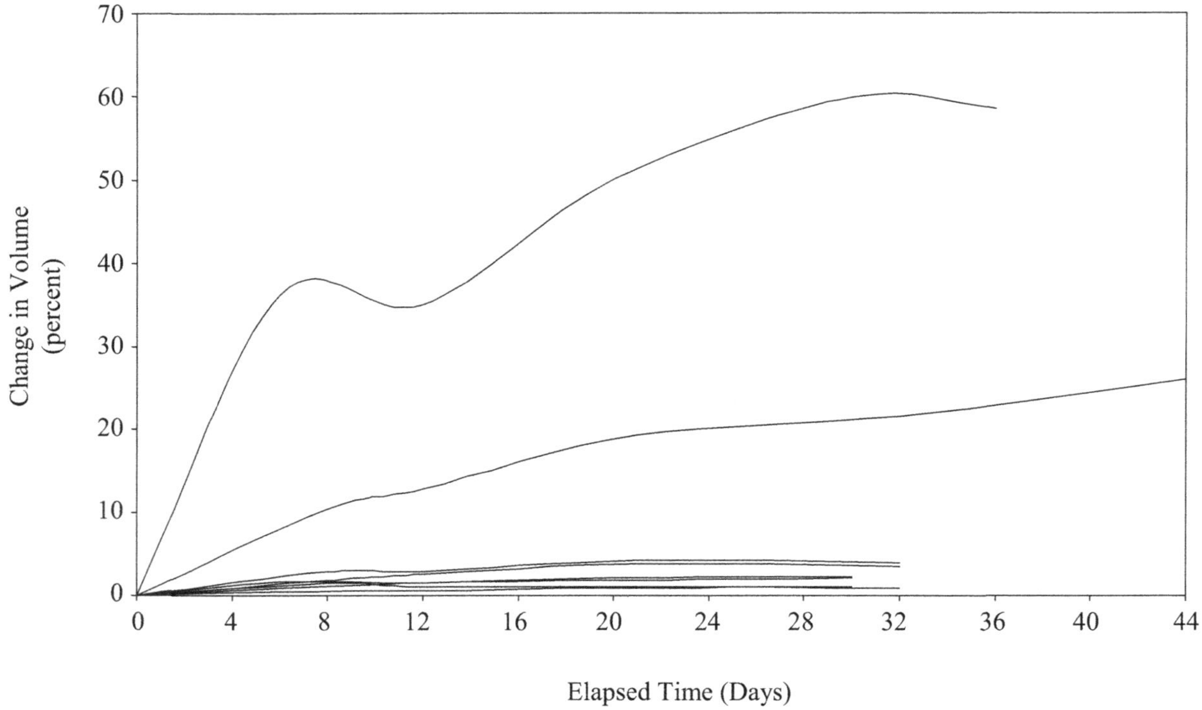

Figure 2.6 EPDM-S density of test samples during rehydration

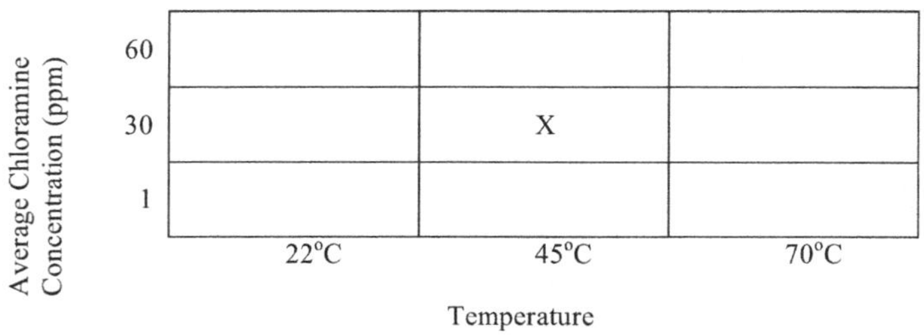

Figure 2.7 Accelerated testing schedule of field samples

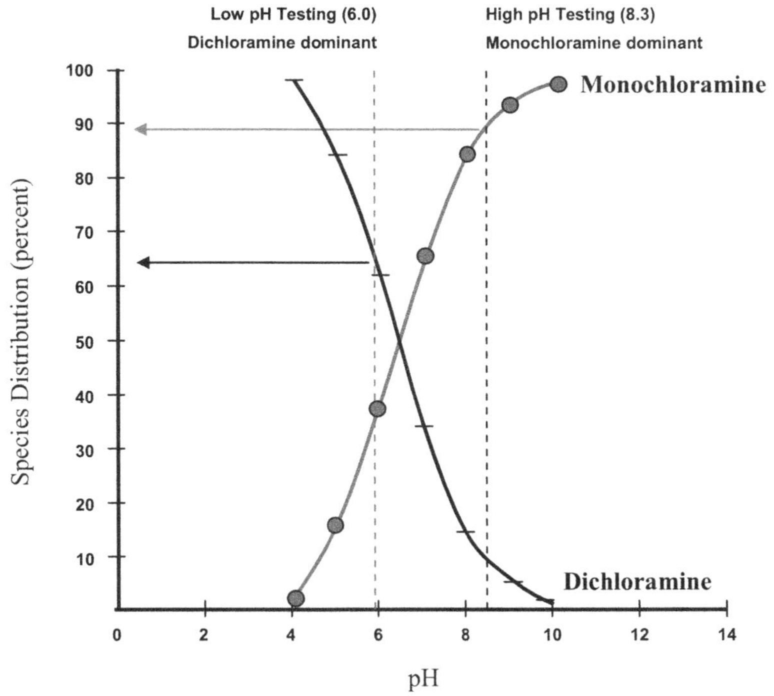

Figure 2.8 Distribution of species as a function of pH (Source: Reiber 1993a)

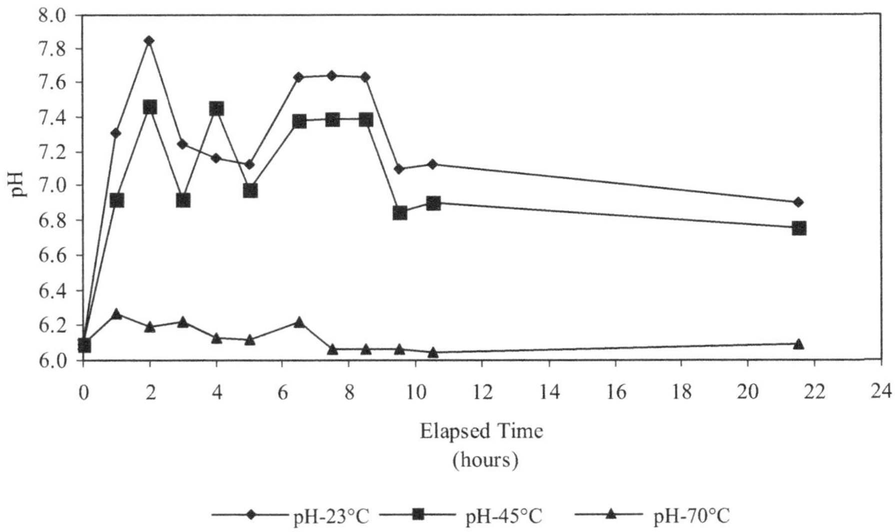

Figure 2.9 pH over 24-hour period at 23°C, 45°C, and 70°C

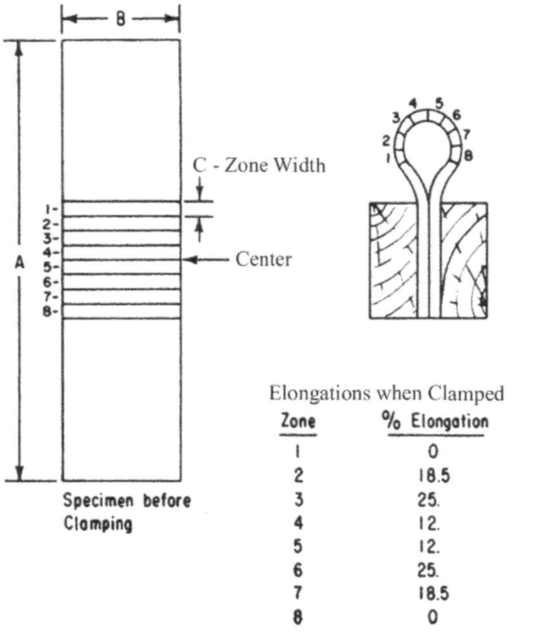

Elongations when Clamped

Zone	% Elongation
1	0
2	18.5
3	25.
4	12.
5	12.
6	25.
7	18.5
8	0

	A	B	C
mm	95	25	3.2
in.	3.75	- 1	0.125

Figure 2.10 Constant strain loop configuration (Source: ASTM 2003. D 518-99 *Standard Test Method for Rubber Deterioration—Surface Cracking*, copyright ASTM International, 100 Barr Harbor Drive, West Conshohocken, PA 19428. Reprinted with permission from ASTM International.)

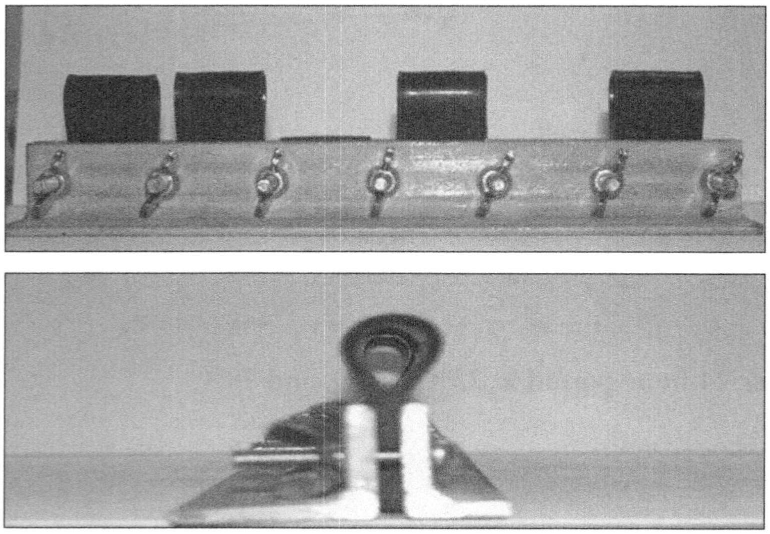

Figure 2.11 Test fixtures used in submersion testing

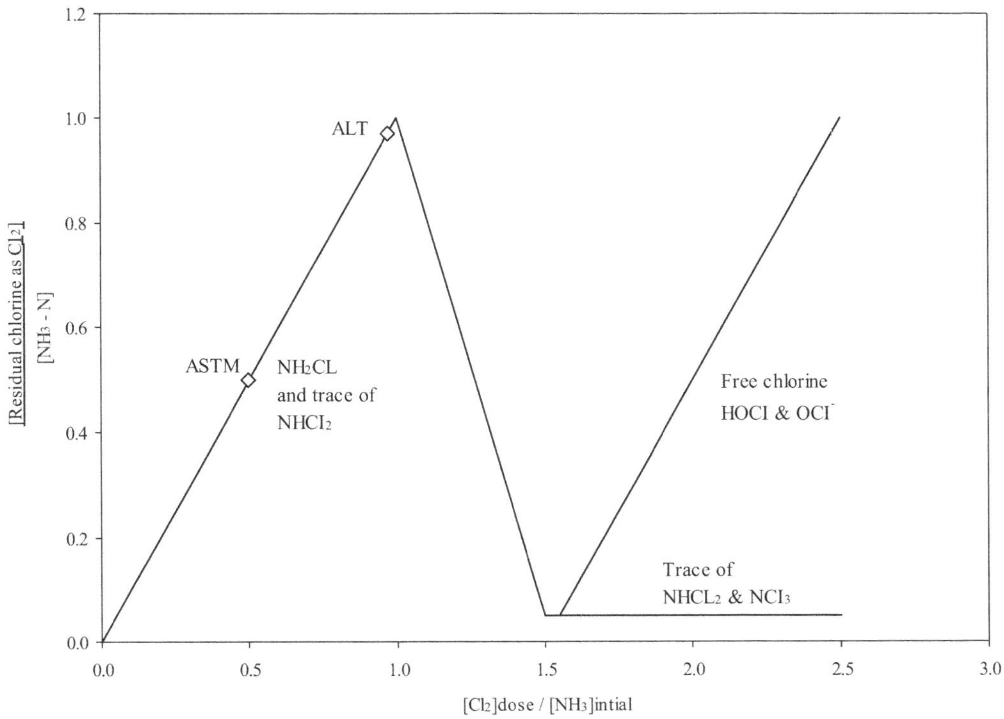

Figure 2.12 Initial molar ratios of the ASTM and ALT methods (Source: V.L. Snoeyink and D. Jenkins, *Water Chemistry*, copyright 1980, John Wiley & Sons, Inc., New York. Reprinted with permission of John Wiley & Sons, Inc.)

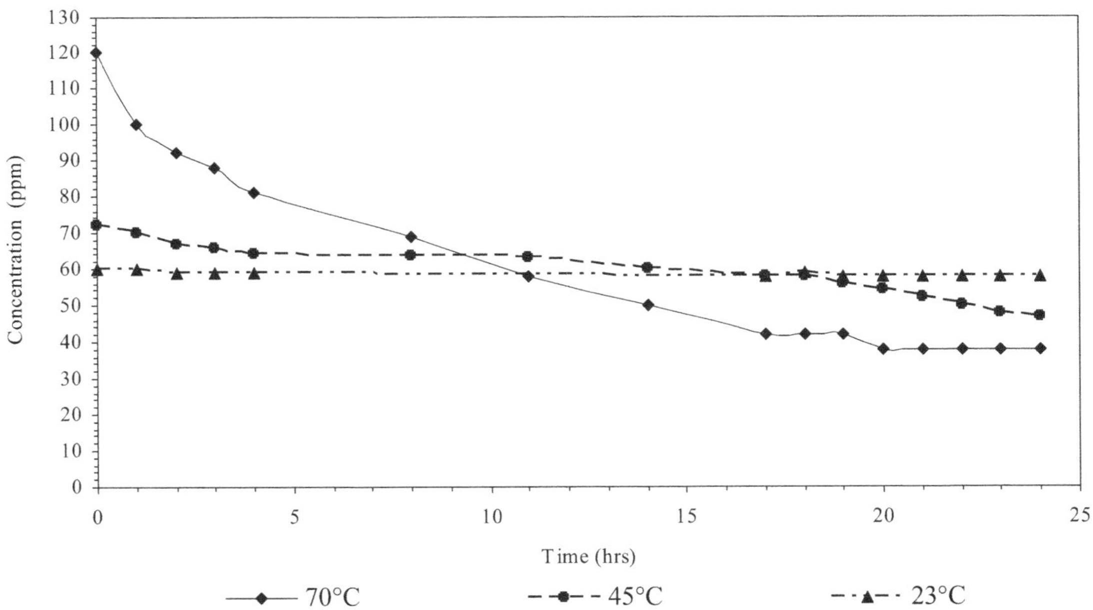

Figure 2.13 Total residual chlorine decay—24-hour interval

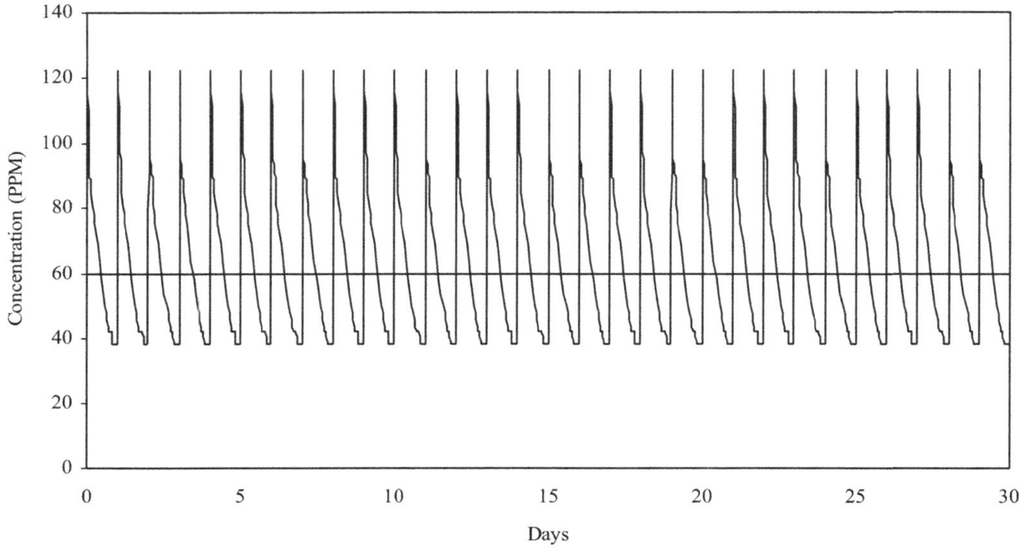

Figure 2.14 Total residual chlorine profile—30-day test

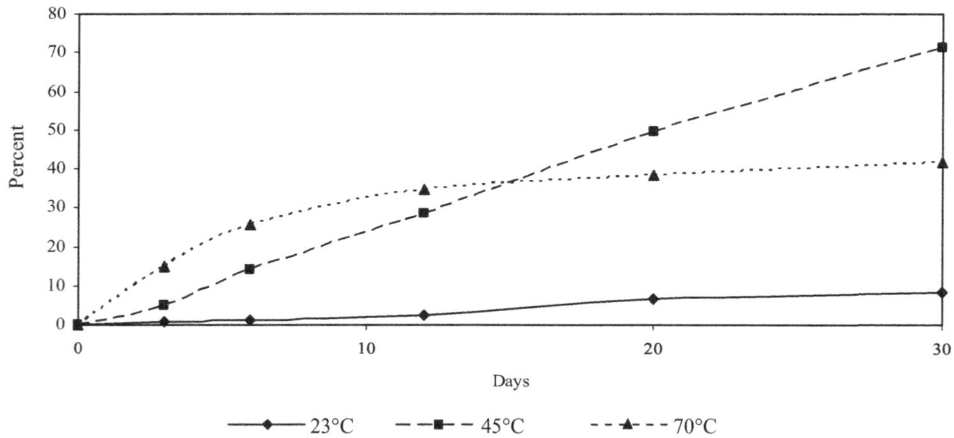

Figure 2.15 Crossover of 70°C curve over 45°C curve of change in mass data during a 30-day test at 30 ppm chloramine concentration

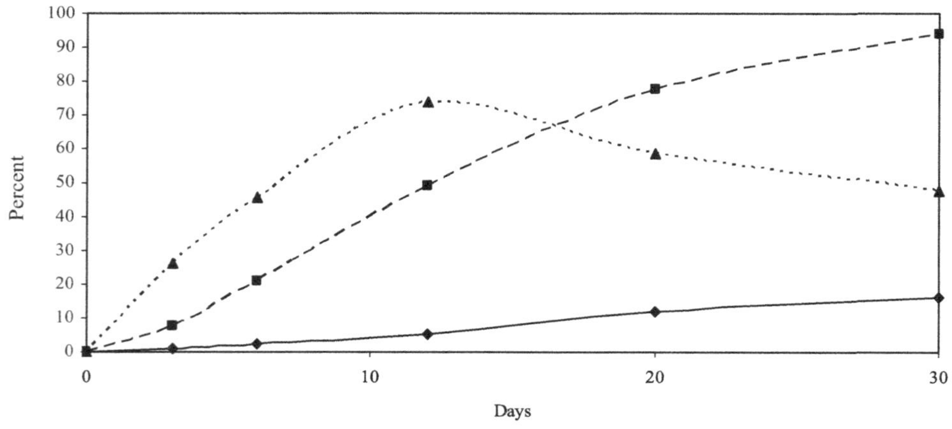

Figure 2.16 Crossover of 70°C curve over 45°C curve of change in volume data during a 30-day test at 60 ppm chloramine concentration

CHAPTER 3
30-DAY TEST RESULTS

Selecting the appropriate elastomer to use for a water utility application is a complex decision that depends on a variety of factors. When using an elastomer in a water system, it is important to consider the elastomer as an engineering material and to match the performance capabilities with the demands required for the particular application. For example, a specific elastomer may be suitable for a gate valve based on hardness performance; however, it may experience an unacceptable volume change that would adversely impact valve performance. The same material, however, may be perfectly suitable for a gasket where volume restrictions are not critical.

To assist utility personnel with selecting the appropriate elastomer for their components, a systematic testing program was created to measure the elastomers' performance with respect to common water utility applications. To achieve this, un-aged elastomer samples were immersed in a solution of water and chloramines, and performance tests were conducted after 3, 6, 12, 20, and 30 days of exposure. To accelerate the elastomer degradation (or aging), the tests were repeated at a variety of high temperatures and high concentrations.

The response variables used to monitor elastomer performance are standard test methods commonly used in the rubber industry. The tests represented a broad range of performance characteristics, including bulk property behavior and surface effects. Specifically, the bulk effects were characterized by weight and volume change, tensile strength and maximum elongation.

In addition to assessing the temperature and concentration effects on elastomer degradation, the degradation influence of chlorine, pH, and strain also were assessed. Throughout the testing, each exposure condition was varied independently for each elastomer. Additionally, it is noted that throughout the test program, the elastomers were segregated from one another, ensuring no possibility of "cross contamination." By varying each parameter independently and measuring the elastomer response with time, a database was populated that was subsequently used to predict long-term elastomer performance.

This chapter presents the individual effects of each parameter during chloramine-induced degradation.

TEMPERATURE EFFECTS

The influence of temperature on elastomer degradation was assessed by systematically blocking other exposure variables so that degradation parameters were only influenced by temperature. A summary of the of six un-aged sample test results is shown in Tables 3.1 and 3.2 for 1 ppm and 60 ppm chloramines concentration exposure and 8.3 pH. It is noted that the summary tables present the total percentage change of the measured performance parameter, at the 30-day test point. In addition, Figures 3.1 through 3.6 are typical plots showing performance versus time through the 30-day test period for one of the most widely used elastomer types, SBR. The temperature-induced degradation curves verses time for all six elastomers at 1 ppm, 30 ppm and 60 ppm concentrations are included in Appendix A.

The effects of temperature-induced degradation were unique with reference to the elastomer materials, exposure concentration and degradation parameter measured. Thus, depending on exposure conditions and test parameters, different materials performed better or worse, and there was not always a consistent preferential order of material performance. However, based on

Table 3.1
Data summary of percent change in property value after 30 days of exposure to chloramines at 1 ppm and pH = 8.3

23°C						
Property change	Natural	EPDM-S	NBR	SBR	Neoprene	EPDM-P
Δ Weight (%)	1	0.2	1	1	3	–0.1
Δ Volume (%)	1	–0.1	1	1	4	0
Δ Hardness (%)	1	2	–3	–4	8	1
Δ Tensile strength (%)	–9	–1	1	–6	–8	2
Δ Elongation (%)	–16	–13	–6	–36	–27	2

45°C						
Property change	Natural	EPDM-S	NBR	SBR	Neoprene	EPDM-P
Δ Weight (%)	4	0.4	2	3	13	0.2
Δ Volume (%)	4	–0.1	2	4	17	0.1
Δ Hardness (%)	0.2	4	–5	2	–10	1
Δ Tensile strength (%)	–17	–3	3	–10	–41	3
Δ Elongation (%)	–29	–24	–14	–43	–48	–2

70°C						
Property change	Natural	EPDM-S	NBR	SBR	Neoprene	EPDM-P
Δ Weight (%)	17	1	4	6	60	1
Δ Volume (%)	19	0.3	5	7	73	1
Δ Hardness (%)	–19	9	–5	2	–44	0
Δ Tensile strength (%)	–65	14	–1	–34	–82	–8
Δ Elongation (%)	–52	–30	–36	–56	–82	–9

the 30-day data shown in Tables 3.1 and 3.2, and the summary figures in the Appendix A, the following general assessments of elastomer behavior can be inferred.

At low exposure concentrations, the temperature effect on the elastomer degradation was less pronounced than for higher exposure concentrations. Of the materials evaluated, both EPDM compounds showed the least amount of performance change due to heat aging when exposed to the 1 ppm solutions. While some performance degradation was observed in the EPDM-S (sulfur cured) material, very little degradation was observed in the EPDM-P (peroxide cured). The SBR and nitrile (NBR) materials exhibited moderate changes in performance, and the neoprene and natural rubber experienced the highest performance degradation as a result of heat aging.

As can be seen in the curves in Appendix A and the SBR performance plots, under all test conditions and with all of the elastomers, performance change with respect to time tended to "saturate" or flatten; i.e., the rate of change of degradation tended to slow as time progressed in the 30-day tests. This provides insight into the expected shape of the "master curves" developed later

Table 3.2
Data summary of percent change in property value after 30 days of exposure to chloramines disinfectant at 60 ppm and pH = 8.3

23°C

Property change	Natural	EPDM-S	NBR	SBR	Neoprene	EPDM-P
Δ Weight (%)	16	14	26	7	12	–1
Δ Volume (%)	19	16	32	8	16	–1
Δ Hardness (%)	–7	–4	–10	–2	–0.5	3
Δ Tensile strength (%)	–31	–25	–40	–30	–40	–11
Δ Elongation (%)	–40	–51	–57	–51	–50	–20

45°C

Property change	Natural	EPDM-S	NBR	SBR	Neoprene	EPDM-P
Δ Weight (%)	73	32	38	85	82	3
Δ Volume (%)	85	36	46	103	94	2
Δ Hardness (%)	–45	–18	–15	–38	–44	5
Δ Tensile strength (%)	–70	–42	–67	–67	–68	–39
Δ Elongation (%)	–47	–54	–61	–68	–55	–41

70°C

Property change	Natural	EPDM-S	NBR	SBR	Neoprene	EPDM-P
Δ Weight (%)	53	70	86	78	107	80
Δ Volume (%)	61	58	105	75	48	94
Δ Hardness (%)	–22	–37	–11	–17	–31	–33
Δ Tensile strength (%)	–81	–63	–86	–87	–87	–72
Δ Elongation (%)	–65	–74	–85	–93	–83	–60

in the project to predict elastomer life. It is also interesting to note that during the low concentration testing, the most sensitive response parameter was the reduction in elongation (at failure).

At the highest concentration (60 ppm), large reductions in performance due to temperature effects were measured in all elastomer types across all performance parameters. While temperature effects appeared to be amplified at high concentrations, EPDM-P exhibited the least degradation, particularly with respect to swelling and change in hardness. However, for EPDM-P materials, there did appear to be a threshold that was not observed at low concentrations. At 45°C and at 60 ppm concentration, all performance parameters were greatly reduced, with bulk parameters changing over 50 percent, and hardness dropping over 30 percent. As can be seen from the tabulated data and the aging curves in Appendix A, temperature-induced degradation effects were highly non-linear. For example, EPDM-P at high concentration experienced relatively small changes in swelling and hardness performance as temperature increased from 23°C to 45°C. But much larger performance changes were observed when temperature increased from 45°C to 70°C.

While most materials followed anticipated trends of increasing degradation with increasing temperature across all concentration levels, some anomalies were observed. For the EPDM-S material, an unexpected 14 percent increase in tensile strength was measured after the 30-day exposure in low concentration at 70°C. In addition to tensile strength improvement, hardness increased (2–9 percent) as temperature was increased. This gain in hardness was in contrast to the dramatic softening measured at the high concentrations.

One explanation for the improved EPDM-S elastomer performance at the higher temperatures is that the additional heat enabled the formation of new cross-links (incomplete curing). To assess the state of curing, the samples were post-cured and "glass transition temperatures" were measured on a differential scanning calorimeter. These values were compared to typical materials commonly used in the water industry. No significant difference was observed between the laboratory samples and typical materials, indicating that the samples were adequately cured. However it is known that some cross-links do continue to form post-vulcanization, and this formation of new cross-links is in opposition to cross-link breakage due to chloramines exposure. Thus, it is possible that EPDM-S materials experienced a net increase in cross-links at high temperature and low concentration, and a net decrease at high temperature and high concentration.

SBR and nitrile (NBR) elastomers showed very little temperature-induced degradation at low concentrations for swelling and loss of hardness. At higher concentrations (30 ppm and 60 ppm) SBR samples exhibited an unusual pattern in weight and volume change over the 30-day test. During the 30-day exposure period, the performance of the SBR at 45°C appeared worse than the performance at 70°C. This "crossover" effect is believed to be an artifact of the test program, driven by severe deterioration (loss of material) under high concentration and temperature. This "crossover" effect is discussed in more detail in Chapter 2.

Natural rubber and neoprene samples generally followed the anticipated trends for temperature changes. Degradation of these elastomers was more severe than the other materials tested. As can be seen in the data summary tables and Appendix A, all performance parameters for these materials showed large degradation.

CONCENTRATION EFFECTS

The same data used to assess temperature effects also can be used to assess the effects of increased concentration on elastomer degradation. The data has been reconfigured in Table 3.3 to show the effects of changes in chloramines concentration on the elastomer samples at constant temperature. Table 3.3 shows changes in material performance due to increasing temperatures after 30-days exposure at room temperature. Table 3.4 shows similar information for changes in material performance due to increasing temperatures after 30-days exposure at 70°C. In addition, typical plots showing performance versus time through the 30-day test period is shown for SBR. A complete set of curves showing concentration effects for all elastomers and all test conditions is shown in Appendix A.

The effects of increased chloramine exposure were reasonably consistent across all temperature levels for the six elastomer materials. In general, as the exposure concentration increased, a greater amount of elastomer degradation was observed during the test window. This trend appeared to be consistent through all performance parameters tested and all elastomer materials. Thus, based on the 30-day data shown in Tables 3.3 and 3.4, and the summary figures in the Appendix A, the following general assessments of elastomer behavior can be inferred.

Table 3.3
Summary table of percent change in elastomeric property after 30 days of exposure to chloramine disinfectant at 23°C and pH = 8.3

1 ppm						
Property change	Natural	EPDM-S	NBR	SBR	Neoprene	EPDM-P
Δ Weight (%)	1	0.2	1	1	3	–0.1
Δ Volume (%)	1	–0.1	1	1	4	0
Δ Hardness (%)	1	2	–3	–4	8	1
Δ Tensile strength (%)	–9	–1	1	–6	–8	2
Δ Elongation (%)	–16	–13	–6	–36	–27	2

30 ppm						
Property change	Natural	EPDM-S	NBR	SBR	Neoprene	EPDM-P
Δ Weight (%)	7	5	15	3	7	–0.1
Δ Volume (%)	8	6	18	4	9	–0.3
Δ Hardness (%)	–3	–0.5	–8	–2	4	3
Δ Tensile strength (%)	–30	–21	–20	–13	–28	–11
Δ Elongation (%)	–35	–36	–33	–35	–38	–10

60 ppm						
Property change	Natural	EPDM-S	NBR	SBR	Neoprene	EPDM-P
Δ Weight (%)	16	14	26	7	12	–1
Δ Volume (%)	19	16	32	8	16	–1
Δ Hardness (%)	–7	–4	–10	–2	–0.5	3
Δ Tensile strength (%)	–31	–25	–40	–30	–40	–11
Δ Elongation (%)	–40	–51	–57	–51	–50	–20

Based on the testing data obtained, the effects of increasing chloramines concentration is less pronounced on EPDM-P and SBR. At room temperature, both showed strong resistance to swelling and change in hardness. At elevated temperatures, all elastomers showed large degradation when exposed to high chloramine concentrations. However, again under all test conditions and with all of the elastomers, performance change with respect to time tended to "saturate" or flatten; i.e., the rate of change of degradation tended to slow as time progressed in the 30-day tests. This phenomenon is illustrated in Figures 3.7 through 3.12.

pH EFFECTS

In addition to temperature and concentration effects, previous research indicates that pH also may be a contributing factor when assessing elastomer degradation (Reiber 1993a). A portion of the testing program was specifically designed to assess the pH effects on elastomer degradation

Table 3.4
Summary table of percent change in elastomeric property after 30 days of exposure to chloramines disinfectant at 70°C and pH = 8.3

1 ppm						
Property change	Natural	EPDM-S	NBR	SBR	Neoprene	EPDM-P
Δ Weight (%)	17	1	4	6	60	1
Δ Volume (%)	19	0.3	5	7	73	1
Δ Hardness (%)	–19	9	–5	2	–44	0
Δ Tensile strength (%)	–65	14	–1	–34	82	–8
Δ Elongation (%)	–52	–30	–36	–56	–82	–9

30 ppm						
Property change	Natural	EPDM-S	NBR	SBR	Neoprene	EPDM-P
Δ Weight (%)	36	31	30	42	88	57
Δ Volume (%)	42	37	37	51	123	64
Δ Hardness (%)	–24	–12	–8	–18	–41	–30
Δ Tensile strength (%)	–78	–50	–67	–74	–81	–66
Δ Elongation (%)	–61	–63	–64	–72	–75	–51

60 ppm						
Property change	Natural	EPDM-S	NBR	SBR	Neoprene	EPDM-P
Δ Weight (%)	53	70	86	78	107	80
Δ Volume (%)	61	58	105	75	48	94
Δ Hardness (%)	–22	–37	–11	–17	–31	–33
Δ Tensile strength (%)	–81	–63	–86	–87	–87	–72
Δ Elongation (%)	–65	–74	–85	–93	–83	–60

for the six elastomer compounds. Results of the testing are shown in Table 3.5. In addition, the SBR figures below are representative plots of performance changes with time during the 30-day test program. Plots for all of the elastomers are shown in Appendix A.

In general, 30-day testing of un-aged samples under the specified test conditions showed that low pH, (dichloramine species predominant), resulted in performance degradation higher than the corresponding monochloramine condition (Figures 3.13 and 3.14). It should be noted that the differences were not extremely large, generally less than 10 percentage points and may be within typical elastomer sampling variability. Variation was observed within the performance parameters; except for EPDM-S, all elastomer materials tended to exhibit higher swelling (weight and volume increase) in the dichloramine environment. All elastomers exhibited more softening and loss of strength as well. SBR and natural rubber tended to be more sensitive to pH than the other elastomers tested. However, EPDM-P showed a relatively large degradation in tensile strength (39 percent) and in elongation (22 percent) when pH was lowered and the environment changed from monochloramine to dichloramine.

Table 3.5
Data summary of percent change in performance after 30 days of exposure to 60 ppm disinfectant in solution at 23°C

pH 6.0—Di-chloramine predominant						
Property change	Natural	EPDM-S	NBR	SBR	Neoprene	EPDM-P
Δ Weight (%)	23	13	23	14	14	3
Δ Volume (%)	26	15	26	16	19	3
Δ Hardness (%)	–13	–8	–13	–8	–9	1
Δ Tensile strength (%)	–42	–30	–42	–47	–40	–50
Δ Elongation (%)	–42	–43	–42	–48	–44	–42

pH 8.3—Monochloramine predominant						
Property change	Natural	EPDM–S	NBR	SBR	Neoprene	EPDM–P
Δ Weight (%)	16	14	26	7	12	–1
Δ Volume (%)	19	16	32	8	16	–1
Δ Hardness (%)	–7	–4	–10	–2	–0.5	3
Δ Tensile strength (%)	–31	–25	–40	–30	–40	–11
Δ Elongation (%)	–40	–51	–57	–51	–50	–20

EFFECT OF CHLORINE VS. CHLORAMINES

Tests were performed to determine the relative impact of chlorine versus chloramines on performance. A set of tests was run at fairly severe test conditions, 70°C and 30 ppm and high pH, to contrast performance with monochloramine predominant. (This is roughly equivalent to the ASTM test condition described in ASTM D 6284.) In addition, tests were run at low pH to check the effect in a dichloramine-predominant solution. The 30-day results are summarized in Tables 3.6 and 3.7. In addition, performance versus time data for SBR are shown for the entire 30-day test period. A complete set of performance curves versus time is in Appendix A.

As shown in Table 3.6, elastomer materials exposed to monochloramine under the test conditions specified were observed to have a much greater swelling and loss in hardness than those exposed to chlorine at the same conditions. Tensile performance loss also was much worse in monochloramine than chlorine, as exhibited by SBR in Figures 3.15 through 3.17. The difference in performance between the two environments is large (10 to 50 percent) for all of the elastomers with the exception natural rubber, which showed a change in performance in the range of 1 to 10 percent.

As shown in Table 3.7, the materials exposed to dichloramine also degraded substantially more than those in chlorine under the same conditions. However, the relative differences between the amount of monochloramine degradation and the amount of dichloramine degradation cannot be compared directly because the test temperature was lower (23°C vs. 70°C), and the concentration was higher. However, the trends are generally consistent in that both monochloramine and dichloramine more aggressively attack elastomer materials than chlorine (Figures 3.18 and 3.19). There are variations within the results. Natural rubber showed a much larger performance difference in all measured parameters, and EPDM-P had a much smaller difference in swelling and loss of hardness.

Table 3.6
Data summary of percent performance change after 30 days, chlorine vs. chloramines, at 70°C and 30 ppm (pH 8.3—monochloramine predominant)

Chlorine						
Property change	Natural	EPDM-S	NBR	SBR	Neoprene	EPDM-P
Δ Weight (%)	30	8	13	23	79	6
Δ Volume (%)	33	8	16	26	109	6
Δ Shore A hardness (%)	–23	–1	–6	2	–39	–2
Δ Tensile strength (%)	–77	10	–18	–32	–85	–26
Δ Elongation (%)	–61	–31	–40	–59	–81	–25

Monochloramine						
Property change	Natural	EPDM–S	NBR	SBR	Neoprene	EPDM–P
Δ Weight (%)	36	31	30	42	88	57
Δ Volume (%)	42	37	37	51	123	64
Δ Hardness (%)	–24	–12	–8	–18	–41	–30
Δ Tensile strength (%)	–78	–50	–67	–74	–81	–66
Δ Elongation (%)	–61	–63	–64	–72	–75	–51

Table 3.7
Data summary of percent performance change after 30 days, chlorine vs. chloramines, at 23°C and 60 ppm (pH 6.0—dichloramine predominant)

Chlorine						
Property change	Natural	EPDM-S	NBR	SBR	Neoprene	EPDM-P
Δ Weight (%)	3	4	3	3	4	–1
Δ Volume (%)	3	5	4	3	6	–1
Δ Hardness (%)	–1	–7	–9	–8	–3	2
Δ Tensile strength (%)	–13	2	–13	–15	–39	–39
Δ Elongation (%)	–7	0.1	–5	4	–35	–22

Di-chloramine						
Property change	Natural	EPDM–S	NBR	SBR	Neoprene	EPDM–P
Δ Weight (%)	23	13	23	14	14	3
Δ Volume (%)	26	15	26	16	19	3
Δ Hardness (%)	–13	–8	–13	–8	–9	1
Δ Tensile strength (%)	–42	–30	–42	–47	–40	–50
Δ Elongation (%)	–42	–43	–42	–48	–44	–42

Table 3.8
Data summary of percent change in performance after 30 days, loop strain vs. no strain, at 45°C and 30 ppm (pH 8.3)

Property change	System	Natural	EPDM-S	NBR	SBR	Neoprene	EPDM-P
Δ Weight (%)	No strain	61	16	19	56	58	4
	Loop strain	35	12	11	33	30	1
Δ Volume (%)	No strain	71	18	24	69	81	4
	Loop strain	38	9	11	36	36	–3
Δ Hardness (%)	No strain	–44	–5	–13	–36	–40	1
Tensile	Loop strain	–72	–47	–34	–57	–67	–28
Compression	Loop strain	–68	–53	–60	–58	–68	–34
Δ Tensile	No strain	–50	–38	–41	–48	–61	–54
Strength (%)	Loop strain	–34	–16	–6	–27	–56	–42
Δ Elongation (%)	No strain	–26	–41	–25	–44	–43	–41
	Loop strain	–32	–28	–9	–31	–39	–26

LOOP STRAIN EFFECTS

The purpose of the strain loop testing was to determine if elastomers under strain would have a different rate of degradation than unstrained samples exposed to identical test conditions. This data could then be used to determine the effect of strain on life expectancy of elastomeric materials within a distribution system. To assess strain influence, identical samples were immersed in elevated concentration and temperature chloramine solutions, but held in a loop configuration. The loop induced tension on the outside of the elastomer and a compression on the inside as described in Chapter 2. The results of the loop testing are shown in Table 3.8 and 3.9. In addition, performance-versus-time curves for the 30-day test period are shown for SBR. A complete set of curves for all of the elastomers is included in Appendix A.

As shown in Table 3.8, loop strain had a measurable impact on performance of the elastomers tested. The presence of loop strain actually *reduced swelling* (less weight and volume change) and *improved tensile performance* for all of the elastomers. Conversely, the presence of loop strain dramatically increased softening (loss of surface hardness), typically 20 to 40 percent. The EPDM compounds showed the least degradation under these conditions, followed by nitrile (NBR). SBR, neoprene, and natural rubber showed the greatest performance change. In addition, for all of the elastomer types except natural rubber, the softening was worse on the compression (inside) surface of the loop versus the tensile (outside) surface. This may be an artifact of the testing method whereby the sample is pressed flat for the hardness testing.

The effect of chlorine versus chloramines in the presence of loop strain is summarized in Table 3.9. In general, the elastomers showed a substantially smaller performance loss in chlorine than in chloramines (Figures 3.20 through 3.22). The exception is EPDM-P, which performed relatively well overall but showed a small (4 percent) increase in swelling and hardness in the chlorine disinfectant versus the chloramine disinfectant.

Table 3.9
Data summary of percent change with loop strain at 30 days, chloramine vs. chlorine, at 45°C and 30 ppm (pH 8.3)

Property change	System	Natural	EPDM-S	NBR	SBR	Neoprene	EPDM-P
Δ Weight (%)	NH_2Cl	35	12	11	33	30	1
	Cl	10	10	3	8	16	5
Δ Volume (%)	NH_2Cl	38	9	11	36	36	–3
	Cl	11	12	3	10	23	5
Δ Hardness (%)	NH_2Cl	–72	–47	–34	–57	–67	–28
Tensile	Cl	–49	–40	–30	–34	–55	–32
Δ Hardness (%)	NH_2Cl	–68	–53	–60	–58	–68	–34
Compression	Cl	–43	–41	–36	–43	–54	–38
Δ Tensile	NH_2Cl	–50	–38	–41	–48	–61	–54
Strength (%)	Cl	–34	–16	–6	–27	–56	–42
Δ Elongation (%)	NH_2Cl	–26	–41	–25	–44	–43	–41
	Cl	–32	–28	–9	–31	–39	–26

TURBIDITY

Turbidity is a measure of the degree to which light is scattered by suspended particulate material. During accelerated-aging testing, material is sloughed off into the test solution as the elastomer degrades, affecting the turbidity of the solution. This loss of material can be an indicator of chemical attack, because as materials are extracted from the elastomer, they will remain in solution in the contact fluid. Measuring turbidity of the test solution on a daily basis and on a cumulative basis provides a qualitative estimate of the extent of chemical attack on rubber.

Results of cumulative turbidity are shown in Tables 3.10 and 3.11 for various test temperatures and concentrations of 30 and 60 ppm chloramines. Each of the six elastomers was aged in separate containers, which allowed measuring the turbidity of the respective test solutions at all the testing conditions. Results are in standard NTU units.

Consistent with the other findings discussed above, the turbidity measurements from EPDM-P indicated the least amount of material extracted from the elastomer during the aging tests. Daily turbidity data are shown for this elastomer in Figure 3.23 at a test concentration of 30 ppm chloramines, with daily readings lower than 2 NTU for the entire 30-day test.

As shown in Figures 3.24 and 3.25, EPDM-S and SBR had daily turbidity readings approximately 10 NTU or less. As shown in Figure 3.26, nitrile (NBR) rubber test solutions had low turbidity at the 23°C and 45°C, but the highest turbidity (79.8 NTU) at 70°C at the end of 30-day testing.

Consistent with other results cited earlier, natural rubber performed poorly in turbidity tests and had the second-highest turbidity readings (Figure 3.27). The natural rubber test solution was a hazy brown color.

Neoprene testing resulted in a solution that had particles of the elastomer material visible in it (Figure 3.28). The highest turbidity measured with neoprene in the test solutions was 17.2 NTU at 70°C and 30 ppm chloramine concentration.

Table 3.10
Cumulative turbidity for the 30-day test period at 30 ppm chloramine and various test temperatures

23°C—Chloramine						
Day	Natural	EPDM-S	NBR	SBR	Neoprene	EPDM-P
0	0.3	0.3	0.3	0.3	0.3	0.3
3	1.5	1.8	1.4	1.6	7.2	0.6
6	0.7	1.5	0.7	0.7	2.6	0.5
12	0.5	0.8	0.5	0.7	2.7	0.4
20	0.5	0.5	0.4	0.5	1.1	0.4
30	1.1	0.5	0.4	0.6	0.9	0.4

45°C—Chloramine						
Day	Natural	EPDM-S	NBR	SBR	Neoprene	EPDM-P
0	0.3	0.3	0.3	0.3	0.3	0.3
3	1.0	4.6	6.3	2.0	6.0	0.8
6	0.9	2.8	4.4	1.6	5.8	0.5
12	1.0	10.4	2.7	1.2	6.1	0.7
20	1.0	5.2	2.8	0.7	4.0	0.4
30	4.7	2.4	2.2	1.4	4.6	0.7

70°C—Chloramine						
Day	Natural	EPDM-S	NBR	SBR	Neoprene	EPDM-P
0	0.3	0.3	0.3	0.3	0.3	0.3
3	13.2	21.9	25.4	8.9	17.2	1.1
6	7.9	12.6	16.2	6.6	10.7	1.1
12	6.2	17.3	15.7	3.9	6.2	1.0
20	2.9	16.3	5.2	2.8	2.5	0.6
30	8.6	11.2	28.9	3.7	4.0	0.7

Cumulative turbidity of the test solutions also was recorded in order to get more insights into the behavior of the degrading elastomers. The sample used to record turbidity daily was stored in an amber bottle, and the turbidity of the stored sample was recorded on days 3, 6, 12, 20, and 30 during the testing. A mathematical average of the daily turbidity and the experimental value were compared to understand the aging phenomenon in terms of turbidity. The conclusion reached by the research team was that daily turbidity values were more useful than the cumulative turbidity and tended to follow the trends of other properties measured during the accelerated-aging tests.

Table 3.11
Cumulative turbidity for the 30-day test period at 60 ppm chloramine and various test temperatures

			23°C—Chloramine			
Day	Natural	EPDM-S	NBR	SBR	Neoprene	EPDM-P
0	0.3	0.3	0.3	0.3	0.3	0.3
3	0.4	0.4	0.5	0.4	3.5	0.4
6	0.5	0.6	0.4	0.5	4.0	0.6
12	0.4	0.8	0.5	0.5	1.8	0.5
20	0.5	1.1	0.6	0.6	1.3	0.6
30	0.5	0.7	0.7	0.5	1.3	0.5
			45°C—Chloramine			
Day	Natural	EPDM-S	NBR	SBR	Neoprene	EPDM-P
0	0.3	0.3	0.3	0.3	0.3	0.3
3	0.5	1.8	9.2	1.0	5.2	0.6
6	0.6	3.5	5.2	1.3	5.4	0.9
12	0.7	1.9	3.0	1.3	4.0	0.5
20	0.8	1.5	2.2	0.9	2.6	0.5
30	1.1	0.9	1.4	0.8	2.5	0.4
			70°C—Chloramine			
Day	Natural	EPDM-S	NBR	SBR	Neoprene	EPDM-P
0	0.3	0.3	0.3	0.3	0.3	0.3
3	1.1	5.9	2.4	2.0	4.9	0.4
6	5.4	8.0	11.2	7.9	4.9	1.1
12	3.5	3.9	6.3	4.2	5.6	0.9
20	3.1	2.5	4.6	3.1	4.6	0.6
30	2.7	2.5	8.0	3.8	3.0	1.0

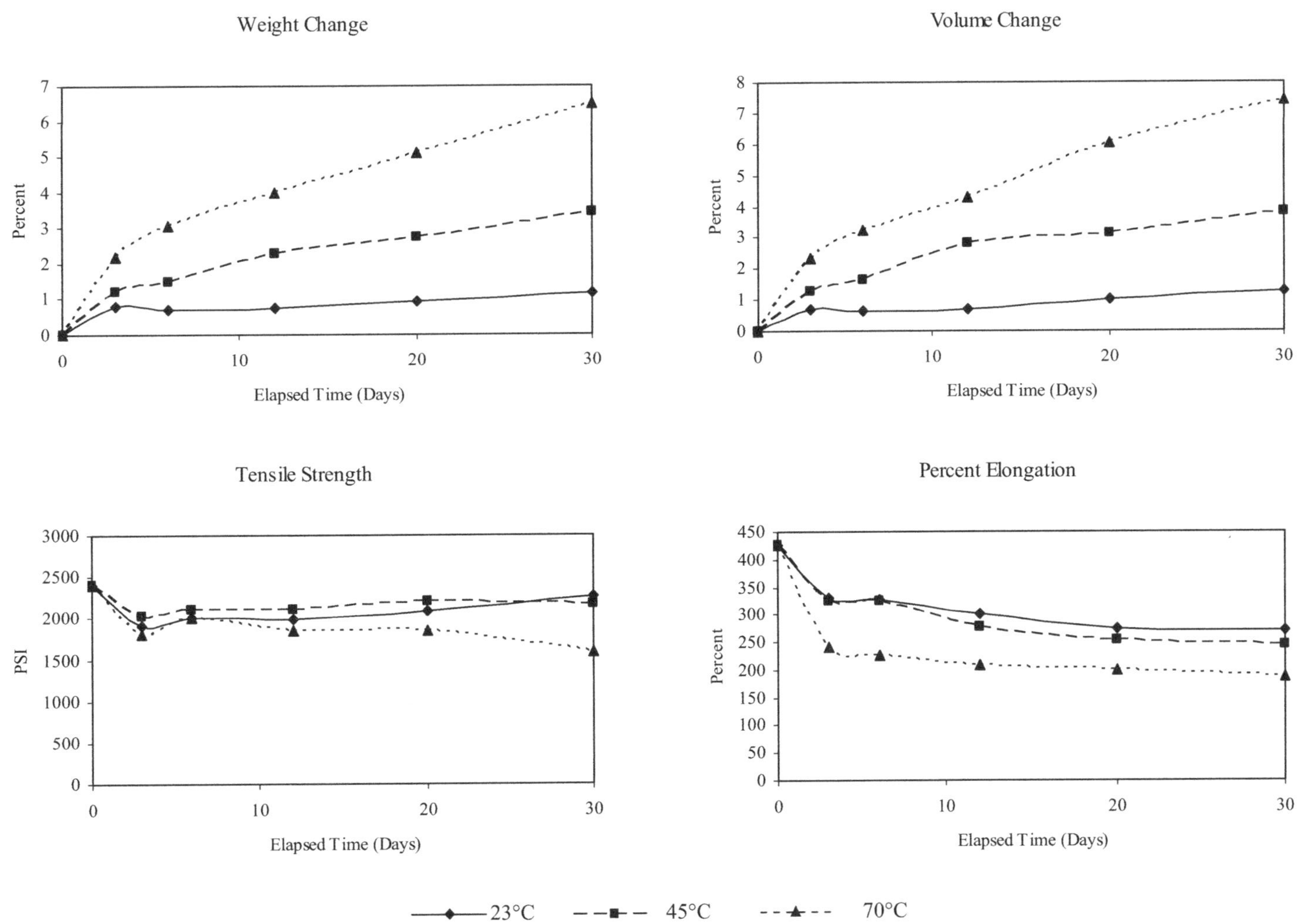

Figure 3.1 Temperature effect on elastomeric properties at 1 ppm chloramine and pH 8.3 of SBR

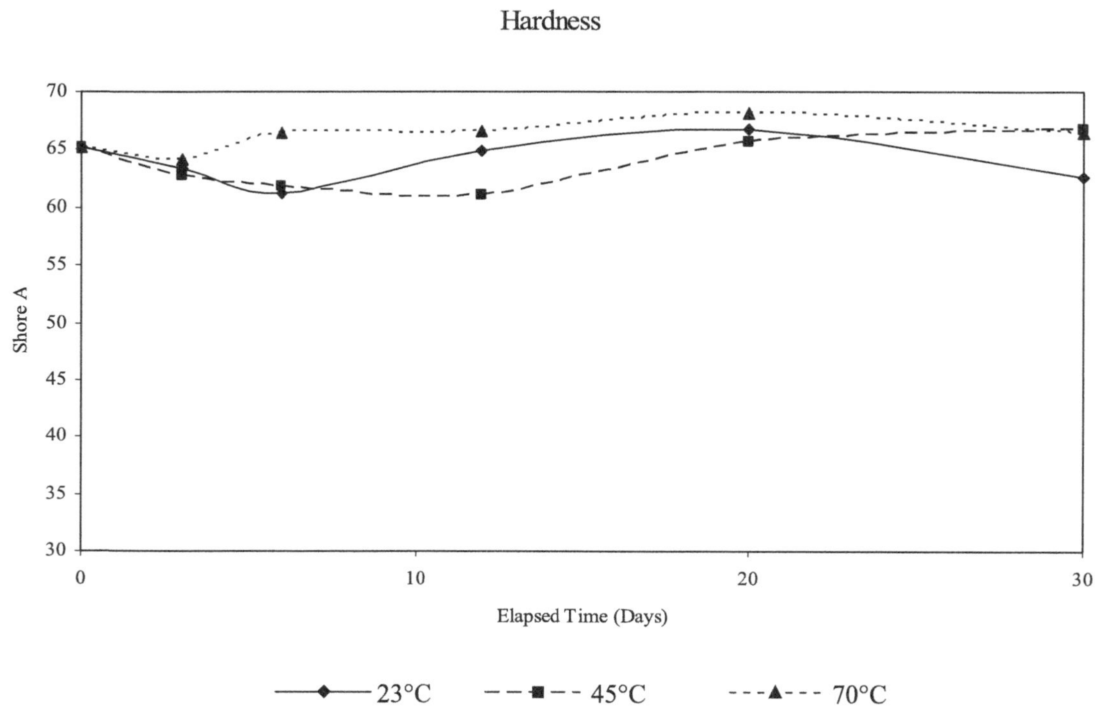

23°C at 30 days

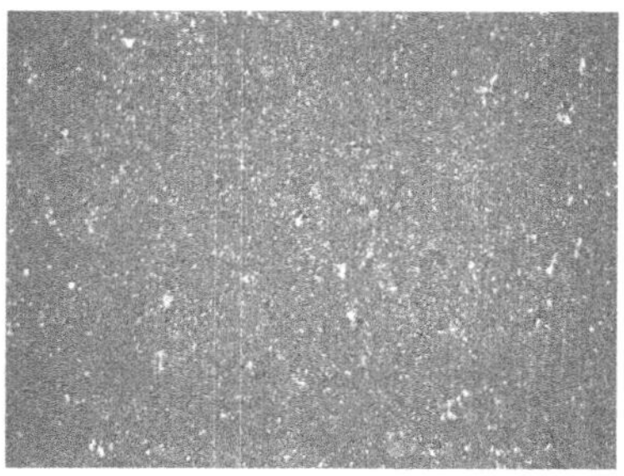

45°C at 30 days

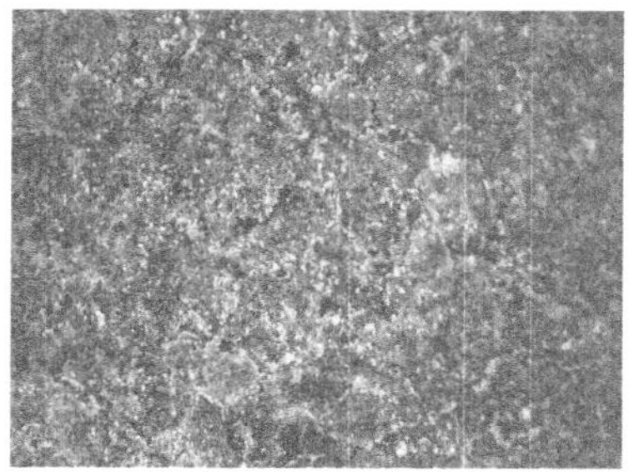

70°C at 30 days

Figure 3.2 Temperature effect on hardness at 1 ppm chloramine and pH 8.3 of SBR with 40× magnified pictographs

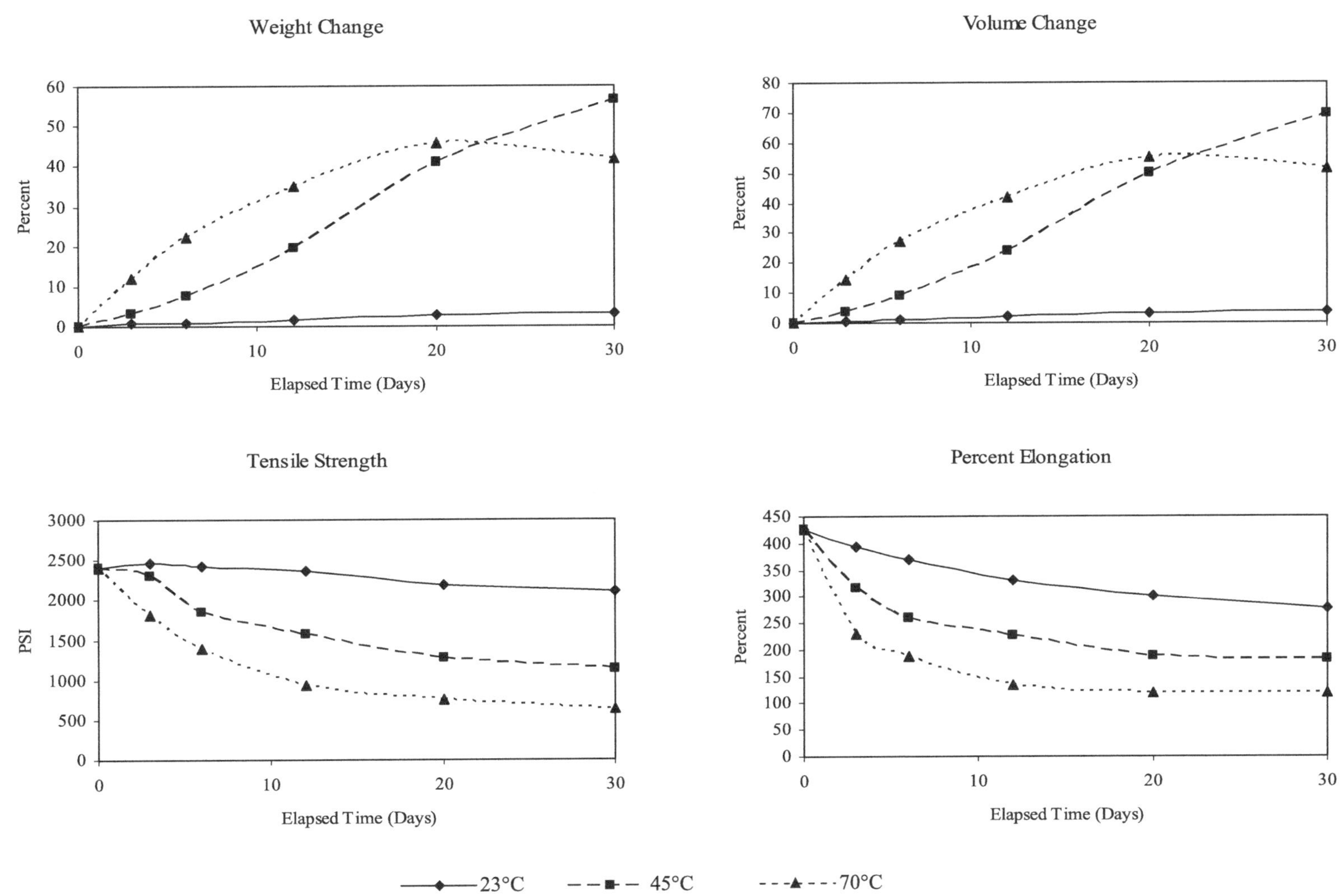

Figure 3.3 Temperature effect on elastomeric properties at 30 ppm chloramine and pH 8.3 of SBR

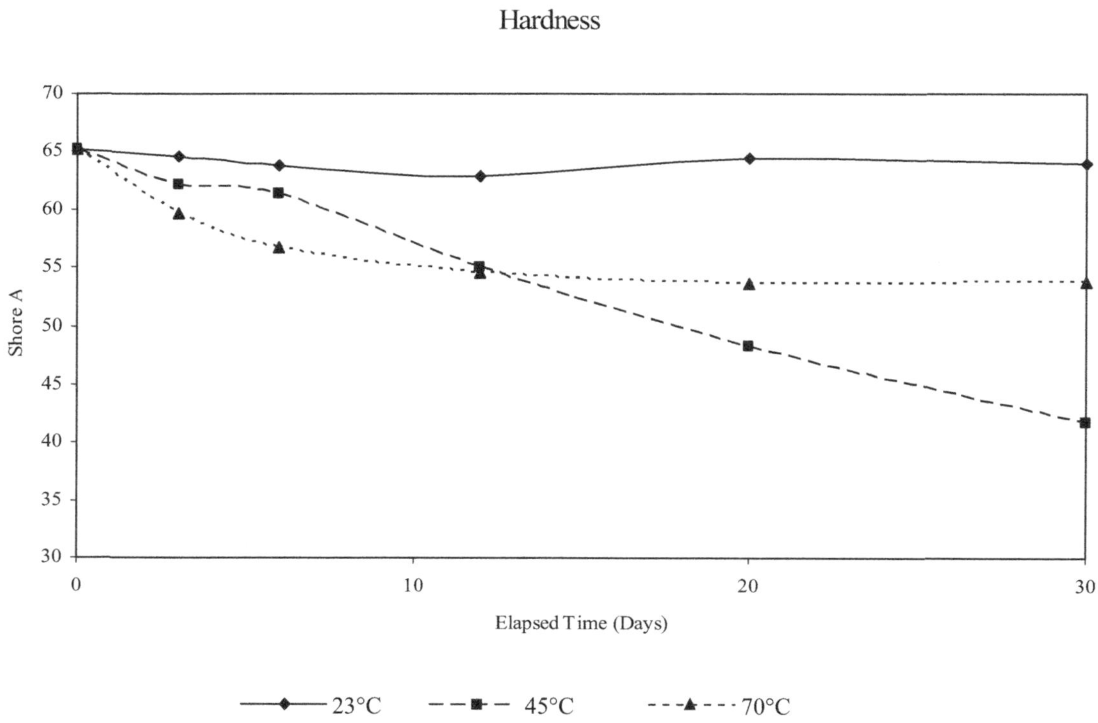

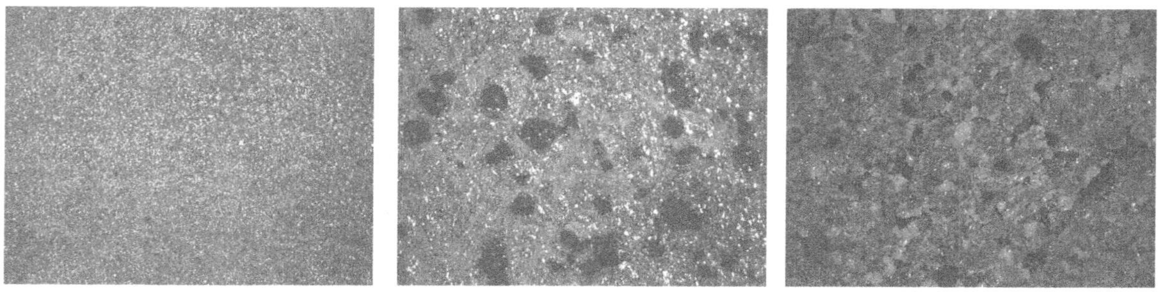

23°C at 30 days | 45°C at 30 days | 70°C at 30 days

Figure 3.4 Temperature effect on hardness at 30 ppm chloramine and pH 8.3 of SBR with 40× magnified pictographs

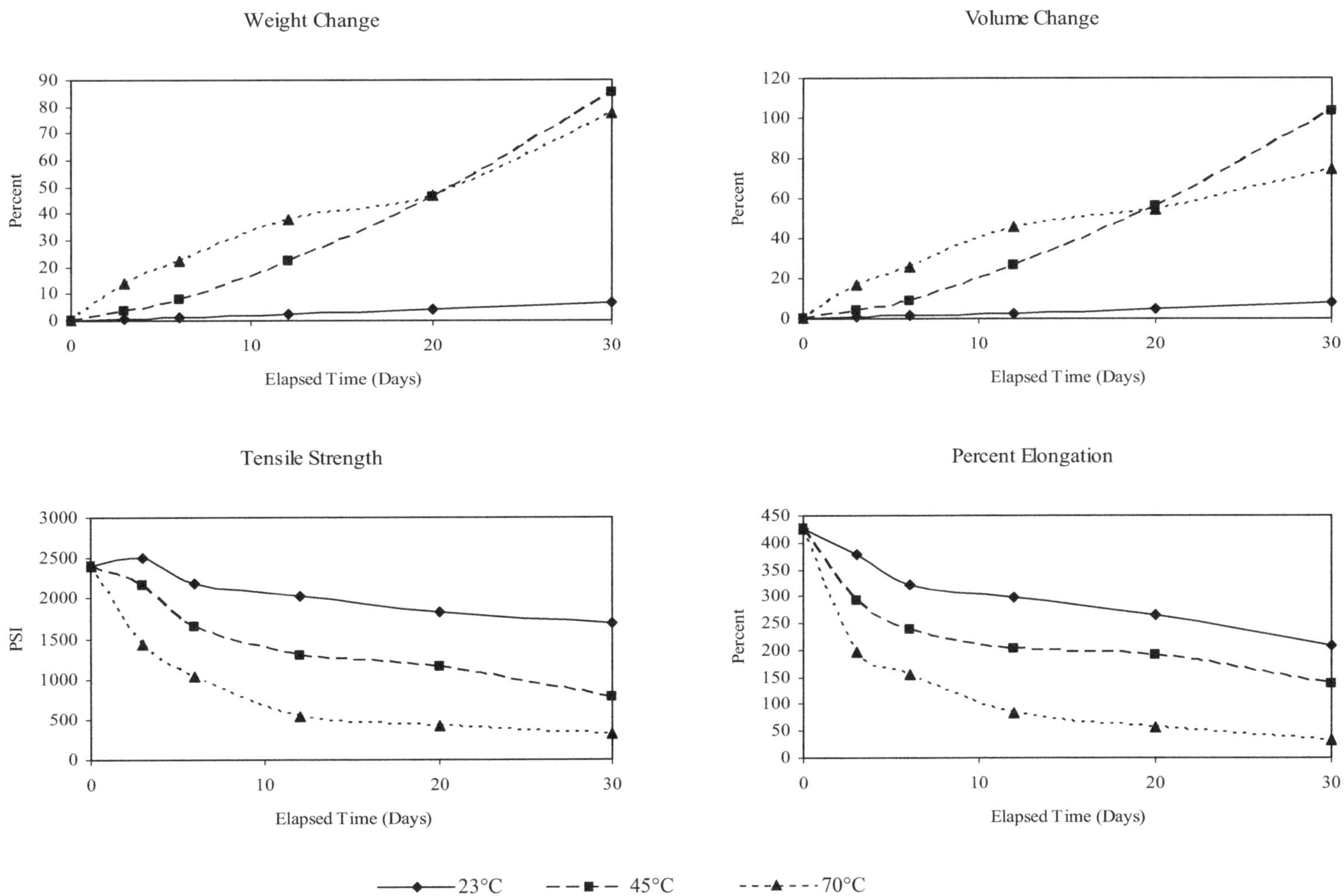

Figure 3.5 Temperature effect on elastomeric properties at 60 ppm chloramine and pH 8.3 of SBR

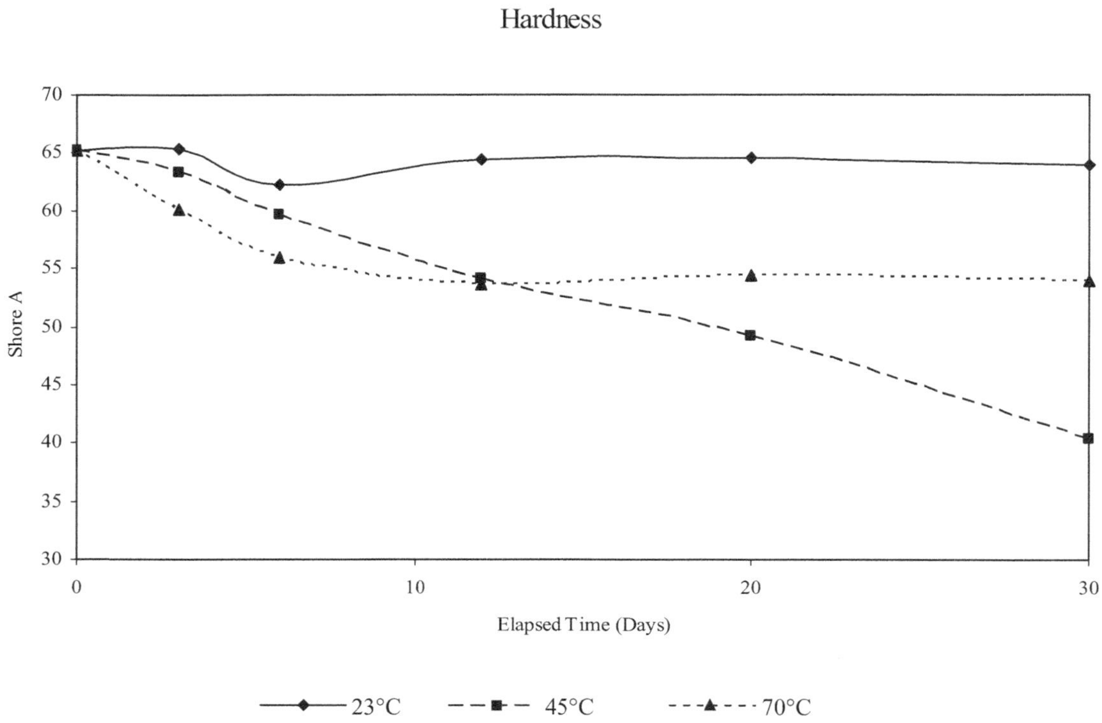

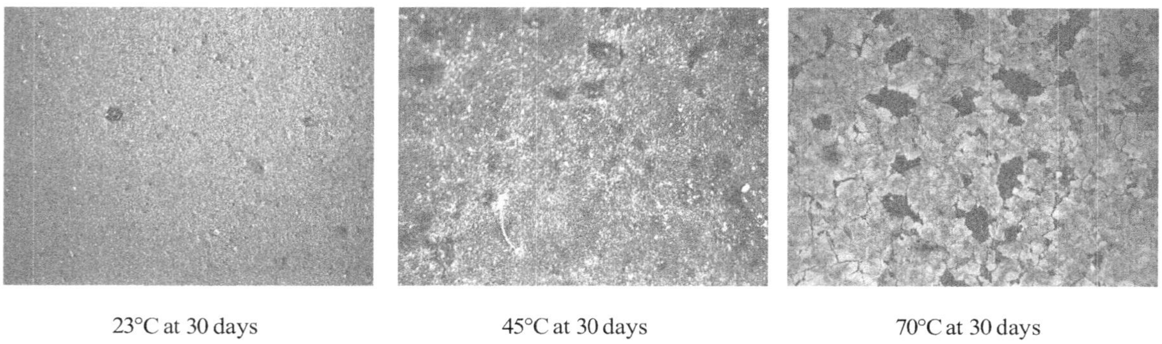

Figure 3.6 Temperature effect on hardness at 60 ppm chloramine and pH 8.3 of SBR with 40× magnified pictographs

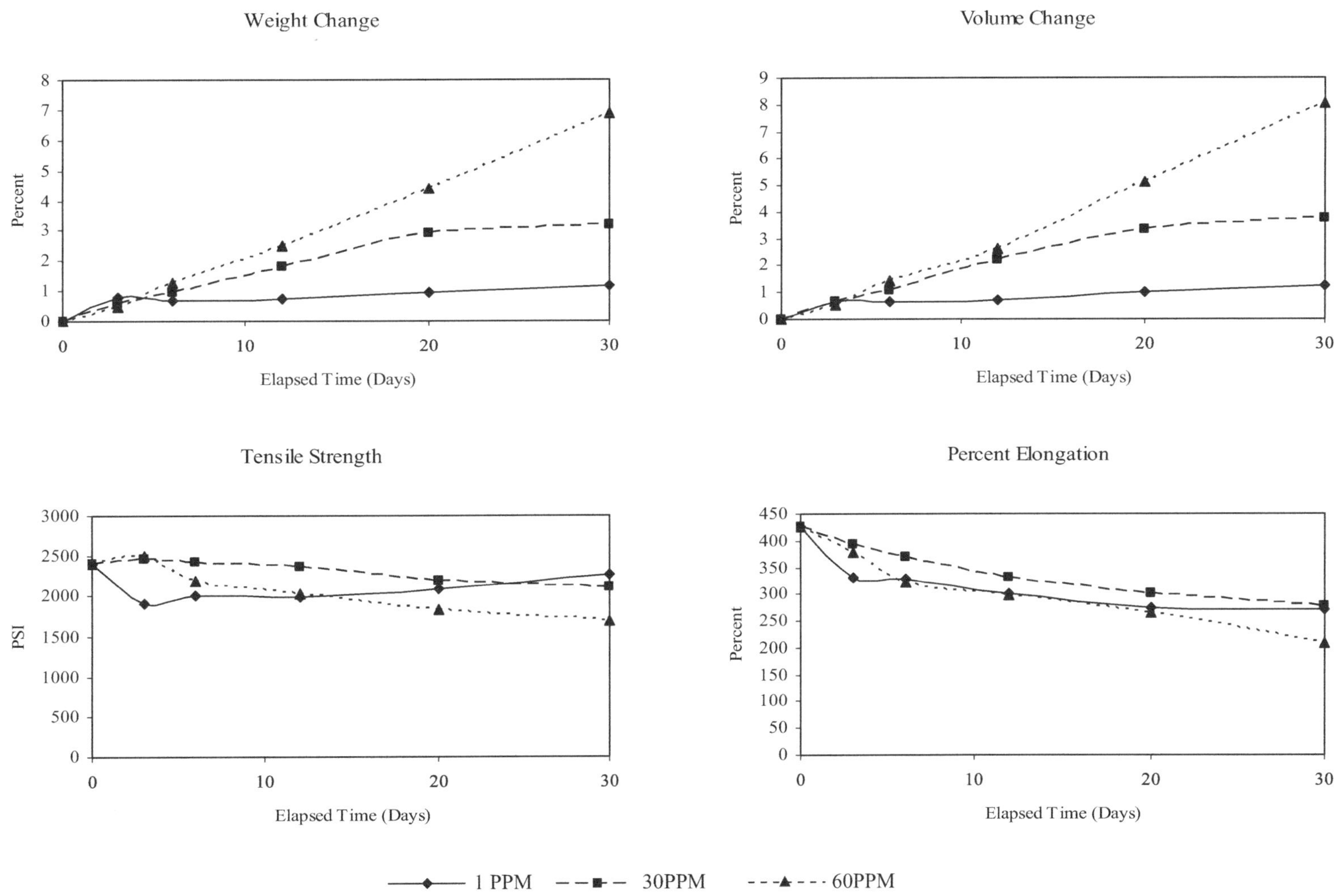

Figure 3.7 Concentration of chloramine effect on elastomeric properties at 23°C and pH 8.3 of SBR

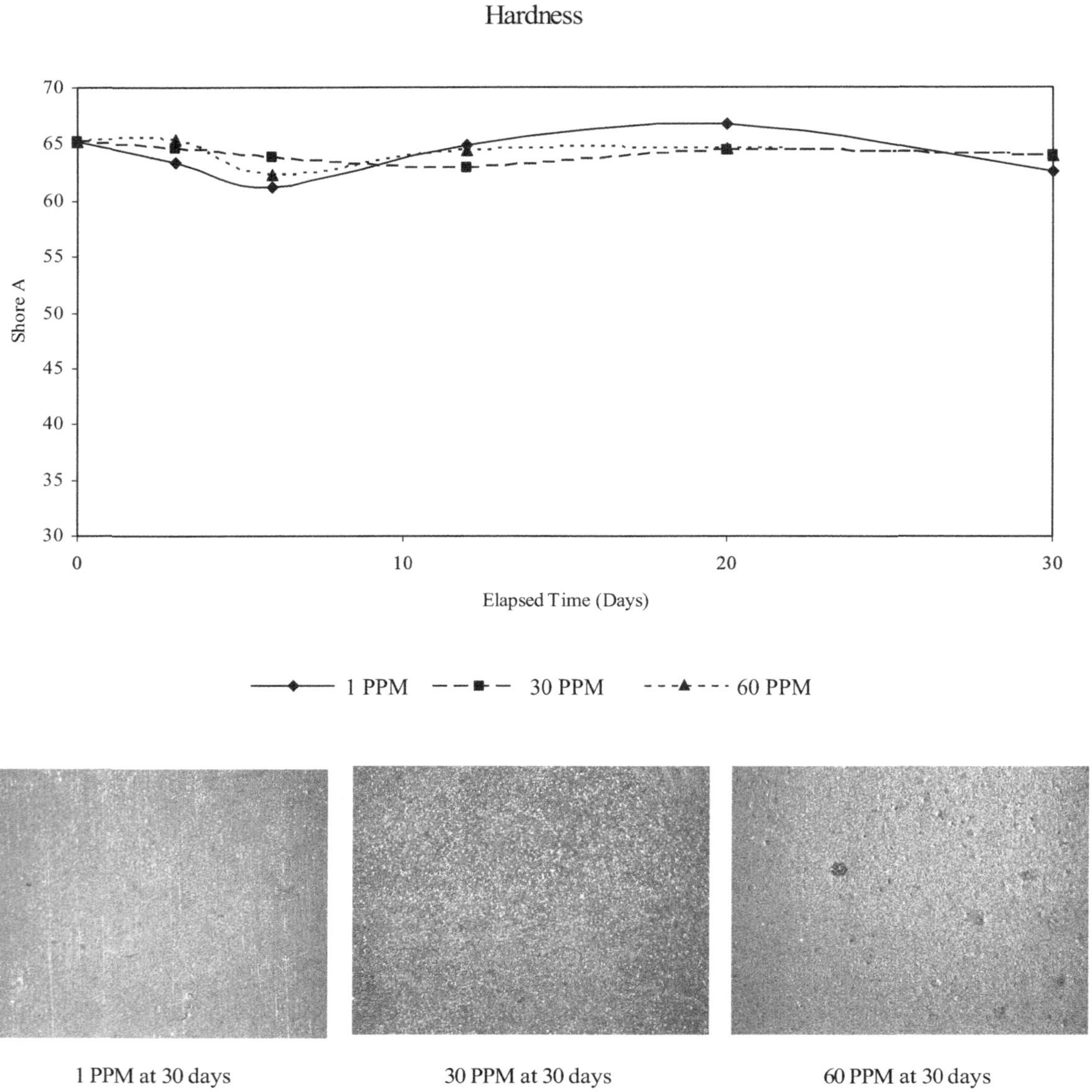

Figure 3.8 Concentration of chloramine effect on hardness at 23°C and pH 8.3 of SBR with 40× magnified pictographs

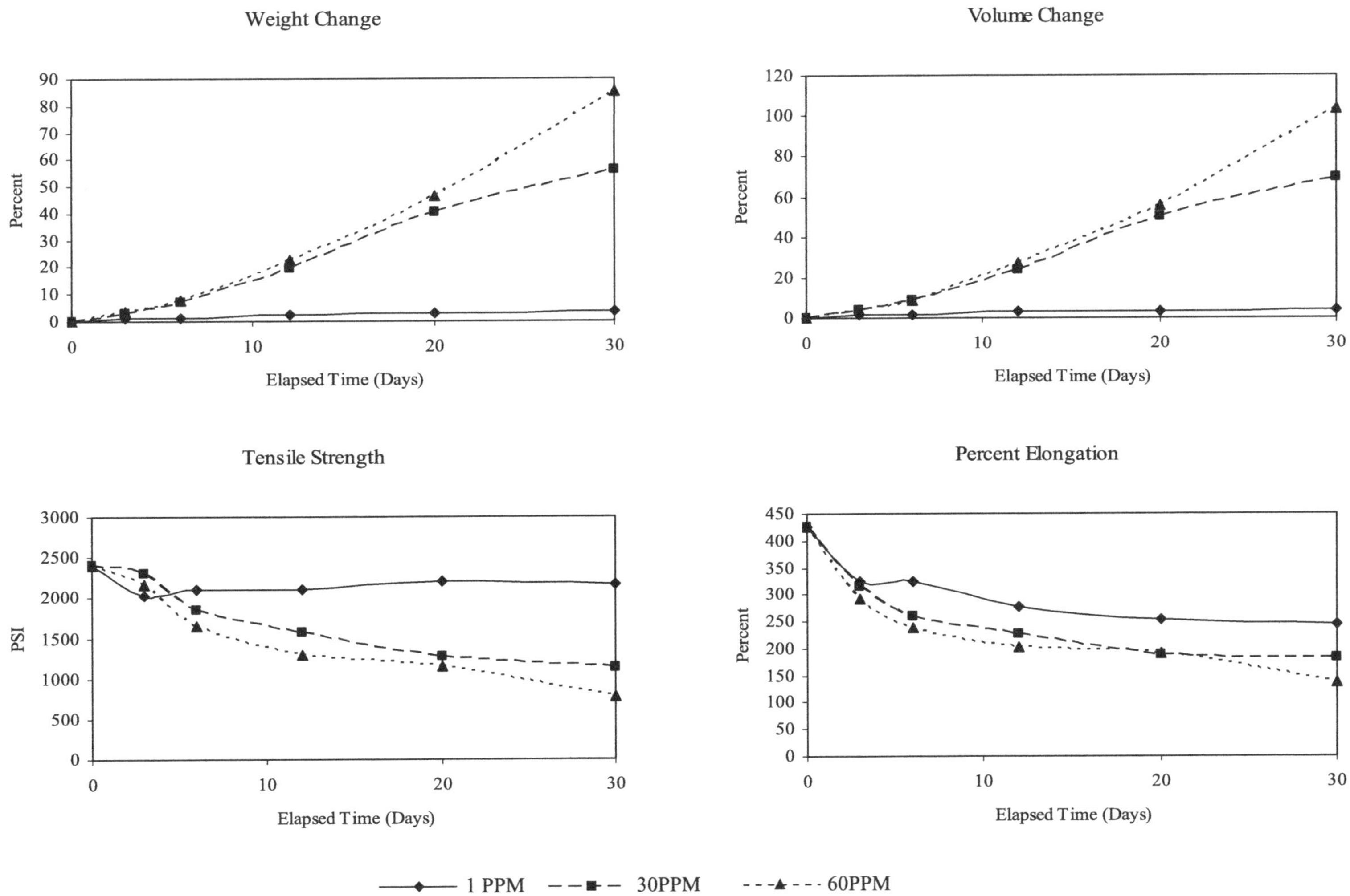

Figure 3.9 Concentration of chloramine effect on elastomeric properties at 45°C and pH 8.3 of SBR

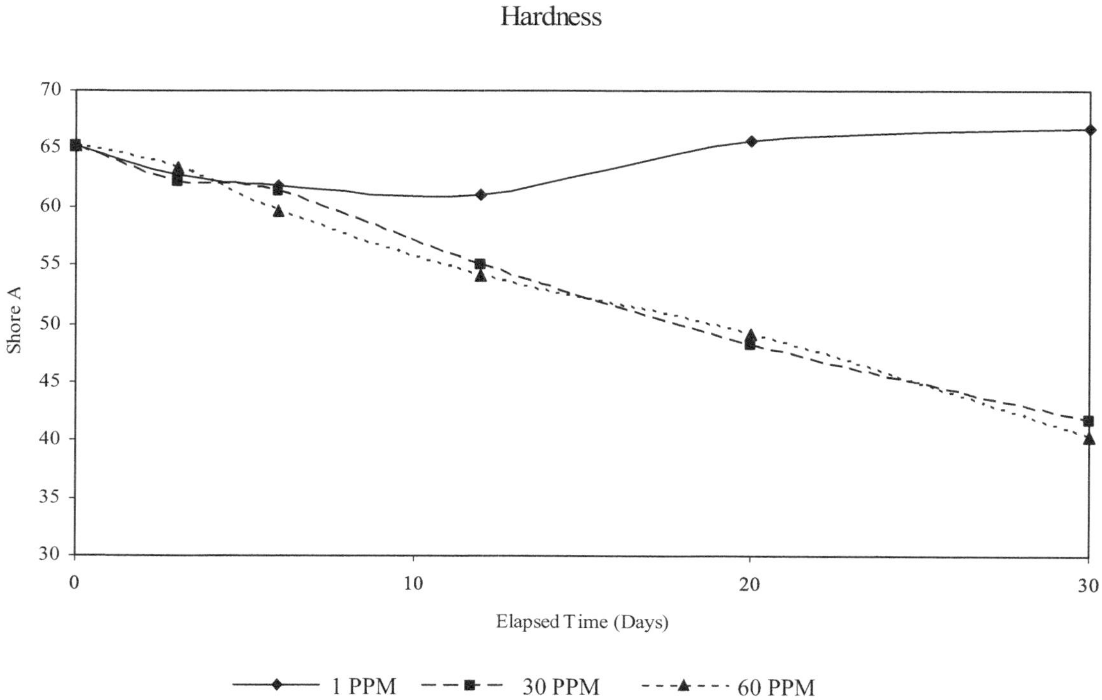

1 PPM at 30 days

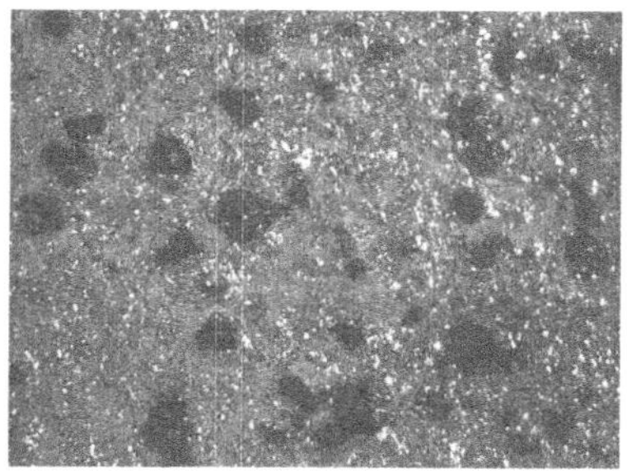

30 PPM at 30 days

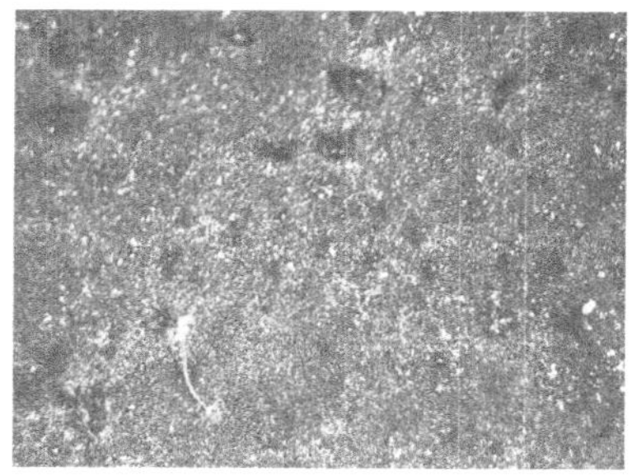

60 PPM at 30 days

Figure 3.10 Concentration of chloramine effect on hardness at 45°C and pH 8.3 of SBR with 40× magnified pictographs

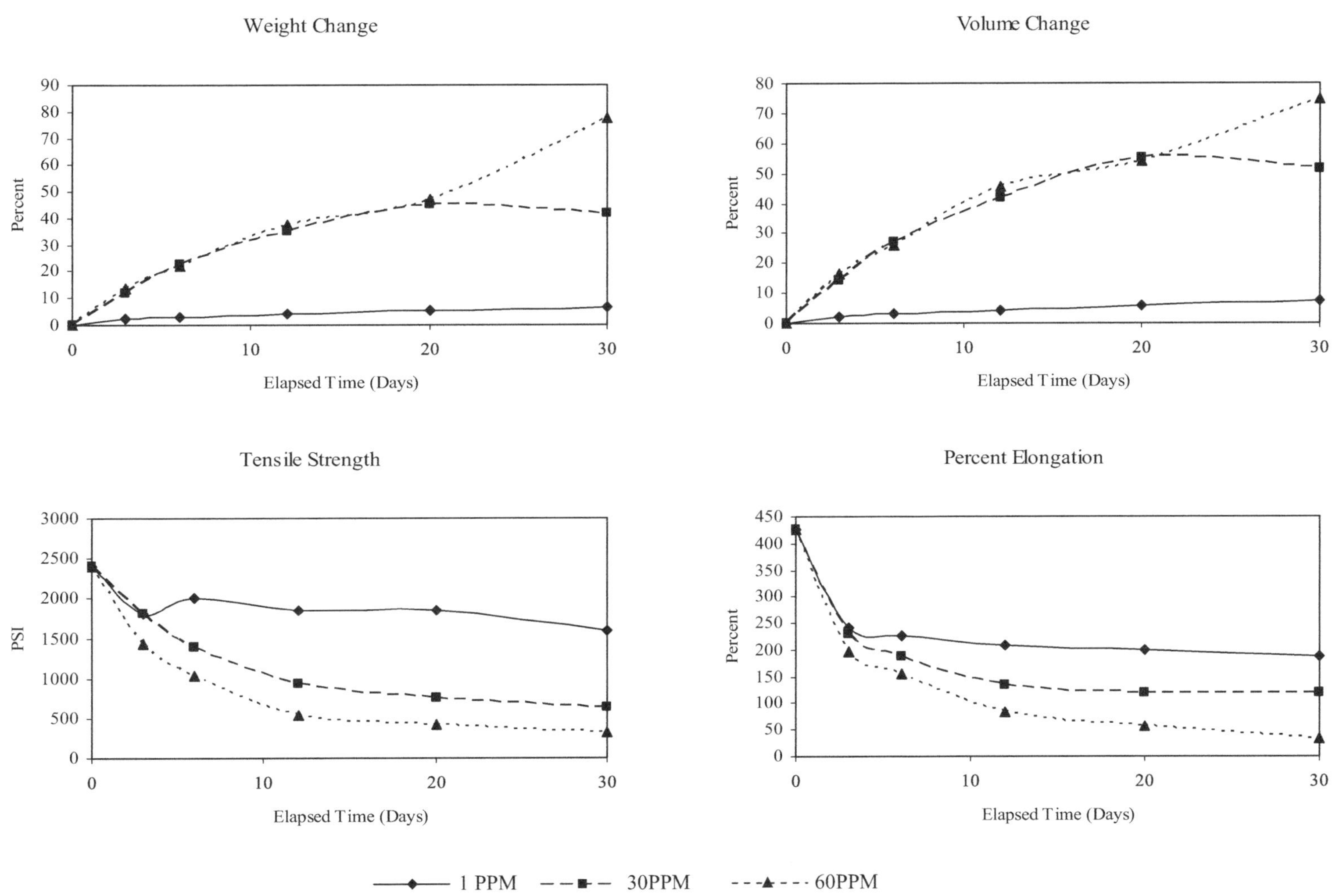

Figure 3.11 Concentration of chloramine effect on elastomeric properties at 70°C and pH 8.3 of SBR

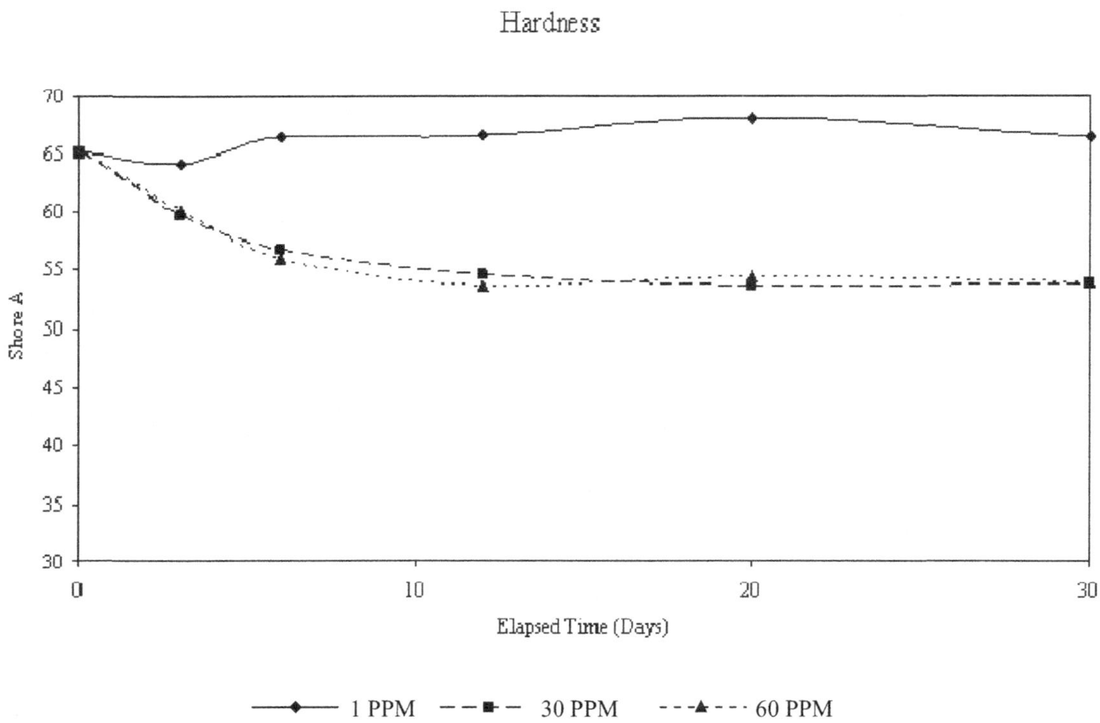

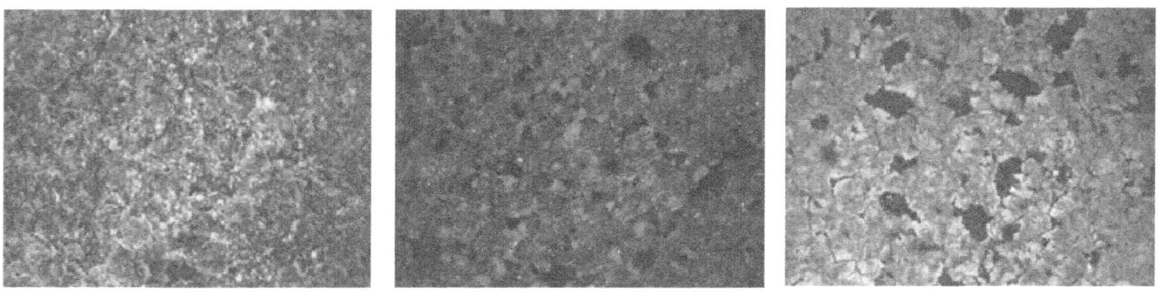

1 PPM at 30 days   30 PPM at 30 days   60 PPM at 30 days

Figure 3.12 Concentration of chloramine effect on hardness at 70°C and pH 8.3 of SBR with 40× magnified pictographs

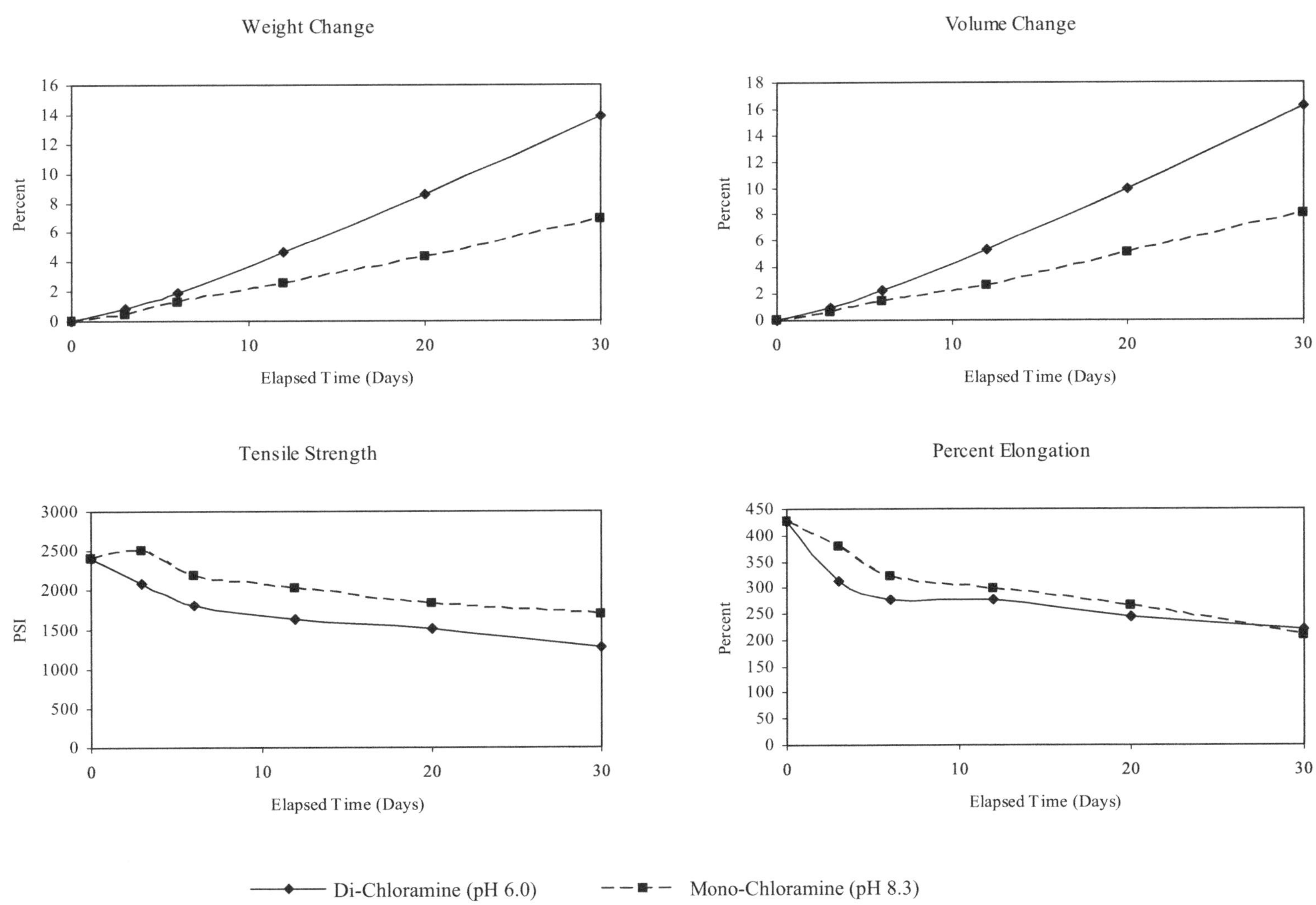

Figure 3.13 pH effect on elastomeric properties at 23°C and 60 ppm chloramine of SBR

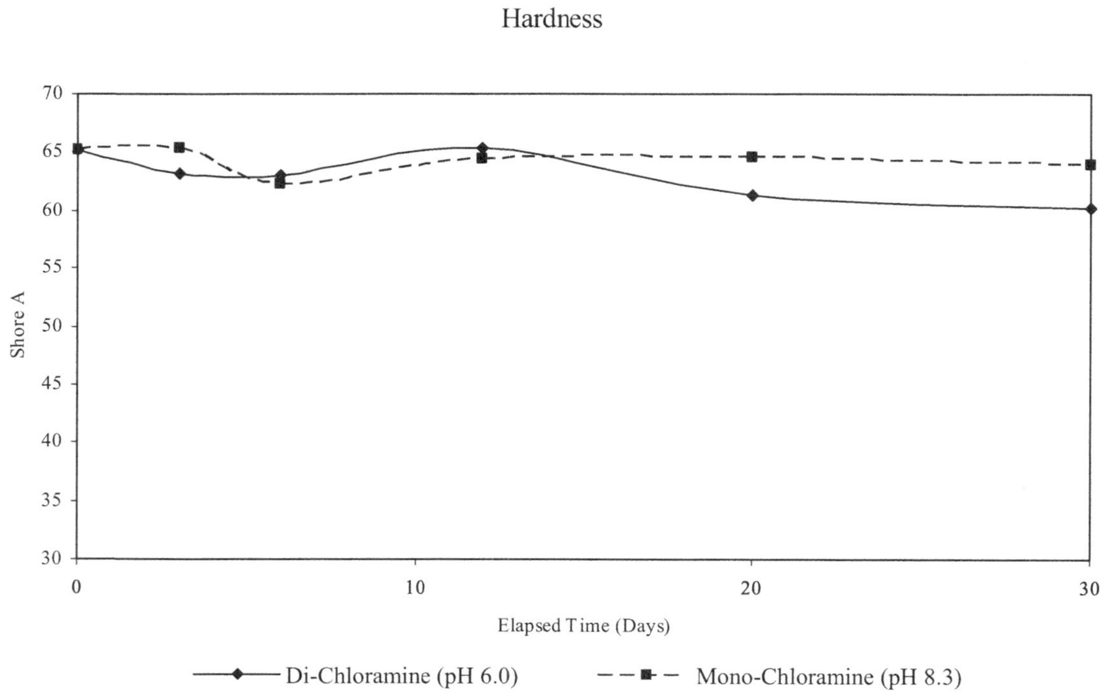

Di-Chloramine at 30 days

Mono-Chloramine at 30 days

Figure 3.14 pH effect on hardness at 23°C and 60 ppm chloramine of SBR with 40× magnified pictographs

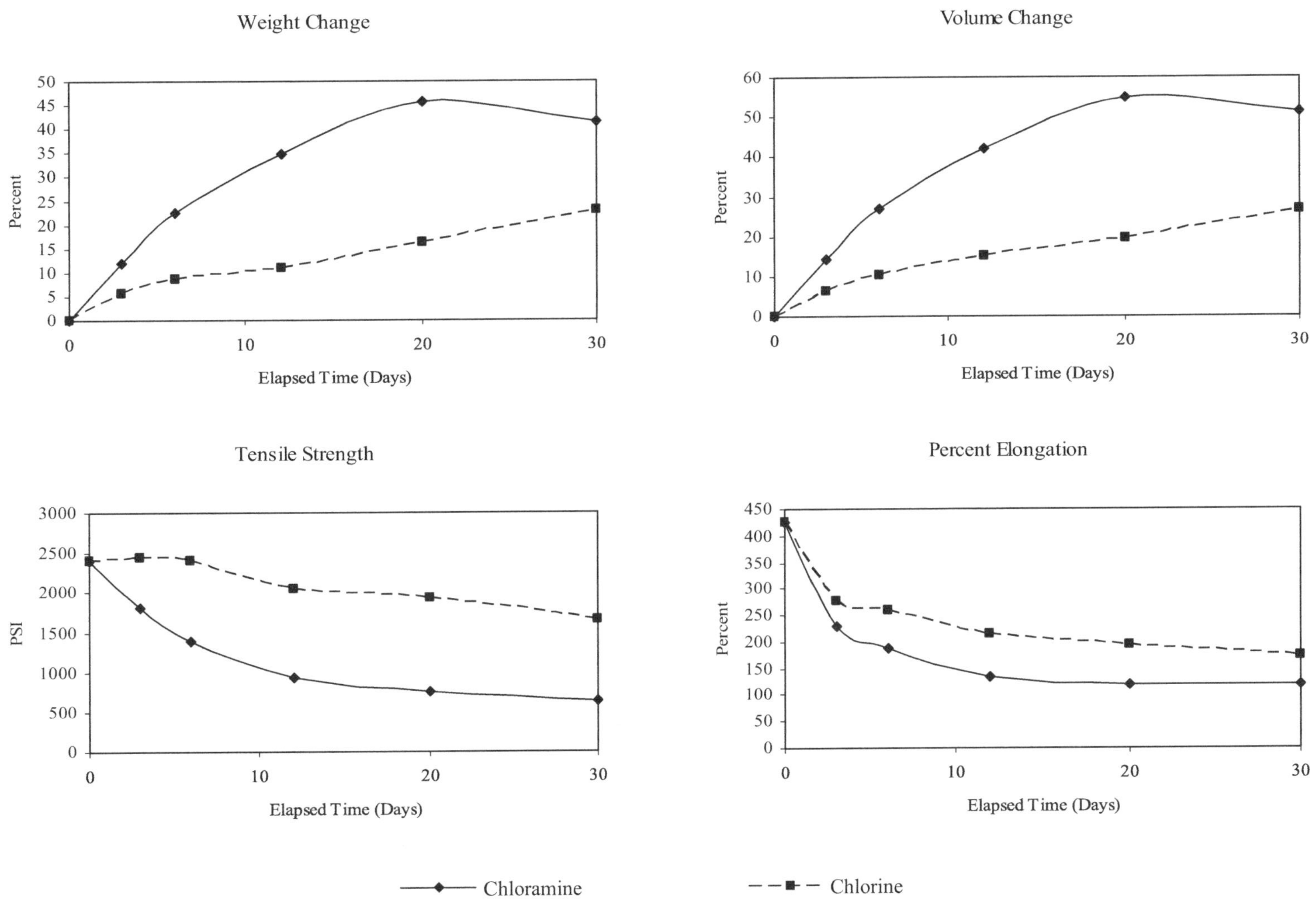

Figure 3.15 Chloramine vs. chlorine effect on elastomeric properties at 70°C, 30 ppm, and pH 8.3 of SBR

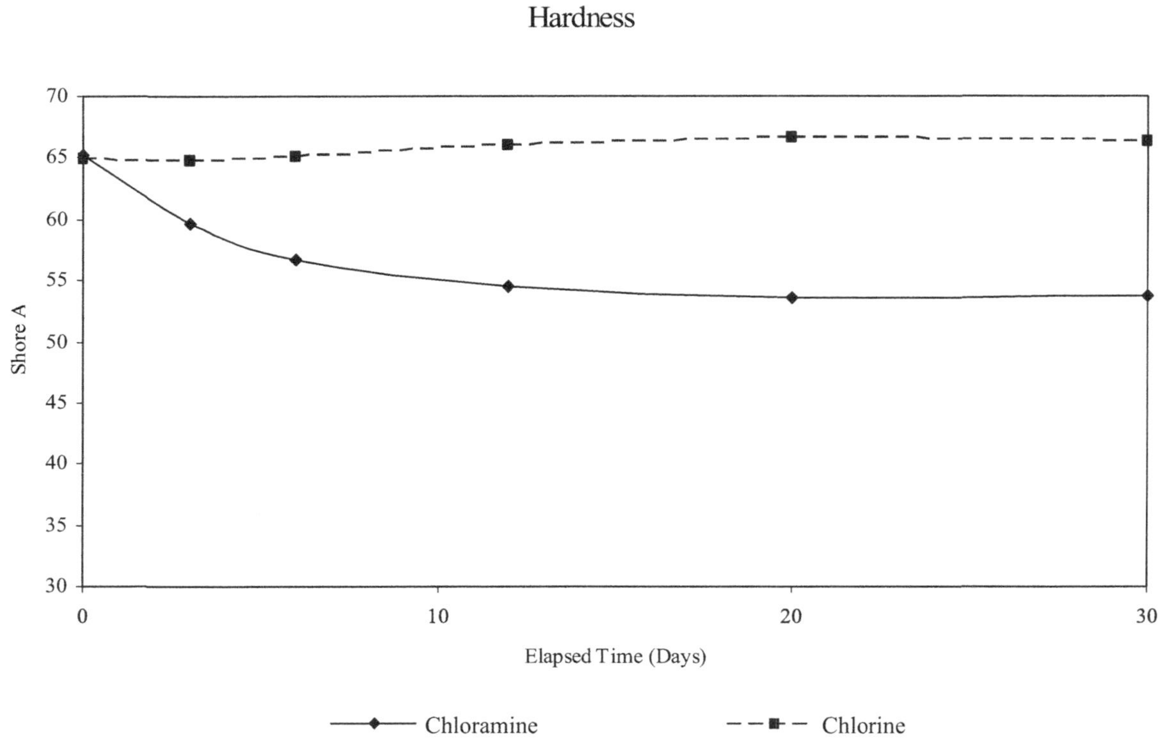

Chloramine at 30 days

Chlorine at 30 days

Figure 3.16 Chloramine vs. chlorine effect on hardness at 70°C, 30 ppm, and pH 8.3 of SBR with 40× magnified pictographs

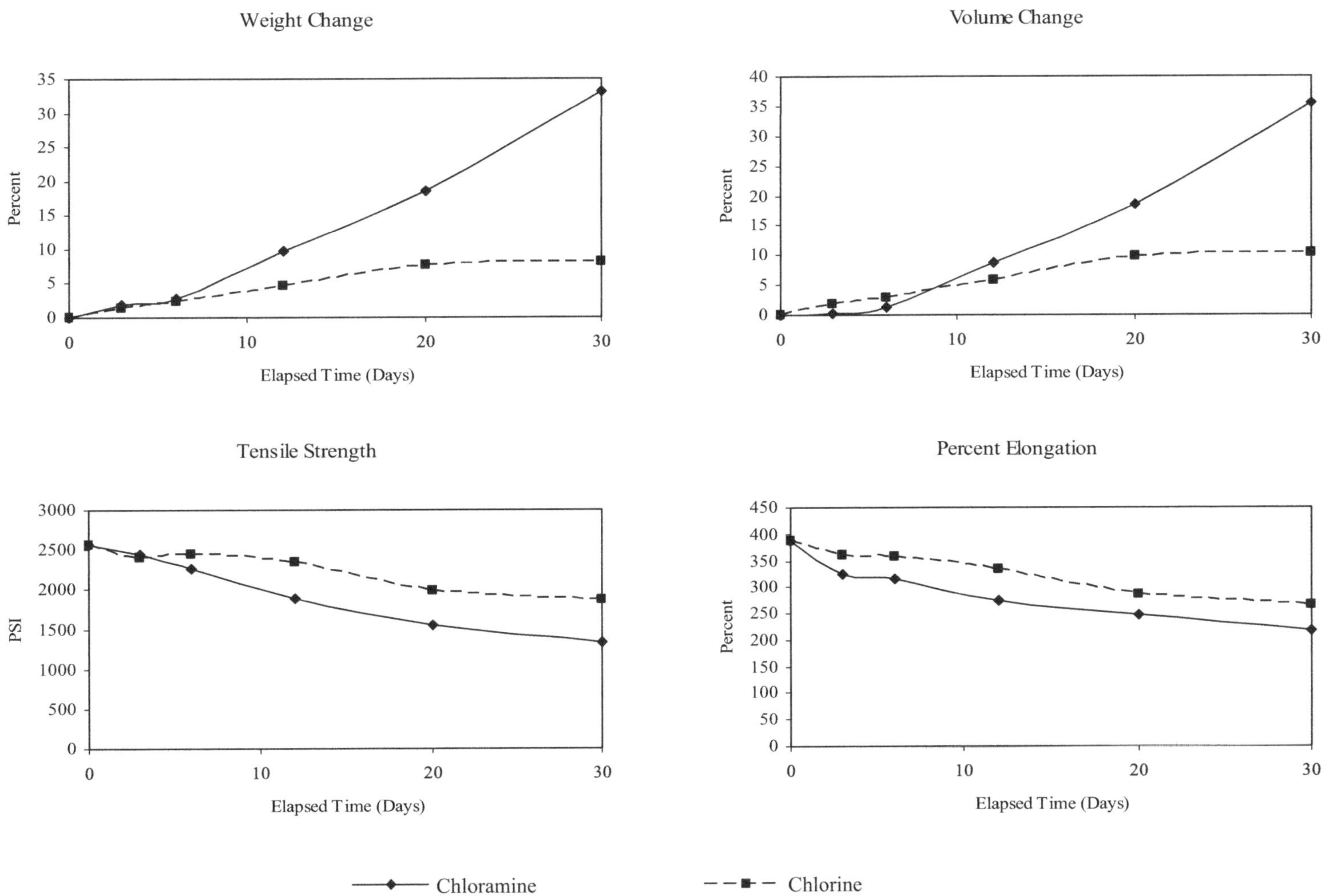

Figure 3.17 Chloramine versus chlorine effect on elastomeric properties at 45°C, 30 ppm, and pH 8.3 of SBR

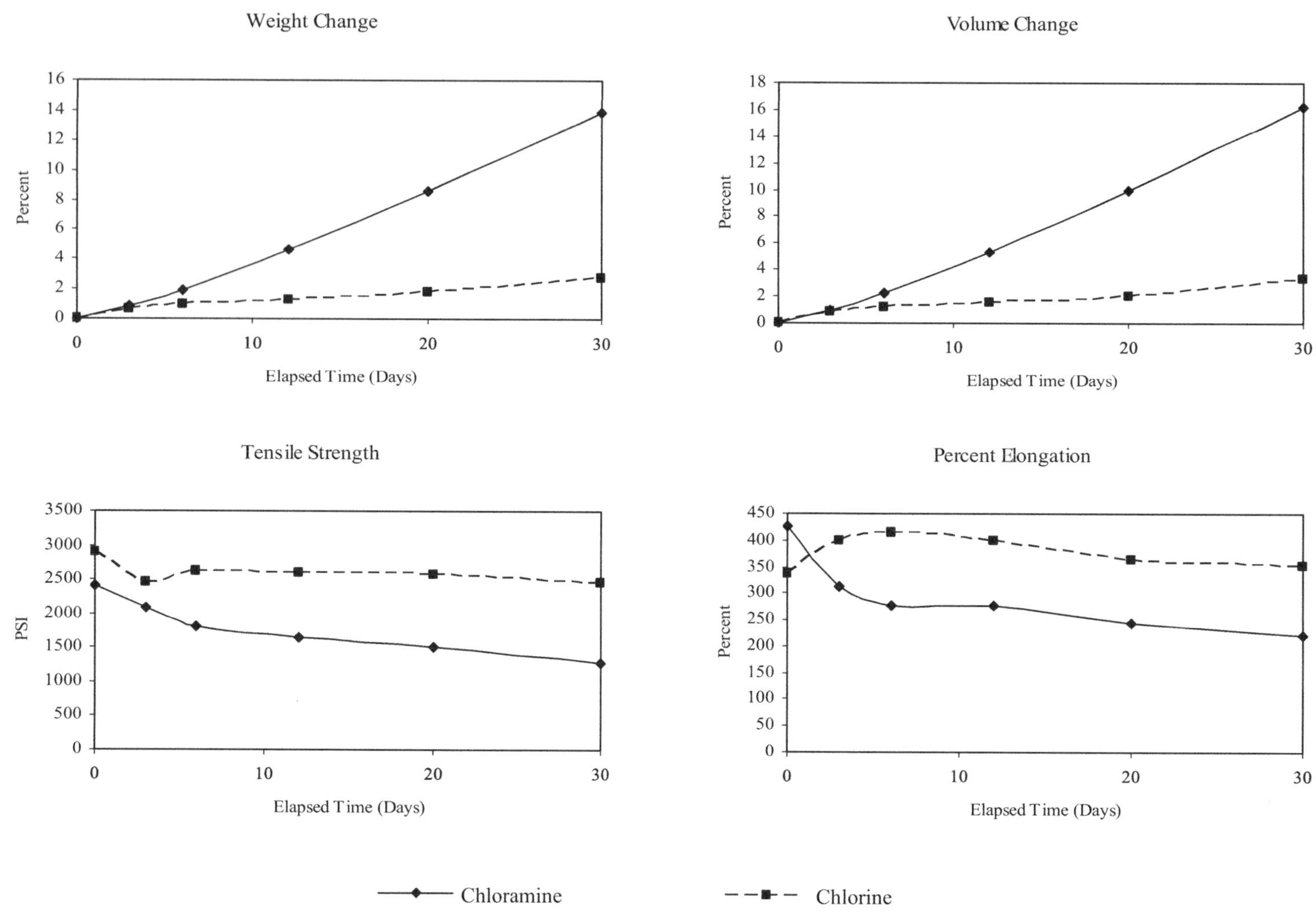

Figure 3.18 Chloramine vs. chlorine effect on elastomeric properties at 23°C, 60 ppm, and pH 6.0 of SBR

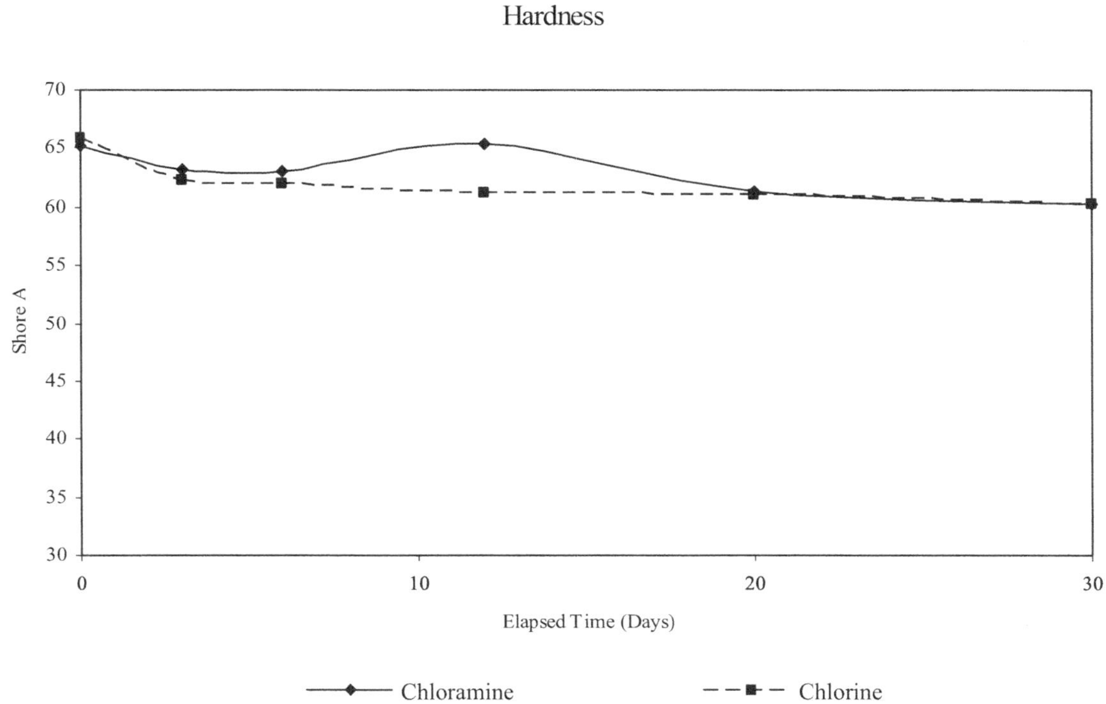

Chloramine at 30 days

Chlorine at 30 days

Figure 3.19 Chloramine vs. chlorine effect on hardness at 23°C, 60 ppm, and pH 6.0 of SBR with 40× magnified pictographs

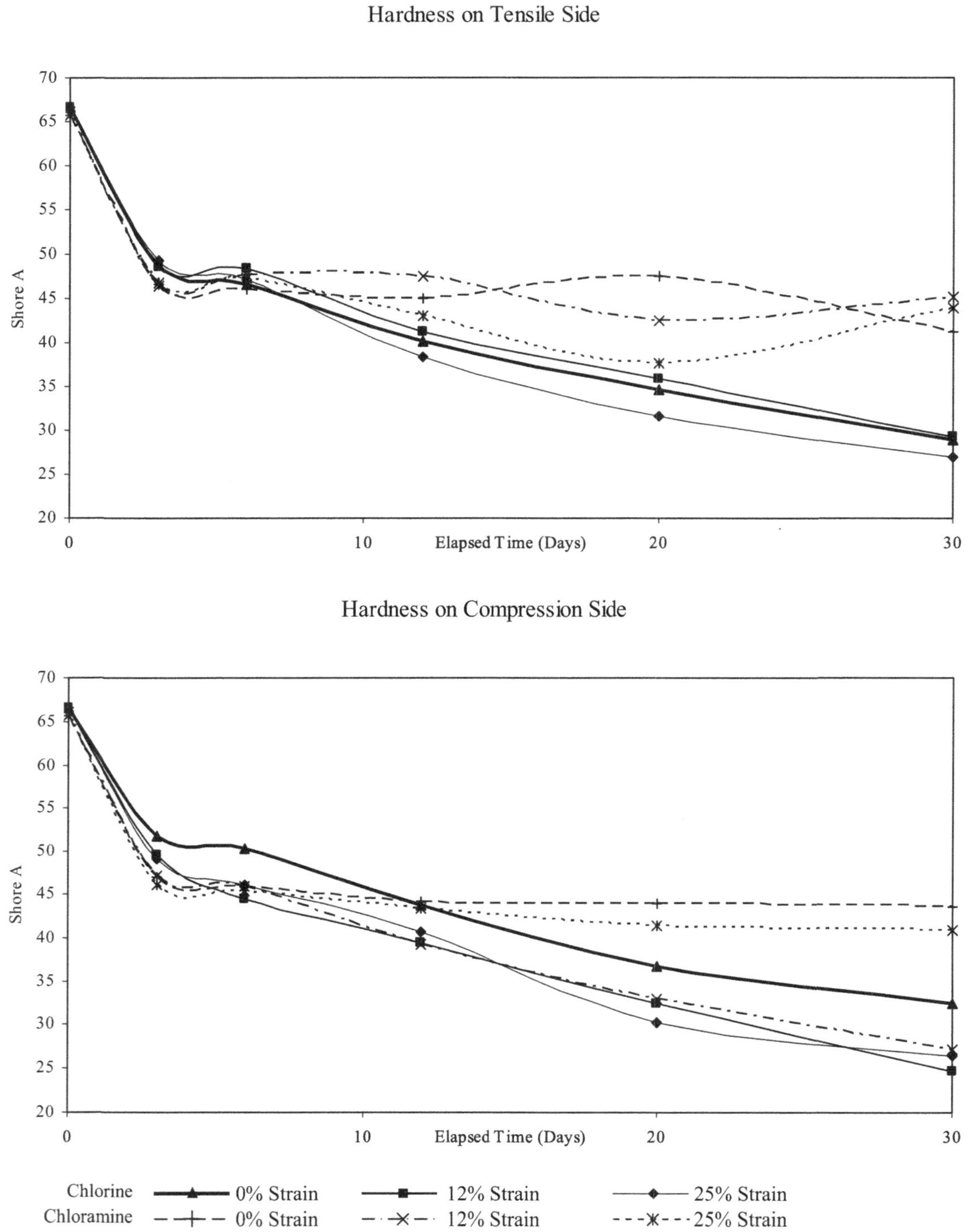

Figure 3.20 Loop strain effects on hardness at 45°C, 30 ppm chloramine compared to 30 ppm chlorine at pH 8.3 of SBR

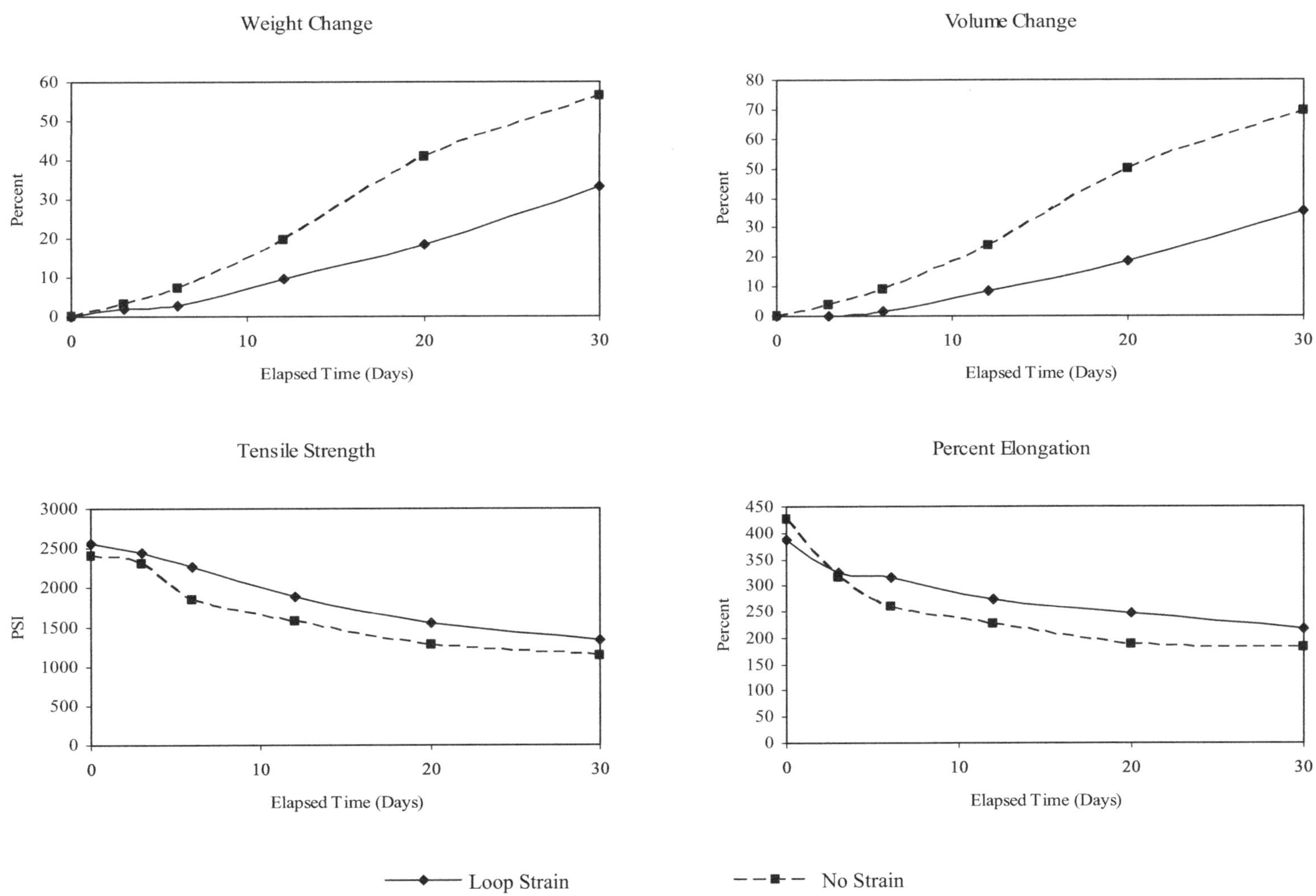

Figure 3.21 Loop strain effects vs. no strain on elastomeric properties at 45°C, 30 ppm chloramine, and pH 8.3 of SBR

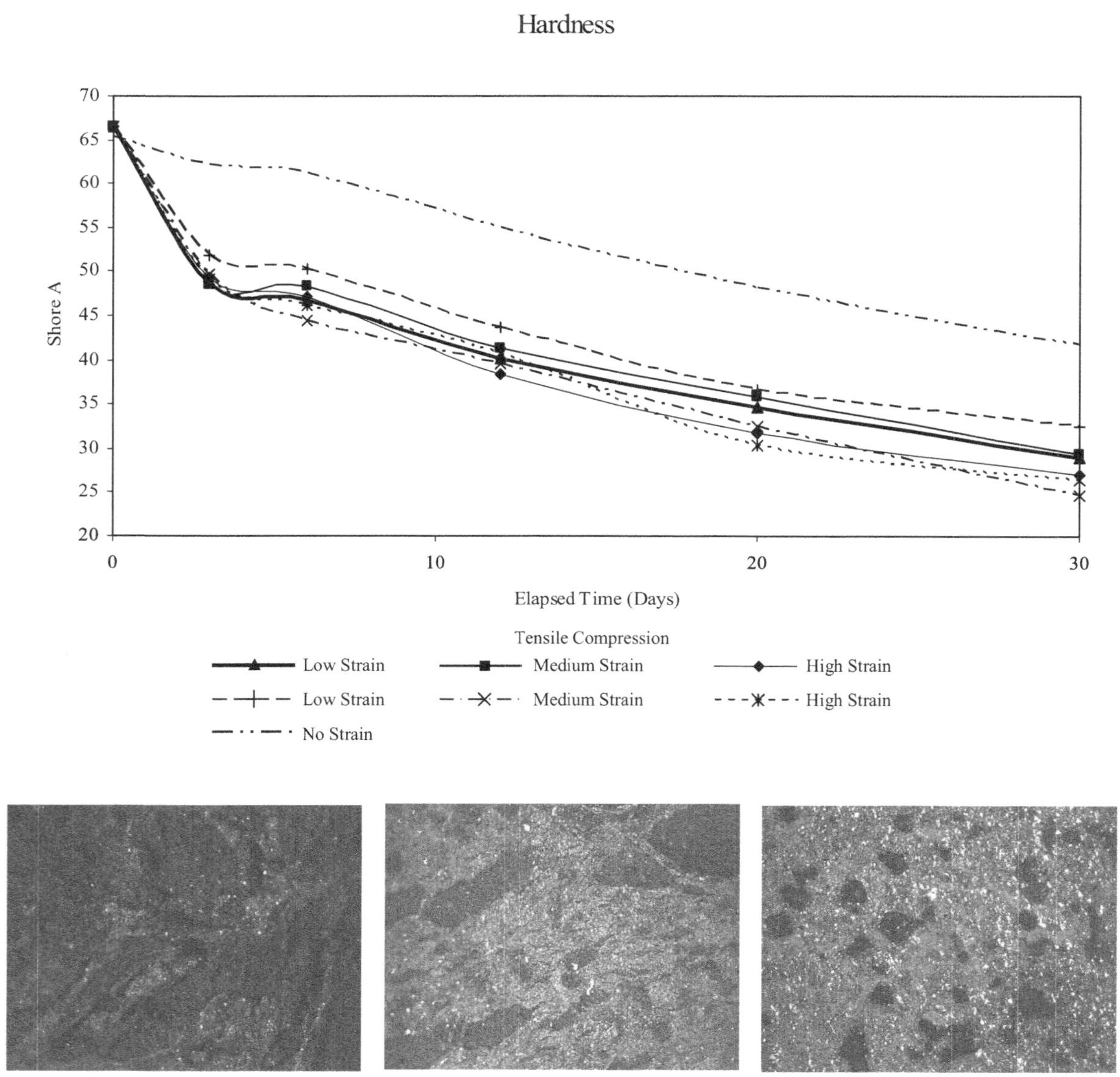

Tensile at 30 days Compression at 30 days No Strain at 30 days

Figure 3.22 Loop strain effects vs. no strain on hardness at 45°C, 30 ppm chloramine, and pH 8.3 of SBR with 40× magnified pictographs

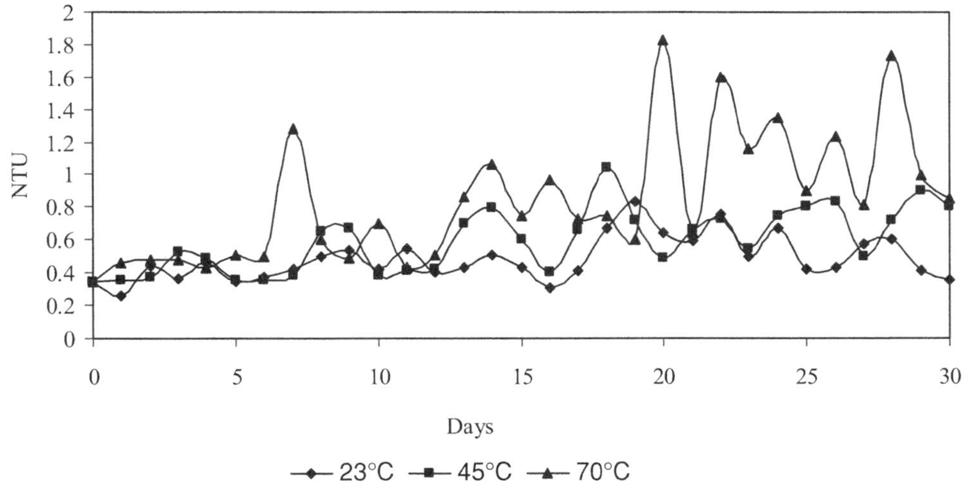

Figure 3.23 Daily turbidity data of EPDM-P at 30 ppm chloramines

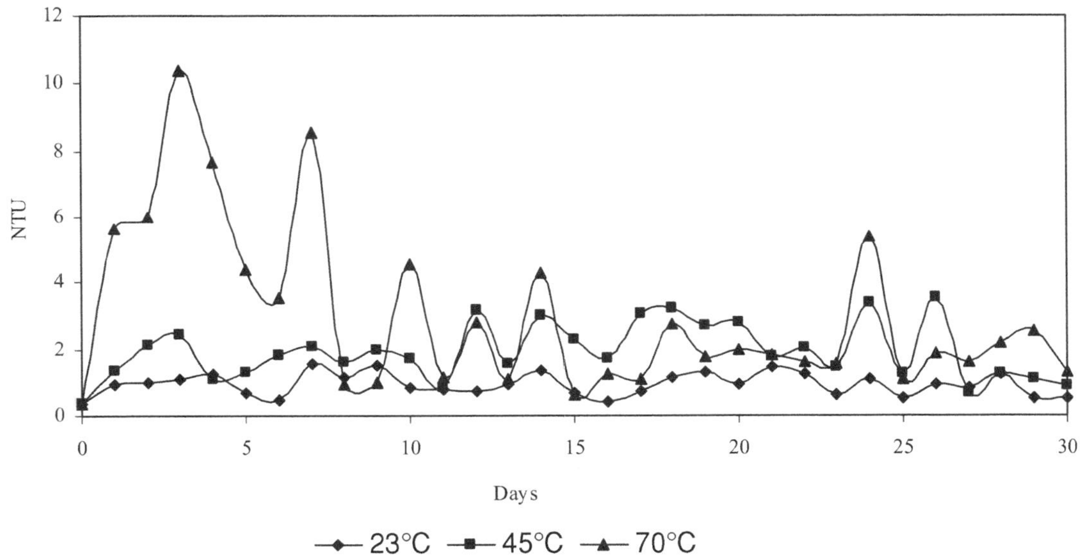

Figure 3.24 Daily turbidity data of EPDM-S at 30 ppm chloramines

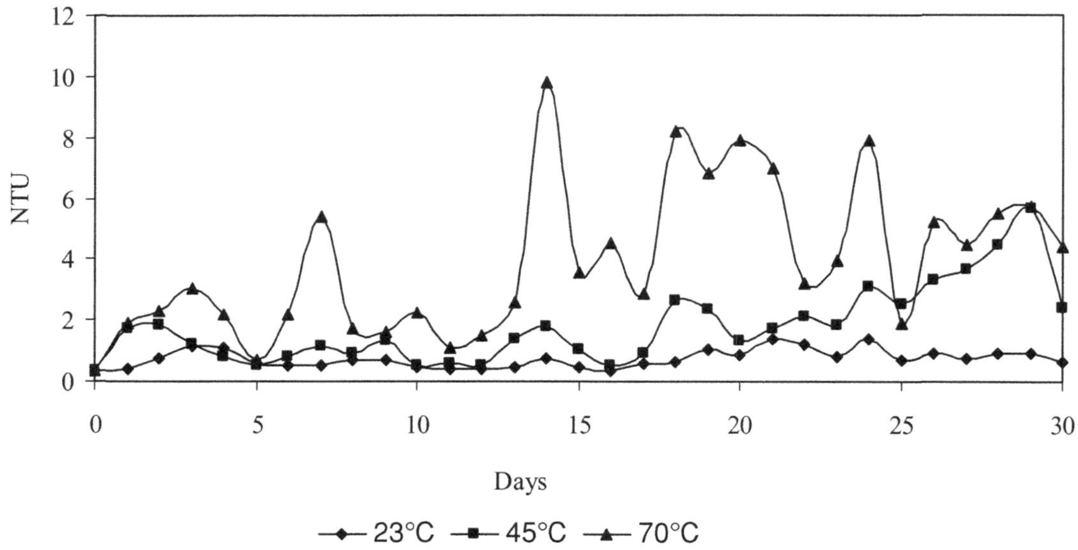

Figure 3.25 Daily turbidity data of SBR at 30 ppm chloramines

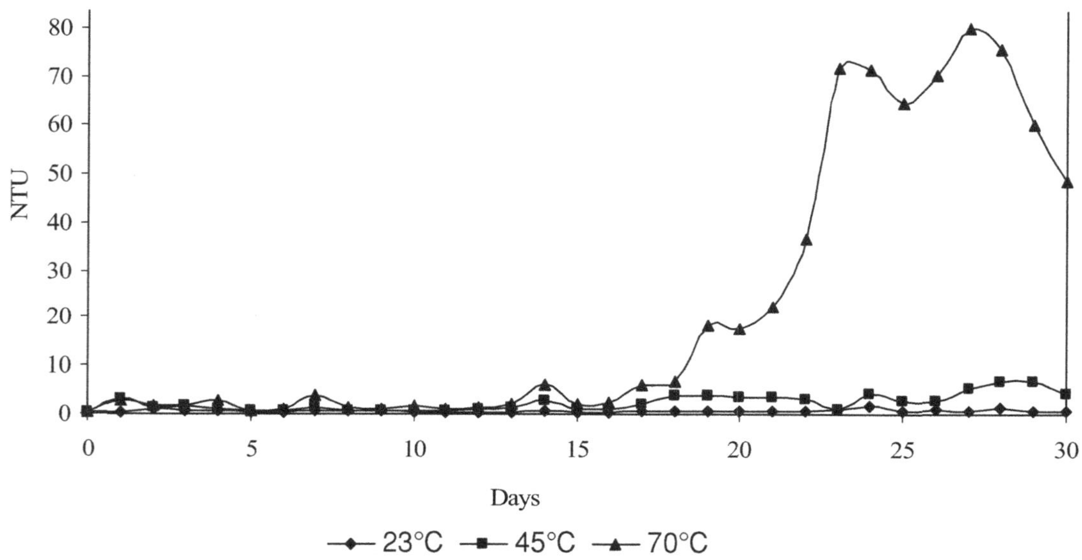

Figure 3.26 Daily turbidity data of NBR at 30 ppm chloramines

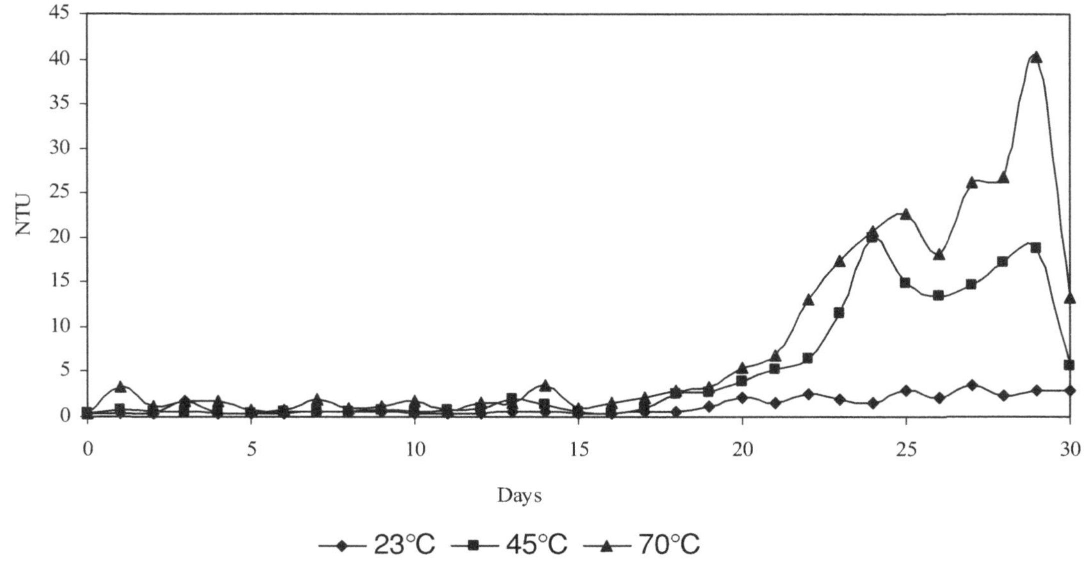

Figure 3.27 Daily turbidity of natural rubber at 30 ppm chloramines concentration test solution

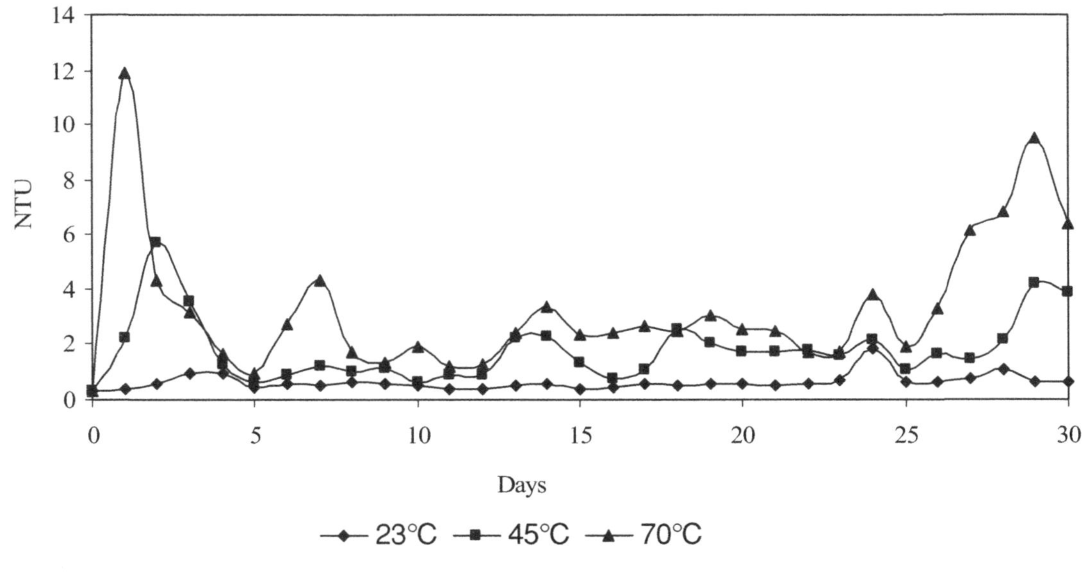

Figure 3.28 Daily turbidity of neoprene at 30 ppm chloramines concentration test solution

CHAPTER 4
DATA ANALYSIS—DEGRADATION PREDICTION

INTRODUCTION

The results presented in the previous chapter give an intriguing picture of elastomer degradation when exposed to chloramines. Over 30 days of exposure, there are many changes that occur in each of the elastomers examined, such as increases in swelling (weight and volume changes) and decreases in the hardness and tensile strength. While this information does help explain what happens to elastomers in situ, the 30-day test window is insufficient to reasonably quantify life expectancy for each kind of elastomer in the field. There are, however, analytical models that can be applied to the data that will effectively extend the range of the tests. By combining multiple sets of degradation data from different test conditions, a better estimate of long-term performance can be obtained. More precise predictions of elastomer life expectancy would allow for better engineering of water works systems and development of a more thorough preventative maintenance plan that could minimize the occurrence of catastrophic failures.

Accelerated Degradation Overview

Predicting long-term performance or life expectancy for materials is one of the most challenging areas of current research. Often, there is insufficient time to perform full-length life expectancy tests on products before they enter the market. But life expectancy is especially important in the water industry, where parts are expected to remain in service in some cases for 100-plus years. Therefore, when long-term tests cannot be performed, researchers look for ways to accelerate degradation in the laboratory. By assessing degradation rates and patterns, it is believed that these relatively short-term tests can be used to reasonably estimate long-term performance.

The automotive industry has generally been the entity that has advanced the study of accelerated-aging tests for elastomers. With the increasing number of elastomeric products in vehicles, it became critical to understand and predict long-term performance. The automotive industry and other researchers have commonly used elevated temperature, pressure, and exposure to rapidly accelerate degradation in the laboratory. It was generally concluded that elastomers degrade in a predictable fashion (Leyden 2003, Osswald and Menges 2003).

While degradation is not likely linear with respect to time, there is generally a "master" degradation curve for each elastomer and test condition so that future performance or remaining life can be reasonably predicted. As shown in Figure 4.1, an elastomer may start with a modulus of 10^{10} dynes/cm^2, but over time this will predictably degrade to a modulus of 10^5 dynes/cm^2. Additionally, it was determined that the elastomer degradation rate can generally be modeled by the Arrhenious rate equation, shown in Equation 4.1, and that temperature and other physical components can be used to effectively speed up the degradation process in a predictable manner.

$$\ln(k)_{Temp} = -\frac{E_a}{RT} + C \tag{4.1}$$

where K = degradation rate
E_a = activation energy
R = universal gas constant
T = absolute temperature in °K
C = constant (material dependent)

For the elastomer degradation study, both elevated temperature and concentrations were used to accelerate the degradation. Thus, when the Arrheneious rate degradation equation, temperature, and concentration effects are taken into consideration, the degradation function was anticipated to take the form expressed in Equation 4.2, with accelerating effects of both temperature and concentration contributing to the degradation.

$$-r_A = f[\text{Temperature-dependent terms, Concentration-dependent terms}] \rightarrow kC_A^a$$

$$= \underbrace{k_o e^{-E/RT}}_{\text{Temperature-dependent term}} \underbrace{C_A^a}_{\text{Concentration-dependent term}} \qquad (4.2)$$

where R_A = degradation parameter
K = degradation rate
E = activation energy
R = universal gas constant
T = absolute temperature in °K
C = constant (material dependent)

Temperature Acceleration

As discussed, elastomer material performance predictions rely to some extent on the degradation, or relaxation, of the individual chains in the polymer matrix. This degradation involves a time component, suggesting that each property will change as a function of time. By increasing the elastomer exposure temperature, researchers can increase the speed at which the polymer chains degrade and effectively move further along the time scale. By measuring property changes over a set time period at several different temperatures, it becomes possible to superimpose all the data sets onto a reference temperature and produce a master degradation curve. The process of combining the individual time and temperature curves through superposition to form a master degradation curve is called "time-temperature superposition."

There are two methods by which a time-temperature superposition can be performed on a data set. The first method (the graphical method) is to plot the data sets onto one graph and physically shift the data sets onto the reference data set (Figure 4.2). This method requires that there be some identifiable "landmark" in each of the data sets that can act as an anchor during the shift. The benefit of this method is that the shift factor can be directly calculated from the degree each of the data sets has to be moved to reach the reference set. It is then possible to determine the material constants for a particular system. While the graphical method is generally suitable for developing the master degradation curve, it is imprecise because the shift value for each curve depends on the reference points selected. However, the graphical method does generally provide a reasonable approximation of the master degradation curve.

The second method to shift the data for time-temperature superposition is through the use of the Williams-Landel-Ferry (WLF) equation (Equation 4.3). The WLF equation defines a shift factor, a_t, in terms of the reference temperature, T_{ref}, and the temperature at which the measurements occur, T, through the following equation:

$$\log a_t = -\frac{C_1(T - T_{ref})}{C_2 + (T - T_{ref})} \tag{4.3}$$

where C_1 and C_2 are material constants. With this method, the validity of the temperature relationship of these data can be confirmed by plotting the inverse of log a_t as a function of the inverse of $(T - T_{ref})$. A linear plot indicates that the property of interest follows the WLF equation and hence can undergo a time-temperature superposition with a reasonable degree of confidence in the predicted values. A non-linear plot indicates that the property of interest may not follow the WLF equation, so any predicted values based on a time-temperature superposition must be utilized with caution. Unfortunately, without advance knowledge of these constants, it is difficult to develop an accurate master curve based on the equation alone.

The time-temperature superposition analysis was performed for each of the measured properties and for each elastomer based on both the graphical method and the WLF equation. In almost every case where a time-temperature superposition could be performed, the breaking stress, breaking strain, and hardness followed predicted trends, and both methods were in close agreement. However, the degree of swelling, as determined by the changes in mass and volume, did not follow expected trends. This was further confirmed by the nonlinear relationship between the inverse of log a_t as a function of the inverse of $(T - T_{ref})$ from the WLF equation (Figures 4.3 and 4.4).

The poor correlation for the weight and volume data does not mean that the graphs do not contain useful information about the changes observed. Poor correlation of weight and volume in certain elastomers may be caused by water or test media extracting or chemically altering such things as plasticizers, extender fillers, and base elastomer polymer gel breakdown. These differences can be clearly observed in the swelling behavior and breaking strain of EPDM-S. The breaking stress shift factor is almost perfectly linear, indicating an excellent agreement with the WLF equation (Figure 4.3), while the shift factor percentage mass change (Figure 4.4) appears almost independent of the temperature difference, indicating a lack of agreement with WLF. This phenomenon indicates other factors at play, which are not detected in the study.

While it may not be intuitive that weight and volume changes do not directly correspond to temperature changes, it is not entirely unexpected. The WLF equation is based on parameters such as the elastic modulus, where individual polymer chains move relative to one another. The degree of cross-linking within a polymer is related to these parameters because the cross-link bonds limit molecular motion. In the case of swelling, the motion of individual polymer chains plays a role in the degree of swelling, but it is not the dominant factor. One can conceptualize this idea by visualizing a pocket within the polymer, which is defined by a series of cross-links. As swelling begins, the pocket will inflate until the cross-links expand to their maximum length and the pocket is fully expanded. As additional liquid enters the pocket, polymer chains will begin to expand to accommodate them, but the expansion will put a strain on the cross-links until they begin to break. Thus, while the motion of individual polymer chains does play a role in swelling, it is local cross-link density and the breaking of those cross-links that determine the swelling behavior of an elastomer. These are different mechanisms than molecular motion and, as such, are not expected to display the same response over time. Given these concepts, it is not surprising that the swelling behavior does not precisely follow the WLF equation.

The three concentrations within the test matrix were blocked so that 1, 30, and 60 ppm data sets could be considered independently. Thus, for each elastomer, degradation curves were generated that represented cumulative time-temperature superposition degradation data for each specific concentration. In almost every elastomer investigated, there was a clear difference between the degradation behavior at 1 ppm and that at 30 and 60 ppm chloramines concentration. Figure 4.5 shows an example of the degradation curves for SBR. These degradation curves were then subsequently used to assess concentration induced degradation.

Concentration Acceleration

The concentration component inducing elastomer degradation may also be characterized as an accelerant, and increasing concentration induces increased rates of degradation. As discussed, the time-temperature superposition analysis resulted in graphs of material degradation at three different concentrations (1, 30, and 60 ppm). These three concentrations could then be superimposed onto one another to develop one master degradation curve that includes the effects of both temperature and concentration as an accelerant.

As presented in Equation 4.2, the concentration effects can be incorporated into the elastomer degradation model through the coefficient $C_A{}^a$. As with the temperature effects, the accelerating effects of concentration can be applied to the model based on the graphical method or through analytical techniques. As previously discussed, the graphical method is imprecise because the shift value for each curve depends on the reference points selected. However, the graphical method does generally provide a reasonable approximation of the master degradation curve.

The analytical model to be used to account for increasing concentration effects in the degradation model is based on Fick's Law of Diffusion. Where the concentration of the disinfectant within the polymer sample at any time and at any location can be estimated based on the steady outside concentration and the diffusion rate of the material, such that:

$$J_i = -D_i\left[\frac{d[c]}{dx}\right] \tag{4.4}$$

$$D_i = f(t) = D_o \cdot e^{-E/(RT)} \tag{4.5}$$

However, in Equation 4.5, the rate of diffusion is also dependent on temperature. The temperature component complicates the characterization of the diffusion rate and makes the analytical understanding of the concentration accelerant very complex and difficult to discern based on the data collected. Therefore, when analyzing the effects of concentration, only the graphical method was used.

Not all the data sets were sufficient to create a master degradation curve of concentration accelerating effects. As discussed previously, when superimposing data based on graphical methods it is critical to have "landmark" or reference points in the data to control the shift. In several of the data sets collected, the parameters monitored did not degrade sufficiently within the testing time window to accurately identify a landmark point. In some cases, material performance actually appeared to slightly improve when exposed to the accelerating temperatures and concentrations. However, this is likely due to the inherent variability with the elastomer materials and within the testing procedures.

MASTER DEGRADATION CURVES

When possible, the effects of temperature and concentration were combined to generate one master degradation curve for each elastomer and physical parameter measured. In some cases, however, it was not possible to combine both effects on the master curve. The primary difficulty was that 1 ppm did not generate sufficient degradation for superposition analysis. Some degradation is necessary in order to have a reference point that corresponds to a similar point on the 30 ppm curve, which would enable a graphical shift. Without this reference point, it is not possible to link the two curves, and the mathematical shift can not be performed. Based on the degradation trends observed, it is likely that months, if not years, would be required before sufficient degradation occurred at 1 ppm and the graphical method would become feasible.

The following graphs present the time-temperature superposition shifts, where sufficient data was available.

Natural Rubber

The degradation curves generated from temperature-accelerated degradation at 1, 30, and 60 ppm were combined using the graphical method to form one master degradation curve that accounted for acceleration based on both temperature and concentration. Due to the observable degradation of the natural rubber in all test conditions, it was possible to construct master degradation curves for all five degradation parameters measured (volume change, weight change, breaking stress, breaking strain, hardness). The master degradation curves for natural rubber are displayed in Figures 4.6, 4.7 and 4.8.

The master degradation curves for natural rubber show some interesting trends. As anticipated, as the material ages measurable parameter performance suffers. With respect to the log of volume change and mass change, the testing indicates a linear change in performance as a function of log time. While this may be assumed for short test windows, the extension of this to infinite swell at infinite time is untenable. This indicates that the testing did not capture the anticipated curve break or saturation point. However, it is noted that over 100 percent swell was observed in the most severely degraded samples, and this would likely correspond to an application failure.

The breaking stress, breaking strain, and hardness degradation curves displayed a much more gradual degradation with respect to log time. Incorporating both the temperature and concentration accelerating effects, the sample degradation time window was extended to between 10,000 and 100,000 days. All parameters were observed to degrade to an extent that would likely correspond to application failure.

SBR

The degradation curves generated from temperature-accelerated degradation at 1, 30, and 60 ppm were combined using the graphical method to form one master degradation curve that accounted for acceleration based on both temperature and concentration. However, sufficient degradation was not observed in all of the SBR rubber degradation parameters to construct master degradation curves. As such, only the master degradation curves for mass change, volume change and breaking strain could be created. The master degradation curves for SBR are displayed in Figures 4.9 and 4.10.

The master degradation curves for the SBR show some interesting trends. As anticipated, as the material ages measurable parameter performance suffers. With respect to the log of volume change and mass change, the testing indicates a linear change in performance as a function of log time. While this may be assumed for short test windows, the extension of this to infinite swell at infinite time is untenable. This indicates that testing did not capture the anticipated curve break or saturation point. However, it is noted that over 100 percent swell was observed in the most severely degraded samples and this would likely correspond to an application failure.

For most of the degradation test window, the breaking strain curve displayed a much more gradual degradation with respect to log time. However, after time, an apparent threshold was achieved and an accelerated degradation rate followed. After the threshold time was achieved, the elastomer quickly accelerated to a point that would likely correspond to application failure.

Nitrile (NBR)

The degradation curves generated from temperature accelerated degradation at 1, 30, and 60 ppm were combined using the graphical method to form one master degradation curve that accounted for acceleration based on both temperature and concentration. However, sufficient degradation was not observed in all of the nitrile rubber degradation parameters to construct master degradation curves. As such, only the master degradation curves for mass change, volume change and breaking strain could be created. The master degradation curves for nitrile (NBR) are displayed in Figures 4.11 and 4.12.

The master degradation curves for nitrile (NBR) show characteristic trends observed in other materials. As anticipated, as the material ages measurable parameter performance suffers. With respect to the log of volume change and mass change, the testing indicates a linear change in performance as a function of log time. While this may be assumed for short test windows, the extension of this to infinite swell at infinite time is untenable. This indicates that testing did not capture the anticipated curve break or saturation point. However, it is noted that over 100 percent swell was observed in the most severely degraded samples and this would likely correspond to an application failure.

For most of the degradation test window, the breaking strain curve displayed a much more gradual degradation with respect to log time. However, after time, an apparent threshold was achieved and an accelerated degradation rate followed. After the threshold time was achieved, the elastomer quickly accelerated to a point that would likely correspond to application failure.

Neoprene

The degradation curves generated from temperature accelerated degradation at 1, 30, and 60 ppm were combined using the graphical method to form one master degradation curve that accounted for acceleration based on both temperature and concentration. Sufficient degradation was observed in measured degradation parameters to generate master degradation curves. As such, only the master degradation curves for mass change, volume change and breaking strain could be created. The master degradation curves for neoprene are displayed in Figures 4.13, 4.14, and 4.15.

The master degradation curves for neoprene show characteristic trends observed in other materials, however, at a much more accelerated rate. As anticipated, as the material ages measurable parameter performance suffers. With respect to the log of volume change and mass change, the testing indicates a linear change in performance as a function of log time. While this may be assumed for short test windows, the extension of this to infinite swell at infinite time is untenable. This indicates that testing did not capture the anticipated curve break or saturation point. However, it is noted that over 100 percent swell was observed relatively quickly and the severely degraded samples would likely correspond to an application failure.

The breaking stress, breaking strain and hardness degradation curves displayed a much more gradual degradation with respect to log time. Incorporating both the temperature and concentration accelerating effects, the sample degradation time window was extended to between 10,000 and 100,000 days. All parameters were observed to degrade to an extent that would likely correspond to application failure.

EPDM-S

The degradation curves generated from temperature-accelerated degradation at 1, 30, and 60 ppm were combined using the graphical method to form one master degradation curve that accounted for acceleration based on both temperature and concentration. However, it was difficult to perform a time-temperature superposition for EPDM-S as there was not enough measurable change in parameter performance over the experimental time scale to support time shifts. Therefore, only the master degradation curves for mass change and breaking strain could be generated. The master degradation curves for EPDM-S are displayed in Figure 4.16.

EPDM-P

The time temperature concentration time shifts could not be generated for the data obtained from the accelerated aging of the EPDM-P samples. Each measured parameter showed very little degradation through the 1 and 30 ppm concentration tests. Thus, even for the more severely stressed samples at the higher temperature and concentrations, there was no "landmark" to which the curves could be shifted. Since the behavior effectively resulted in a horizontal line across the time scale, there was no definitive method for performing the graphical shift. This limited the potential for finding the material constants for the system for that particular variable.

Field Sample Verification of Degradation Curves

To verify the reasonableness of the master degradation curves, researchers compared them to the real-world degradation values of the field samples. The exposure time and degradation state of the field samples were known, so one would expect a close relationship between the master degradation curves generated in the lab and the degradation-time points obtained from the field samples.

However, as discussed previously, it was not practical to test all parameters for the field samples because the parts were irregularly shaped. Only weight, volume, and hardness degradation parameters were obtained. Two of these parameters—weight and volume—did not have a common reference point with the lab data. Weight and volume were measured in percent change, and field samples and lab samples had different initial values. The field sample data began from its as-received condition—not from its original condition. The lab sample data were based on original condition of un-aged, unexposed materials.

Hardness provided a discrete value that was not based on rate of change. Therefore, hardness data could be correlated to the time-temperature-concentration superposition curves of the lab data, illustrated in Figures 4.17 and 4.18.

Because of the slow degradation of some elastomers at 1 ppm, as discussed above, hardness time-temperature-concentration superposition curves were available only for natural rubber and neoprene. The hardness values from the field samples were over-plotted on these curves.

When reviewing the field data and the master degradation curves together, many of the hardness values for the returned field samples were higher than the initial hardness value for the prepared test coupons. Since it was not possible to know the specific formulation of the returned field samples, it was difficult to know how close they were to the test coupons utilized in the experiments. What is important to note, however, is that the field samples were all to the right of the master curve. In other words, the field samples had degraded at a slower rate than predicted by the master curve. This is most likely due to the fact that only a portion of the field samples had been exposed to chloramines, which would limit the degradation rate of the piece. Therefore, the master curve can be considered a conservative estimate of elastomer life in a chloramines system.

Although data are not available for all elastomers and field samples, the close correlation between the two data sets is encouraging. This correlation indicates that the test procedures reasonably model degradation caused by field exposure for other parameters as well.

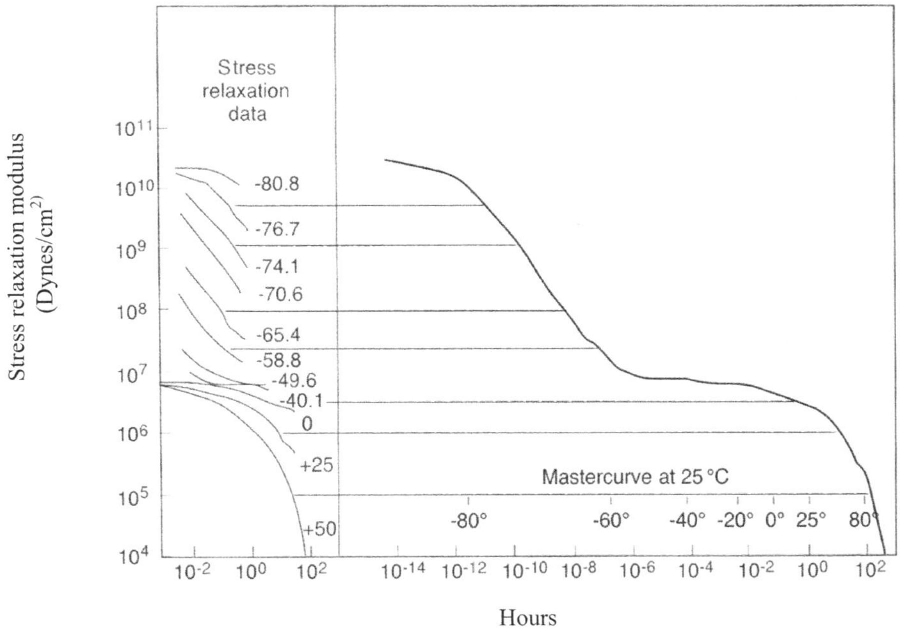

Figure 4.1 Example of 25°C master curve for the relaxation modulus of polyisobutylene (Source: Osswald and Menges 2003)

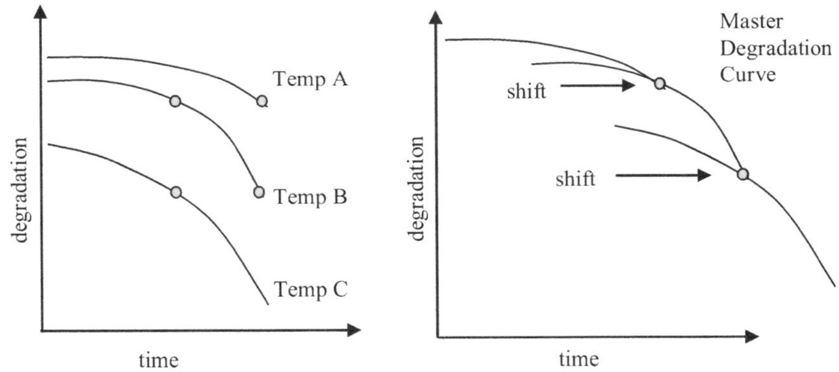

Figure 4.2 Graphical method of time-temperature superposition

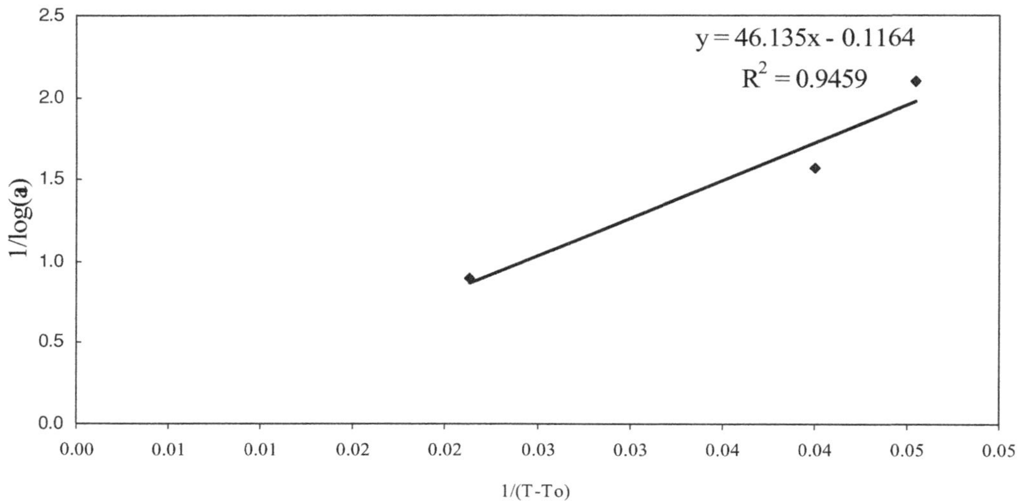

Figure 4.3 Plot of the inverse of log a_t vs. the inverse of ($T-T_{ref}$) for the breaking strain of EPDM-S. The degree of linearity indicates the level of agreement with the Williams-Landel-Ferry equation for time-temperature superposition.

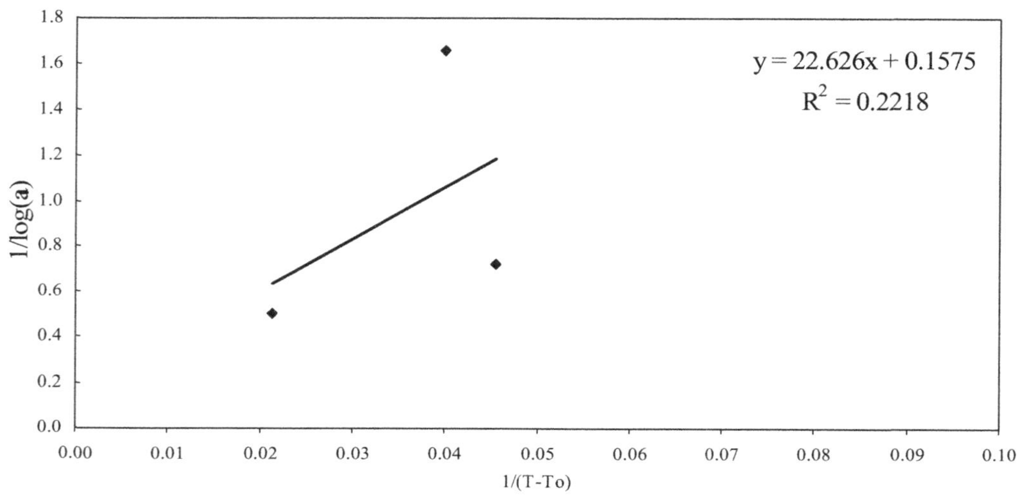

Figure 4.4 Plots of the inverse of log a_t vs. the inverse of ($T-T_{ref}$) for the percent mass change in EPDM-S on exposure to chloramines solutions. The degree of linearity indicates the level of agreement with the Williams-Landel-Ferry equation for time-temperature superposition.

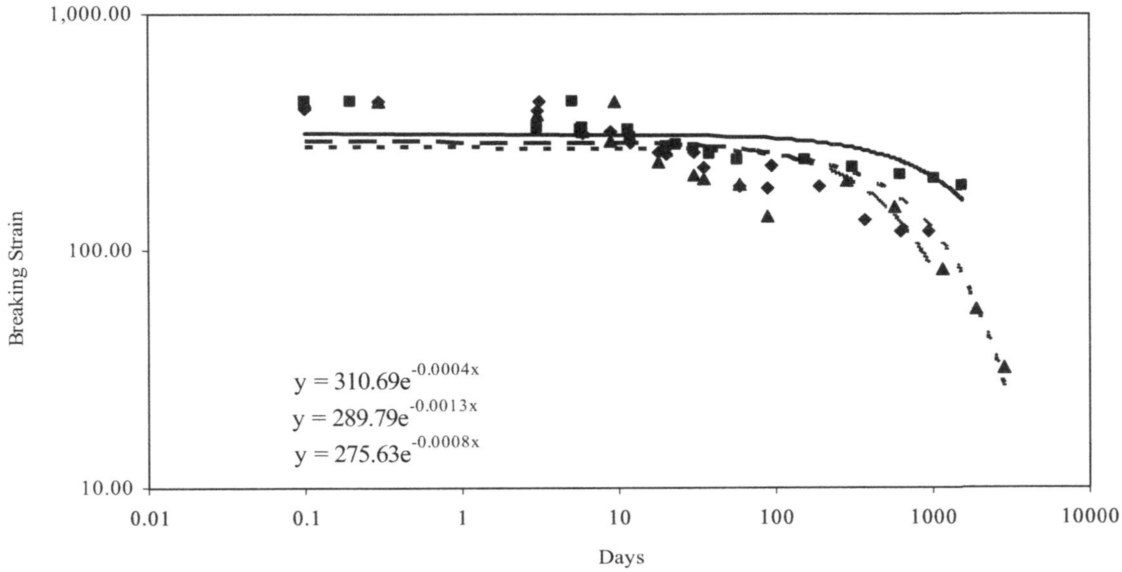

Figure 4.5 Time-temperature superposition curve for breaking strain for SBR

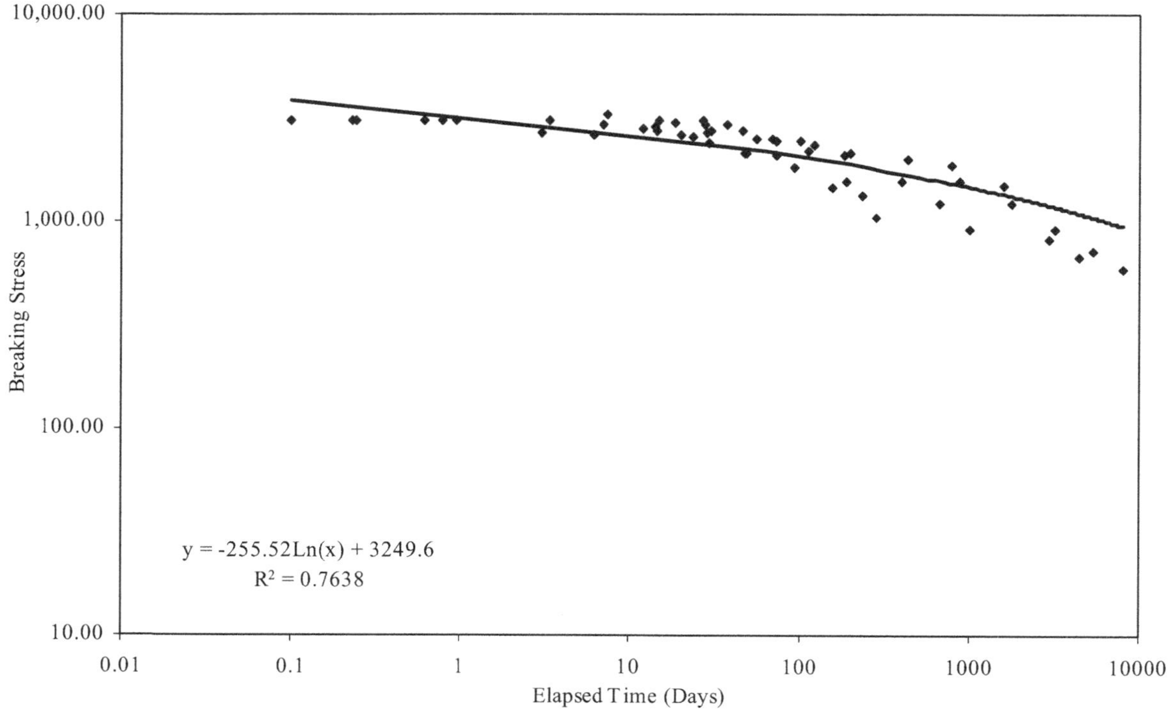

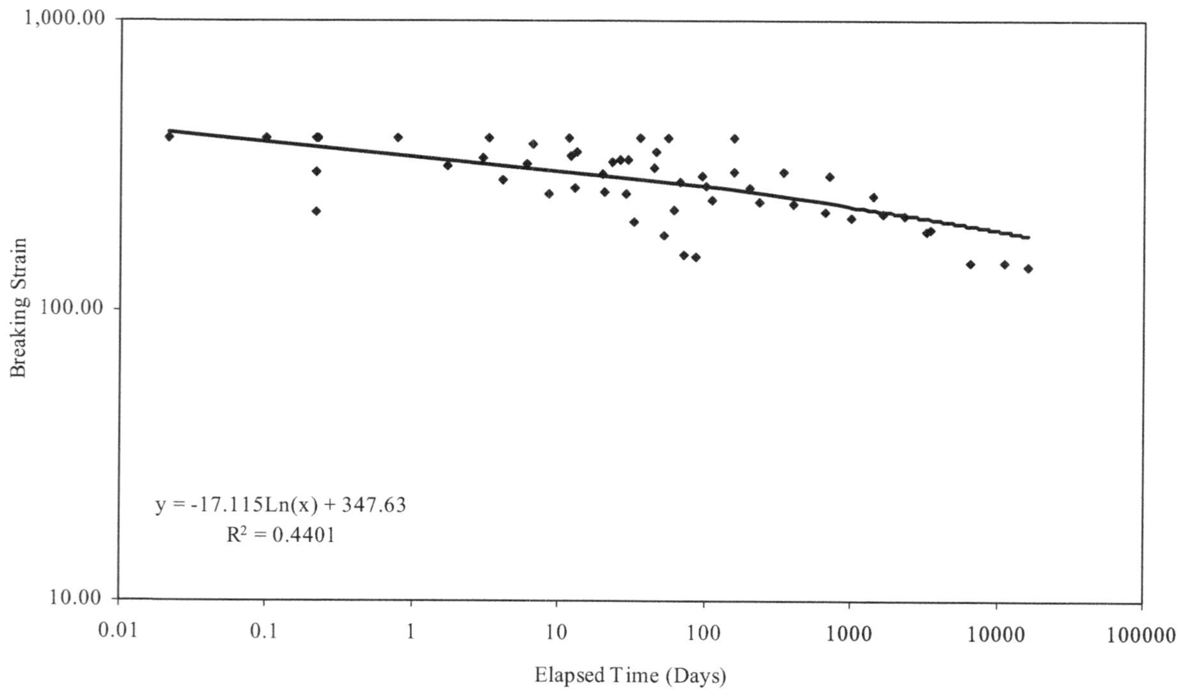

Figure 4.6 Master degradation curves of breaking stress and breaking strain for natural rubber

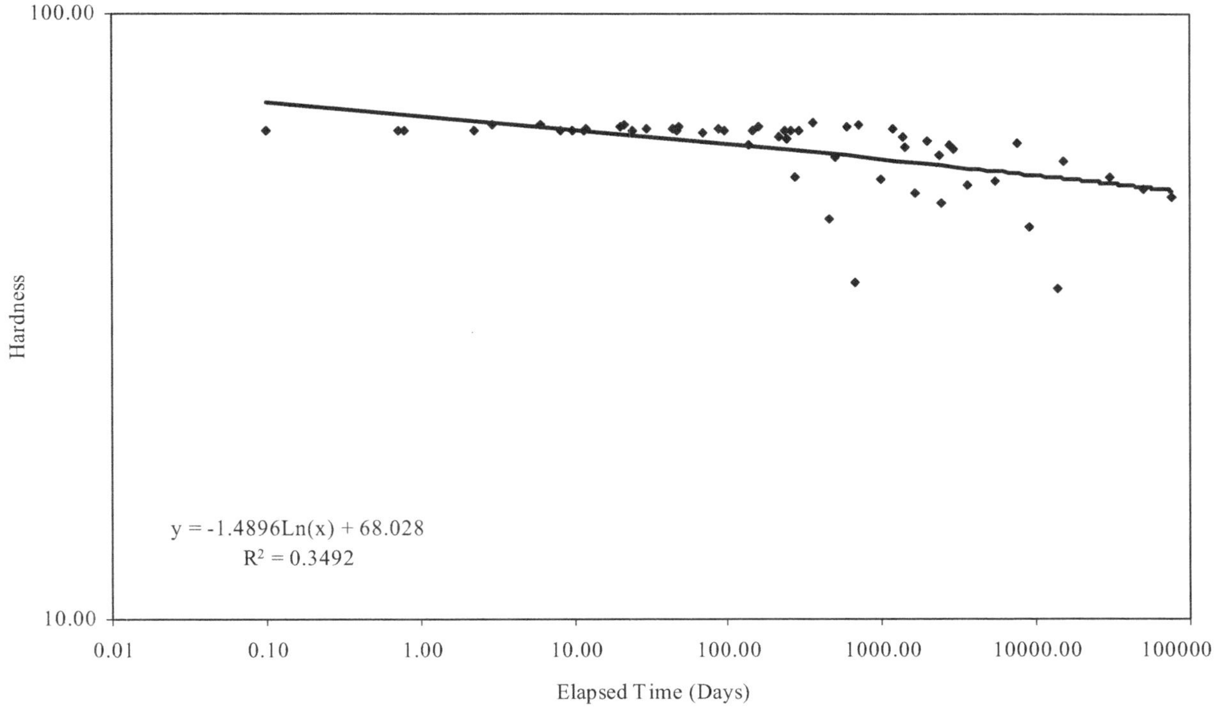

Figure 4.7 Master degradation curves of hardness for natural rubber

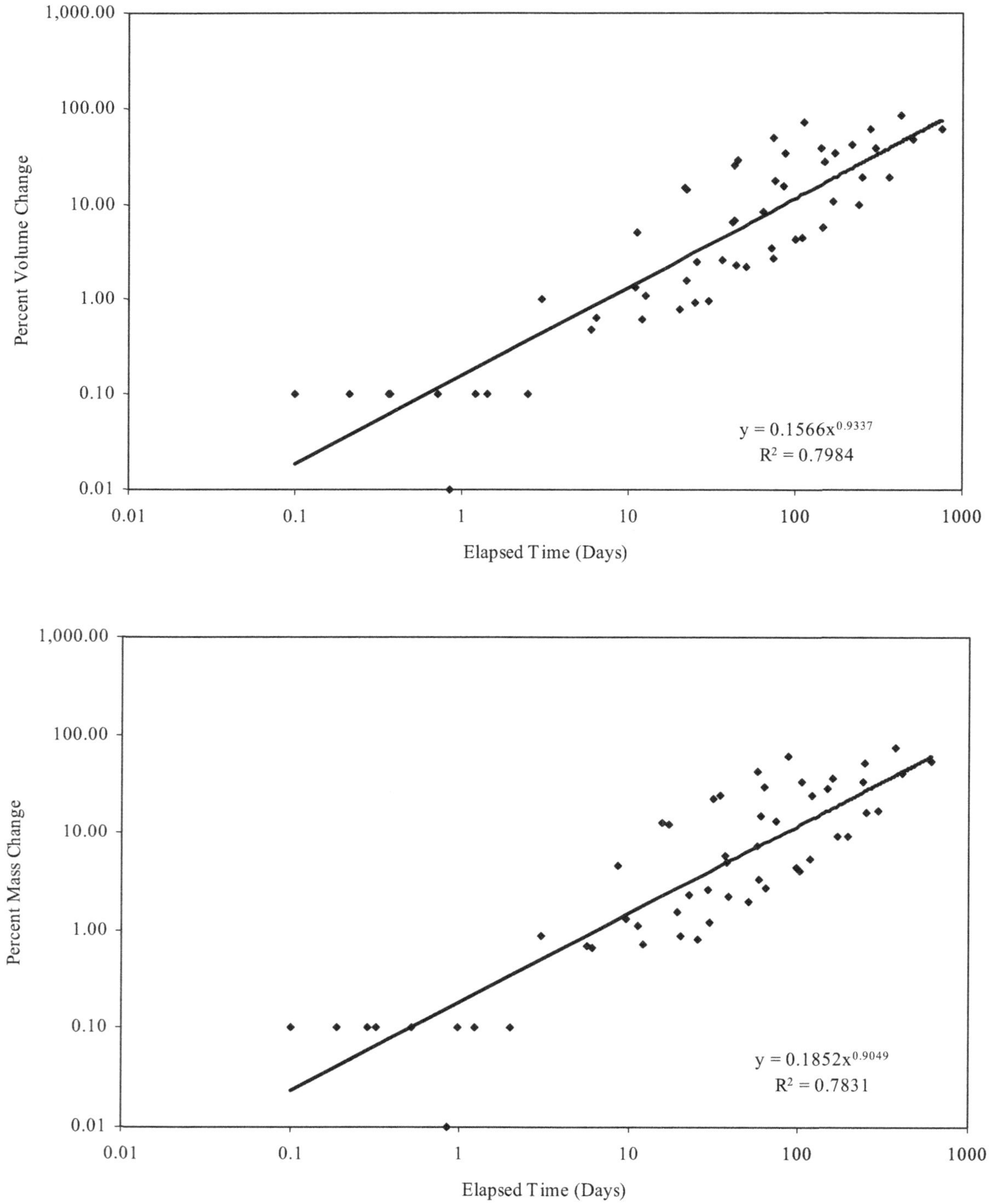

Figure 4.8 Master degradation curves of percent mass change and percent volume change for natural rubber

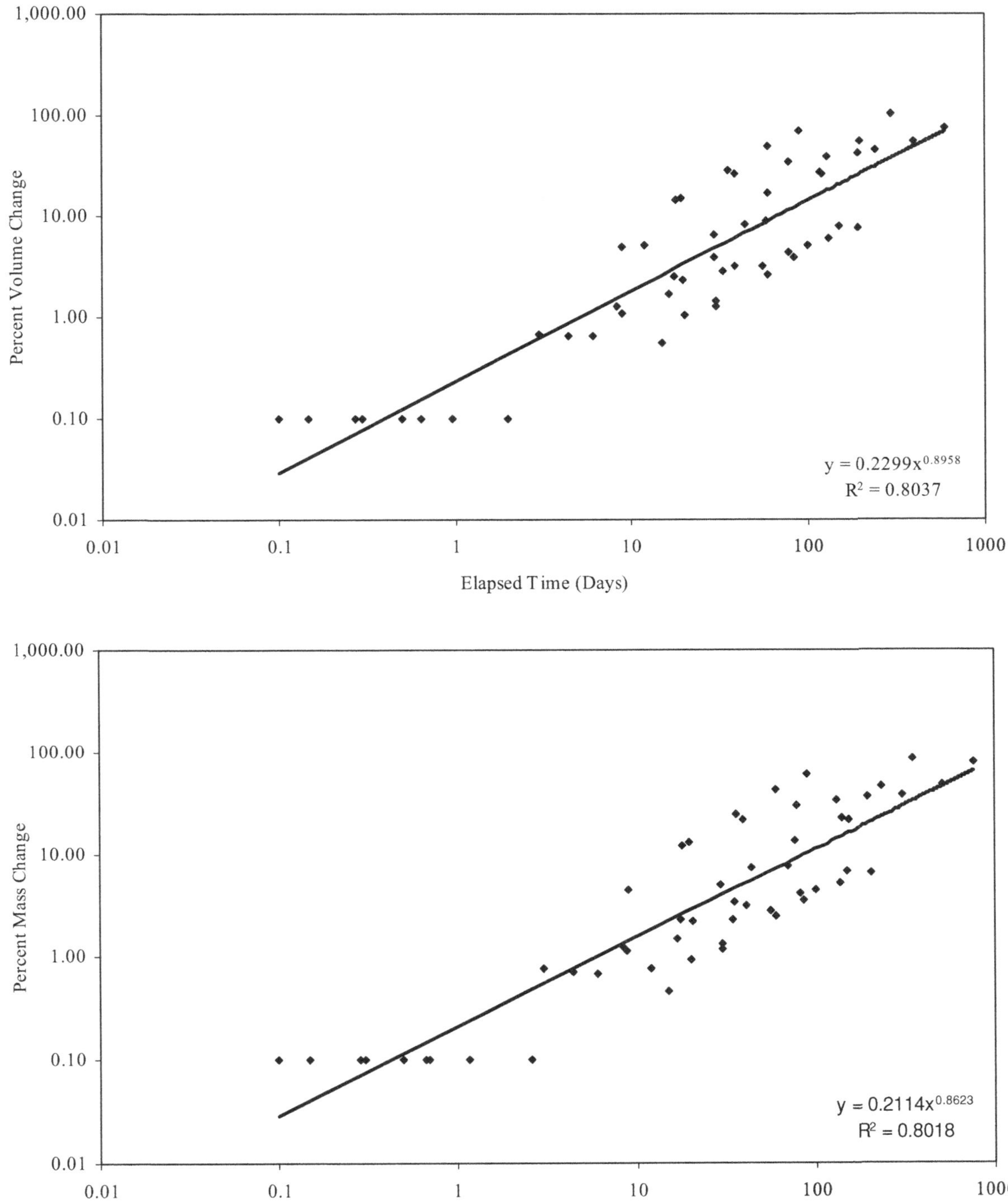

Figure 4.9 Master degradation curves of percent mass change and percent volume change for SBR rubber

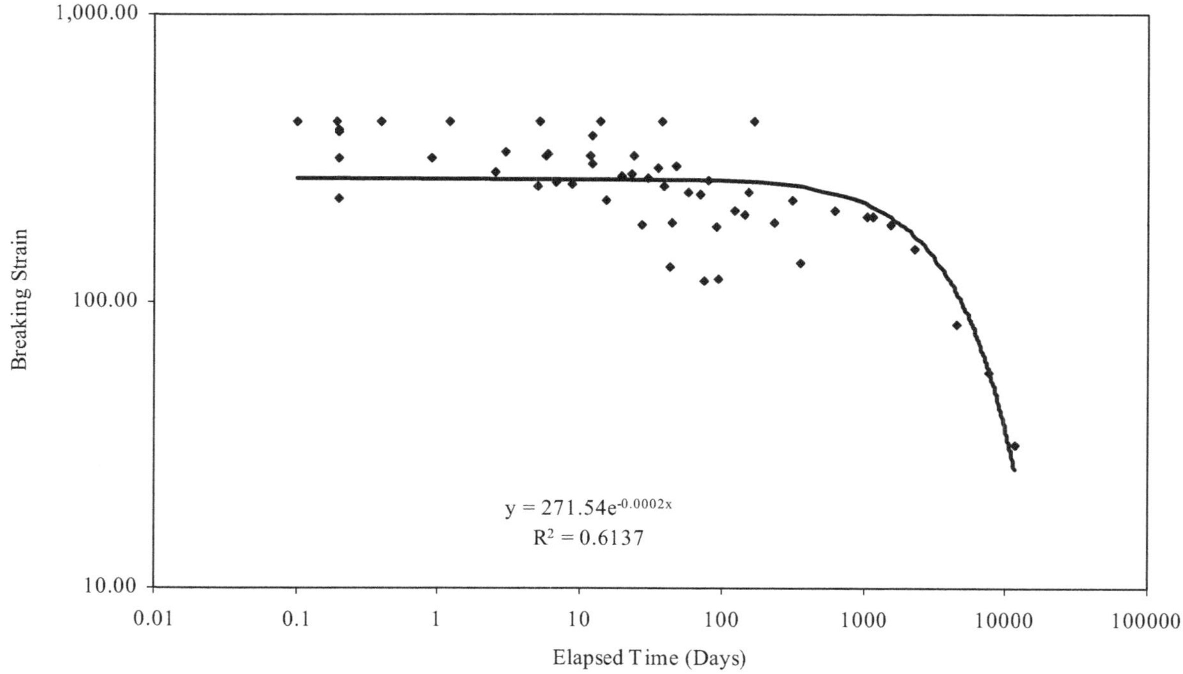

Figure 4.10 Time-temperature superposition curves for breaking strain for SBR

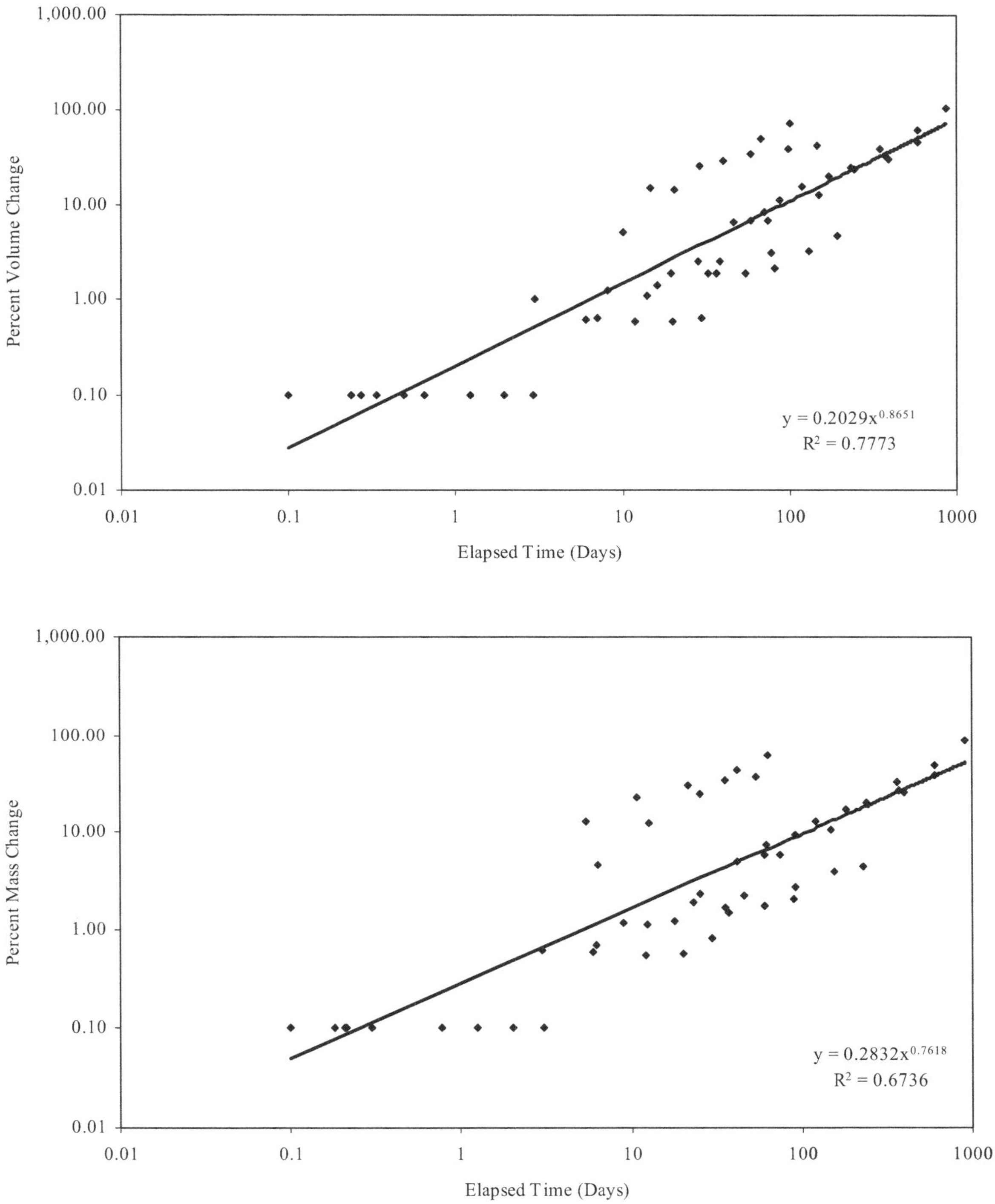

Figure 4.11 Master degradation curves of percent mass change and percent volume change for nitrile (NBR)

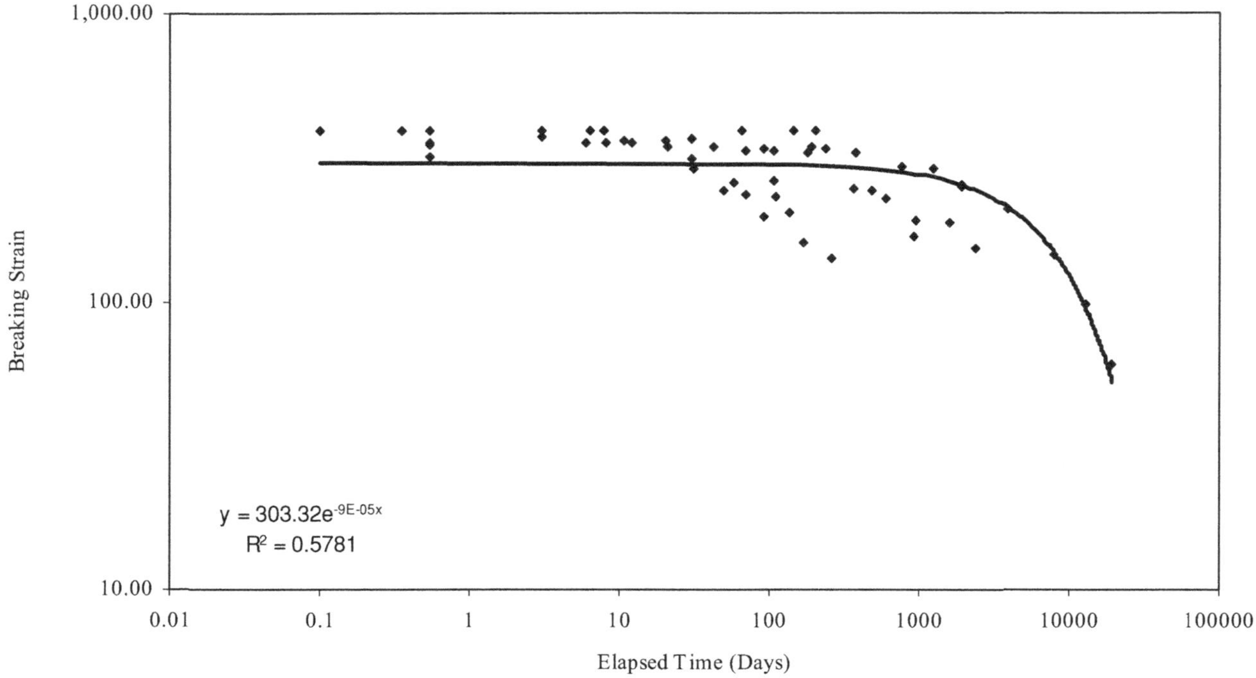

Figure 4.12 Master degradation curves of breaking strain for nitrile (NBR)

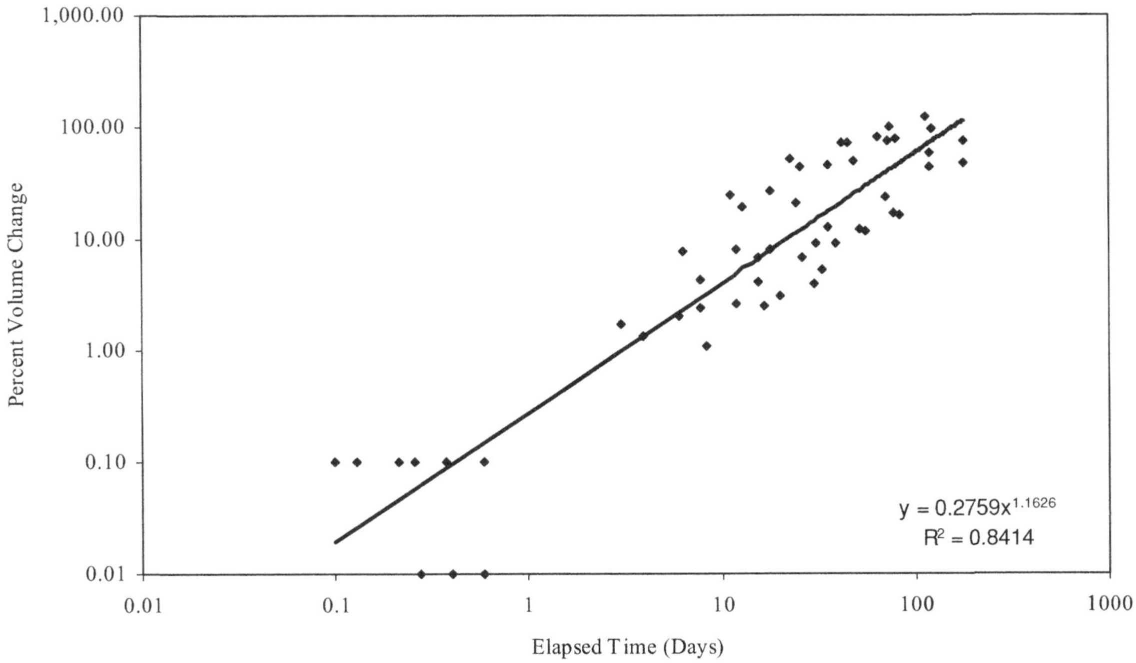

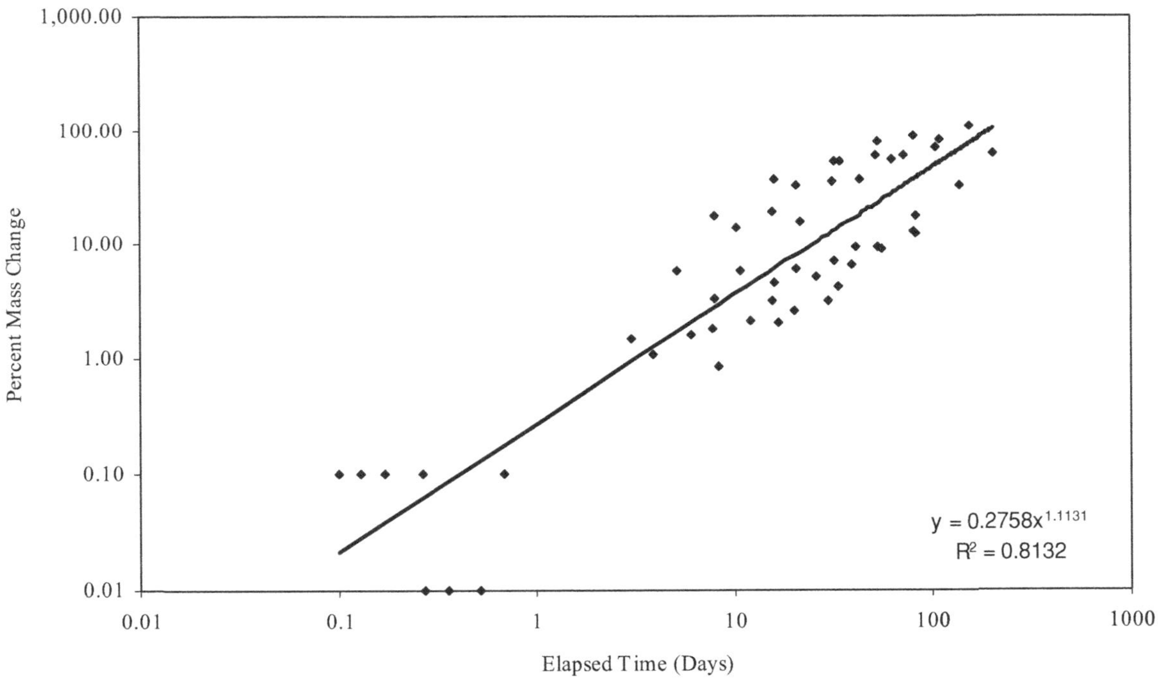

Figure 4.13 Master degradation curves of percent mass change and percent volume change for neoprene

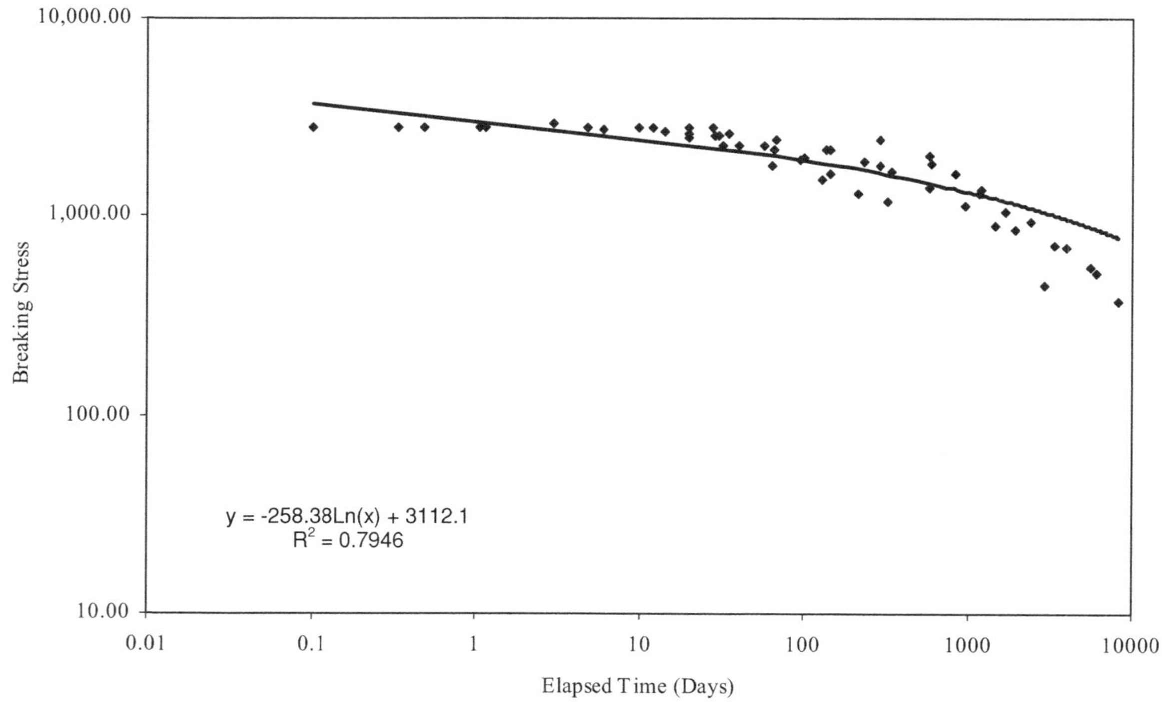

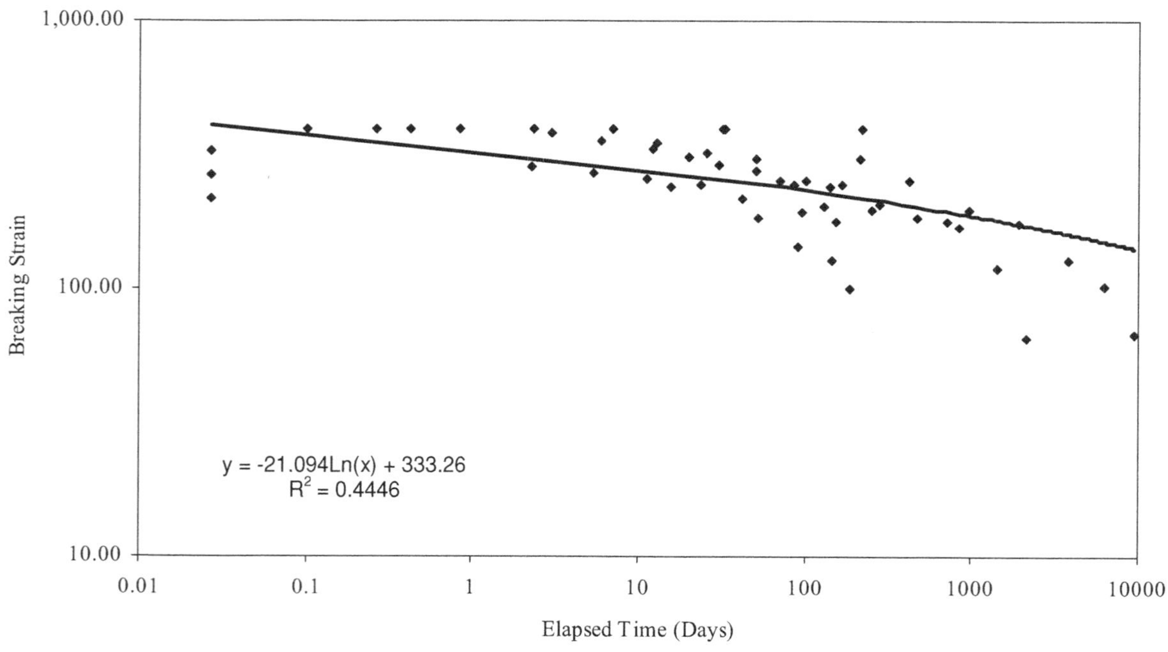

Figure 4.14 Master degradation curves of breaking stress and breaking strain for neoprene

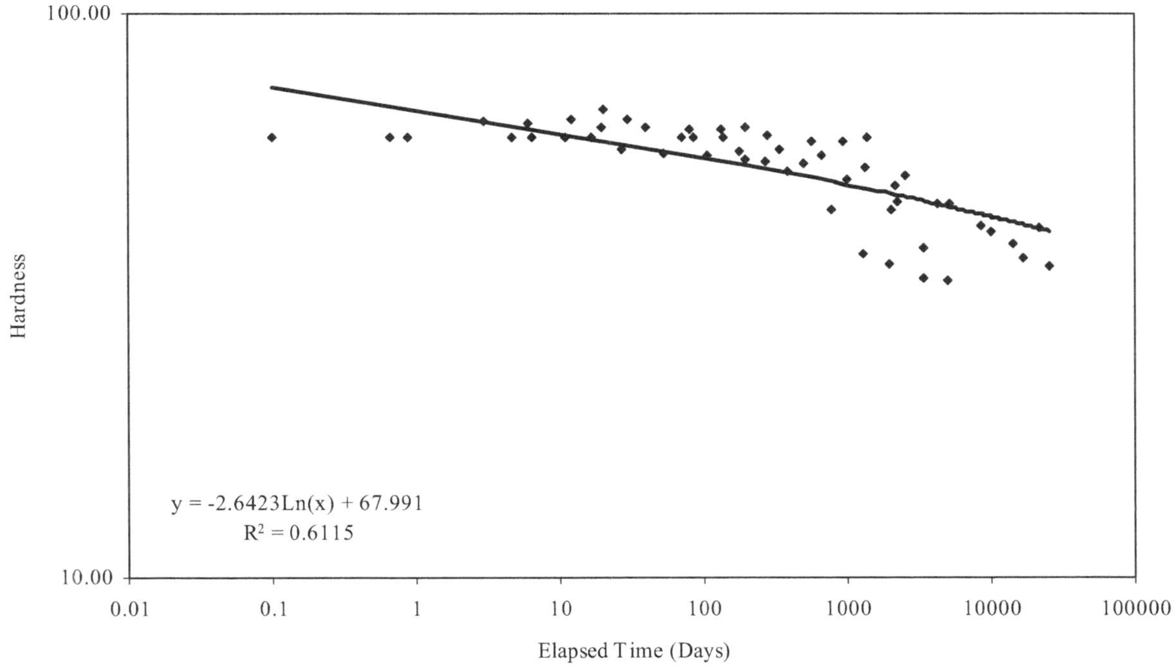

Figure 4.15 Master degradation curves of hardness for neoprene

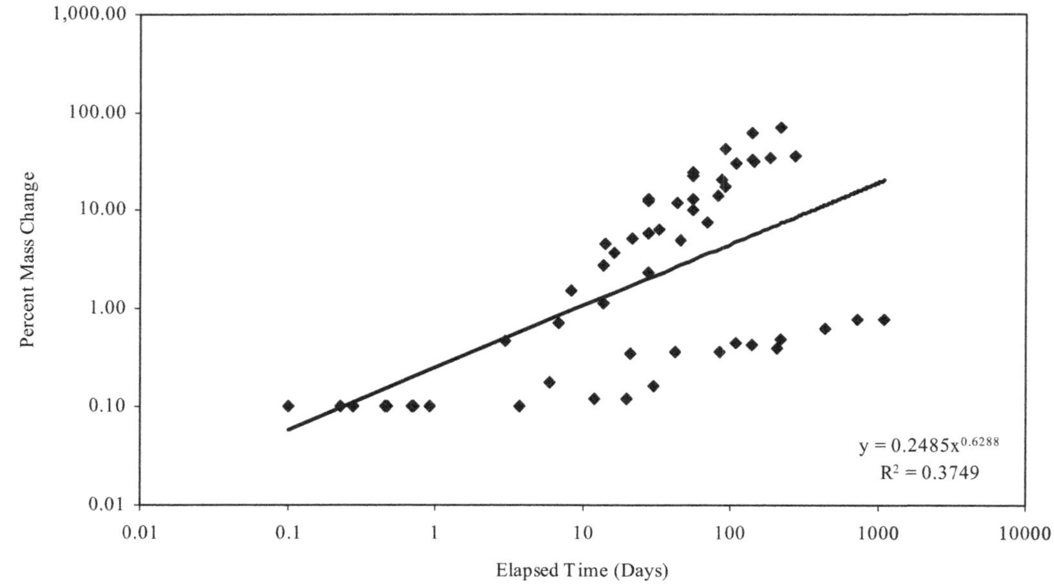

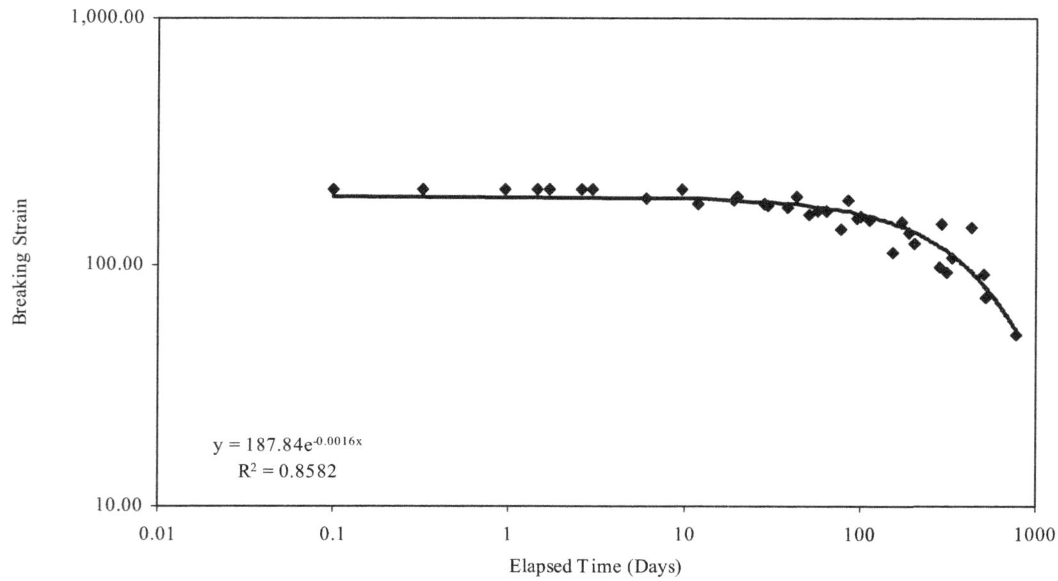

Figure 4.16 Time-temperature superposition curves of percent mass change and breaking strain for EPDM-S

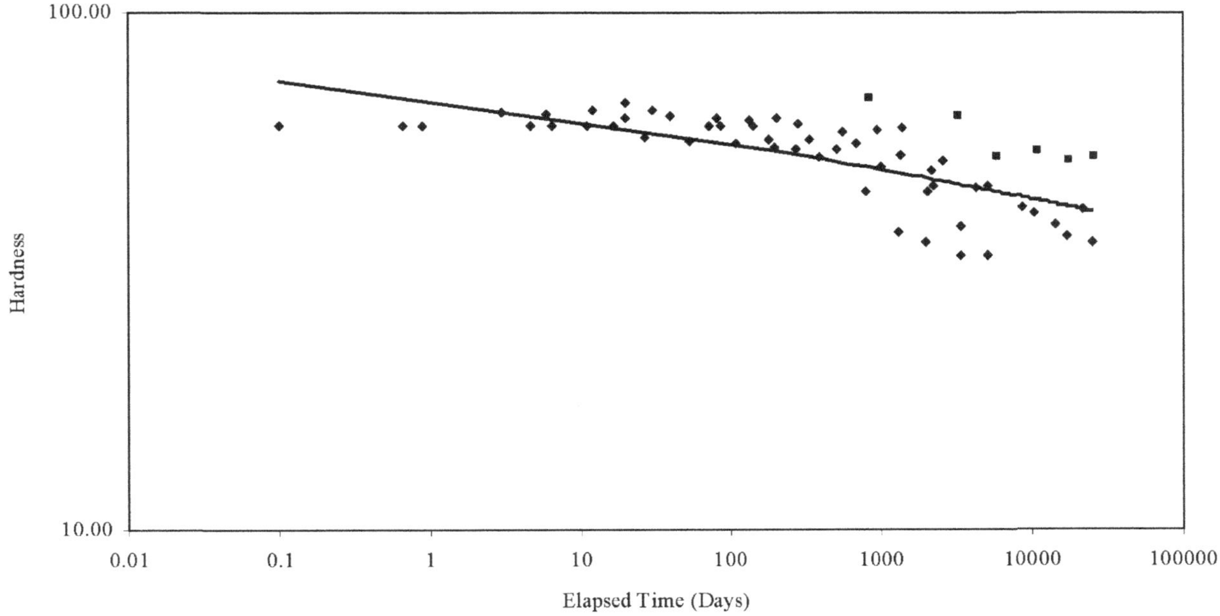

Figure 4.17 Master degradation curve of hardness for neoprene with shifted field return accelerated test results

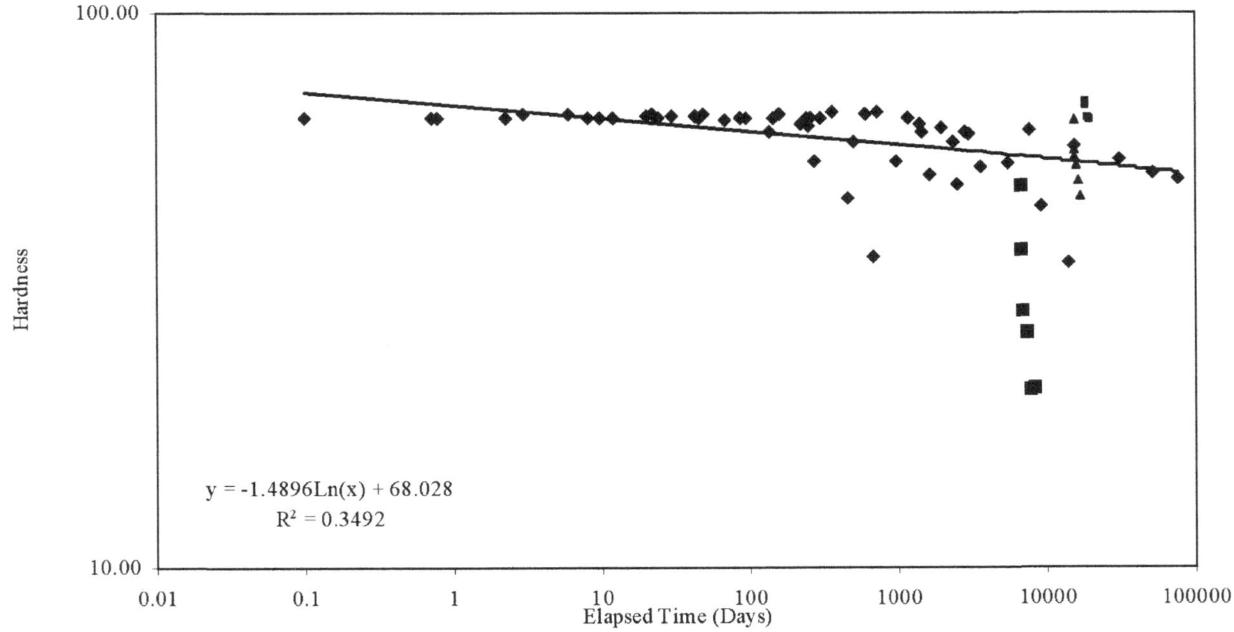

Figure 4.18 Master degradation curve of hardness for natural rubber with shifted field return accelerated test results

CHAPTER 5
SURFACE-AREA-TO-MASS RATIO EFFECTS ON DEGRADATION

INTRODUCTION

The results discussed in the previous chapters provide a unique insight into the degradation of a variety of rubbers utilized in the water utility infrastructure when exposed to chloramines. The effects of concentration and temperature on the degradation have been used to provide conservative estimates of the lifetime of elastomers. There is one additional variable that will, in part, determine the lifetime of these components, namely, the surface-area-to-mass ratio of the part in question.

The overall amount of chloramines absorbed by the elastomer in question is a strong function of both the time of exposure and the area exposed. This is easily seen by returning to Fick's First law, which states that the diffusional flux (J) is equal to the diffusion coefficient (D) multiplied by the concentration gradient across the part:

$$J_i = -D_i\left[\frac{dC}{dx}\right] \tag{5.1}$$

It is also important to note that the diffusion flux is equal to the mass that diffuses through the rubber (M) divided by the time (t) over which the diffusion takes place and the area (A) across which the diffusion takes place:

$$J_i = \frac{M}{t \cdot A} \tag{5.2}$$

The result of this is that as either the time of diffusion or the area across which diffusion occurs is increased, there is a corresponding increase in the amount of chloramines that the rubber will uptake, and hence, the amount of degradation that will occur.

As the surface-area-to-mass ratio is decreased, the amount of material present within the part is increased. This should have a marked effect on the lifetime of a given component based on specific design parameters. It has been discussed in previous chapters that elastomer degradation is primarily driven by surface phenomena. Because of this, it is expected that surface-specific properties, such as hardness, should not be affected by the surface-to-mass ratio, but bulk properties, such as the mass and volume change, could change dramatically. This would primarily be due to the fact that while the diffusion of chloramines into the bulk will most likely not increase, the ultimate degree of swelling will increase due to the increase in material present in the bulk of the part. The degradation in mechanical properties is more difficult to predict as it would be expected that as the mass is increased, the rate of degradation should decrease, but it is also known that as the volume of a part is increased, so too is the average defect size within the part increased. This is why larger parts are typically weaker than smaller parts of the same material. It is unclear what effect the defect size within the part will have on the chloramines-induced degradation.

To more fully investigate the influence of the of the surface-area-to-mass ratio on the elastomer degradation, the testing protocol discussed previously was applied to a series of newly synthesized elastomer samples where the surface-to-mass ratio was decreased from the original 600 in^2/lb to 300 in^2/lb and 150 in^2/lb by increasing the elastomer sheet thickness from 0.080 in to 0.160 in and 0.320 in respectively. Since this was not meant to be an exhaustive study but only provide additional information on which material choices can be made, it was decided that these new samples would be exposed to the 1 and 30 ppm chloramines concentrations and 23 and 70°C temperatures used previously. Additionally, the elastomers chosen for the study were limited to natural rubber as a baseline material, SBR as the most common elastomer used, and EPDM-S and EPDM-P as elastomers formulated to resist chloramine attack. The results of this study are presented within this chapter and combined with data from prior measurements, discussed earlier within this text, for a more detailed discussion of degradation behavior and the influence of surface-area-to-mass ratio.

Natural Rubber

As predicted, the mass and volume change for the thicker samples (lower surface-area-to-mass ratio) did not occur at a faster rate than the original samples (Figure 5.1). This should not be surprising as the overall flux of chloramines into the sample is limited by the surface area of the part, not by its mass. Since there is little change in the surface area from the 0.080 in samples to the 0.320 in samples, one would expect little change in the flux. Increasing the chloramines concentration (Figure 5.2) increased the flux by increasing the driving force, i.e., the concentration gradient, across the sample. Predictably, temperature showed the greatest influence on the mass and volume change due to elastomer chain relaxation between the cross-links (Figure 5.3). It is interesting to note that, as predicted, the ultimate degree of swelling did increase as the part gained more mass, as seen in Figure 5.4 for the high temperature, high chloramines results. It eventually leveled off at a mass increase of ~60 percent and a volume increase of ~70 percent for the 0.320 in samples as compared to the ~30 percent mass and ~40 percent volume increase for the 0.080 in samples. This means that an increase of mass by a factor of four led to only a doubling of the mass increase within the part over the extended testing period.

Like the change in mass and volume, the hardness for the thicker samples followed the same general trend as those of the thinner samples (Figure 5.5), though the rate of decrease increased with concentration (Figure 5.6) and more so with temperature (Figure 5.7). At the highest temperature and concentration (Figure 5.8), the hardness could be seen to decrease to a plateau level of 25 percent of its original hardness. It is important to note that in each of the cases, the hardness of the thicker samples (0.160 inch and 0.320 inch) is less than that for the original rubber samples (0.080 inch). This could simply be a matter of the presence of larger defect sites in the surface regions of elastomer samples, which would make it easier for the chloramines to attack the surface region. Another possibility could be that this is an effect of the manufacturing method as the original 0.080 inch samples were molded in a laboratory environment whereas the 0.160 inch and 0.320 inch samples had to be created on the manufacturing line. The two different methods may have led to subtle microstructure differences between the samples, which would have had a great influence on how the surface was attacked by the chloramines.

The bulk scale mechanical properties in the form of both the tensile strength and percent elongation for the thicker samples followed almost identical trends to the thinner samples. The tensile strength for the 0.080 inch samples was always higher than that of the 0.160 and 0.320 inch samples (Figures 5.1 through 5.4). As discussed, this trend was related to the manufacturing process producing fewer defects at the thinner sample size, which leads to an increase in strength. Additionally, the percent elongation was always lower for the 0.080 inch samples, which most likely relates to the increase in the amount of material, which allows for an increase in the amount that the bulk part can stretch before it fractures. It is interesting to note that at the highest temperature and concentration (Figure 5.4), both the breaking stress and percent elongation seemed to decrease to similar plateau values for all of the thicknesses. These values were similar to those expected for the rubber near the end of its lifespan. Since we can assume that there had been a large amount of damage done to the cross-linked structure by this time, it is reasonable to assume that these values related to the overall stress and strain behavior of the elastomer chains, which is typically much lower than the cross-linked, carbon black filled rubber.

EPDM-S

The results for EPDM-S showed very little variation in either the rate of degradation in or the absolute values of the mass and volume change with respect to the surface-area-to-mass ratio (Figures 5.9 through 5.12). This is a bit surprising as this result seems to indicate that part longevity is independent of its dimensionality. It is also interesting to note that at 30 ppm chloramines concentration, the values for both variables seemed to plateau. At room temperature (Figure 5.10), the values seemed to be 8 percent mass gain and 9 percent volume change, though these values may have been a temporary plateau as both figures seemed to have a step-like appearance. At the higher temperature (Figure 5.12), these values were much higher (~90 percent mass gain and ~100 percent volume increase) which in part results from the change in the mechanical properties of the polyethylene/propylene backbone with temperature. These materials are semi-crystalline thermoplastics and will vary much more in terms of physical properties as those crystalline regions are broken down. While it would be highly unlikely that potable water infrastructure would be subjected to such temperatures during its lifespan, continuous degradation of the elastomer will eventually cause the same degree of degradation. These values have to be considered the high end of expansion, although it is difficult to determine when they would actually reach this state.

In terms of mechanical properties, the breaking stress and breaking strain varied little over the time of the experiment. The only set of conditions that showed any significant decrease in properties was at the highest temperature and concentration (Figure 5.12). Like natural rubber, the 0.080 inch samples always had the highest breaking stress and lowest percent elongation due to the reasons already discussed. Hardness also varied little over the course of the experiment (Figures 5.13 through 5.16) and, as with the breaking stress, the only major decrease in properties came at the highest temperature and concentration (Figure 5.16). Unlike natural rubber, the hardness of the 0.080 inch samples was always lower than those of the thicker samples. The reason for this is unclear, but is most likely due to the change in manufacturing method, perhaps leading to a higher degree of crystallinity at the surface of the thicker samples, which would lead to a higher hardness.

SBR

The increase in weight and volume for SBR was lower for the thicker samples than for the thin ones, although the rate of increase appeared to be consistent, especially at the lower temperature (Figures 5.17 and 5.18). At the higher temperature the thicker samples began to gain in more mass and volume in the period between 20 to 30 days (Figures 5.19 and 5.20). As stated previously, as the area over which diffusion of chloramines can occur does not change between the samples, one would not expect the rate of degradation to vary significantly. What should vary is the ultimate magnitude of the mass and volume increase. SBR proved this to be the case as the thinner samples, for the most part, appeared to reach a plateau value that was lower than that for the thickest (0.320 inch) sample.

The degradation of the bulk mechanical properties as a function of surface-area-to-mass ratio of SBR varied from those of the previously discussed elastomers. While the percent elongation of the 0.080 inch sample was always lower than those of the other two thicknesses, the breaking stress was only higher for the entire experimental time at the lower temperature (Figures 5.17 and 5.18). At the higher temperature, the breaking stress for the 0.080 inch sample started out higher than the thicker samples, but dropped below their breaking stress around day 10 (Figures 5.19 and 5.20). After this point, all three thicknesses showed a very similar rate of decrease in the breaking strain. As was shown in the previous chapter, SBR seemed to lose carbon black from its interior during the initial stage of degradation, which may have been the cause of this behavior. The thicker samples had a larger cross-sectional area that provided more material to resist this initial carbon black loss, although the dynamics of degradation eventually caught up with the thicker samples and the rate of loss between the various thicknesses began to equalize. This explains the similar slopes for all three thicknesses in the breaking stress graphs, especially at the higher temperature and concentration (Figure 5.20).

The hardness results also showed very similar trends in the bulk mechanical properties. At the lower temperature (Figures 5.21 and 5.22), the hardness seemed to be a function of thickness, with the thinnest sample having the lowest hardness values and the thickest sample having the highest values over the course of the experiment. This is again can most likely be attributed to the manufacturing process as opposed to some inherent material property. The trend of the thinnest sample having the lowest hardness value also seemed to hold at the high temperature/low concentration conditions (Figure 5.23). It is at the high temperature/high concentration conditions (Figure 5.24) that there was a major change in this behavior. At these conditions, there was a sharp decrease in the hardness values for both of the thicker samples after day 10, which was not seen in the 0.080 inch sample. Physical observation of the thicker samples at these conditions showed that they took on an almost spongy texture with the outermost layer being easily wiped away with any mechanical contact. While this is somewhat surprising, it can once again be linked to the increase in material, as more material leads to a larger degree of swelling. The decrease of interior carbon black, which provided a certain degree of hardness to the whole sample, led to a softening of the whole sample, as opposed to a softening of just the outermost portion of the sample as seen with the other elastomers. Again, this should be taken as a worst-case scenario of the condition of the elastomer near the end of its lifespan.

EPDM-P

As discussed in previous chapters, EPDM-P appeared to be fairly resistant to attack by chloramines, and decreasing the surface-area-to-mass ratio did not change this assessment. For both concentrations at the lower temperature, there appeared to be an initial decrease in mass and volume over the course of the experiment. Both properties appeared to level off or increase slightly after 60 days of accelerated aging (Figures 5.25 and 5.26). This initial loss of mass an volume may have been due to a loss of carbon black, as shown by the turbidity data, or slight losses of the extending oils used in the EPDM-P manufacturing process. This trend did eventually reverse itself as the higher temperature/lower concentration conditions results show (Figure 5.27). This result is logical if one considers the process that is occurring. As the chloramines attack the elastomer, chlorines become incorporated into the EPDM-P. Polyethylene/propylene is typically hydrophobic, but as chlorine is added to the matrix the material would become less hydrophobic. This is due to the fact that chlorine would act as a hydrogen bonding site within the rubber matrix, allowing for the greater infiltration of water into the rubber. Eventually, the EPDM-P sample did swell like the other elastomers, as can be seen for the high temperature/high concentration conditions (Figure 5.28). The ultimate extent of swelling was ~65 percent by mass and ~75 percent by volume, although the time needed to reach this degree of swelling would be presumably much longer than for the other elastomers tested.

The bulk mechanical properties seemed to vary little over the course of the testing, with only the high temperature/high concentration conditions showing any appreciable decrease in the values over time (Figure 5.28). As with most of the other elastomers, the breaking stress for the 0.080 inch samples was always higher than the other two samples, due mostly to the defect size in the final part. It is interesting to note that the percent elongation did not seem to depend on sample thickness as the values were nearly identical for every thickness at all conditions (Figures 5.25 through 5.28). These results were mirrored in the results of the hardness testing over the course of the accelerated testing (Figures 5.29 to 5.32). As with the other elastomers, the hardness of the 0.080 inch samples was lower than the hardness for the thicker samples over the course of the testing. The only time this changed was at the high temperature/high concentration conditions where the thicker samples showed a rapid decrease in hardness after 30 days. The plateau value for the 0.320 inch sample was much lower than the apparent plateau value for the 0.080 inch sample. Like SBR, this may have been the result of a loss of carbon black in the interior of the sample, which was compounded by the increase in the amount of material in the sample itself. Given the nature of EPDM-P, it cannot be ruled out that this was the result of a preferential decomposition of crystalline regions near the surface, the loss of which would also be compounded by the increased mass within the sample itself.

Surface to Mass Ratio Conclusions

Based on the results presented in this chapter, one can conclude that the surface-area-to-mass ratio of a given part has very little influence on the chloramine degradation process. This, in part, stems from the fact that the diffusion of chloramines into the rubber component is a strong function of the area over which diffusion can occur.

Without increasing the exposed surface area, the added mass did not change the rate at which degradation occurred, but it did change the ultimate extent of the degradation based on the added mass of the component. Specifically, the rates at which the mass and volume increased remained fairly constant for the three different thicknesses of the elastomers investigated, but in each case, the thicker samples (lower surface-area-to-mass ratio) had a much higher degree of swelling at the end of the testing procedure. The rate of change in the bulk mechanical properties was also consistent among the three thicknesses. The breaking stress was always highest for the 0.080 inch samples because of the decreased defect size in the manufactured component. The percent elongation was always highest for the 0.320 inch samples, except in the case of EPDM-P where the percent elongation was apparently independent of sample thickness. The hardness values, like the bulk mechanical properties, seemed to degrade at the same rate regardless of the sample thickness, except at the higher temperature/higher concentration conditions where the thicker samples of SBR and EPDM-P samples seemed to undergo a sharp decrease in hardness. This behavior was possibly due to the loss of carbon black from the interior of the sample or, in the case of EPDM-P, the loss of crystalline regions. In all cases, except natural rubber and the case just discussed, the thicker samples had higher hardness values through the course of the testing as compared to the original 0.080 inch sample. Because of this, one can surmise that the ultimate lifetime of a part in service will change very little based on the surface-area-to-mass ratio of the part, since the rate of degradation is nearly identical for each of the sample thicknesses tested. It is expected that reducing the exposure area would have a greater effect on the ultimate lifetime of the part, as it would limit the flux of chloramines into the component. Further research is needed to fully investigate this particular phenomenon.

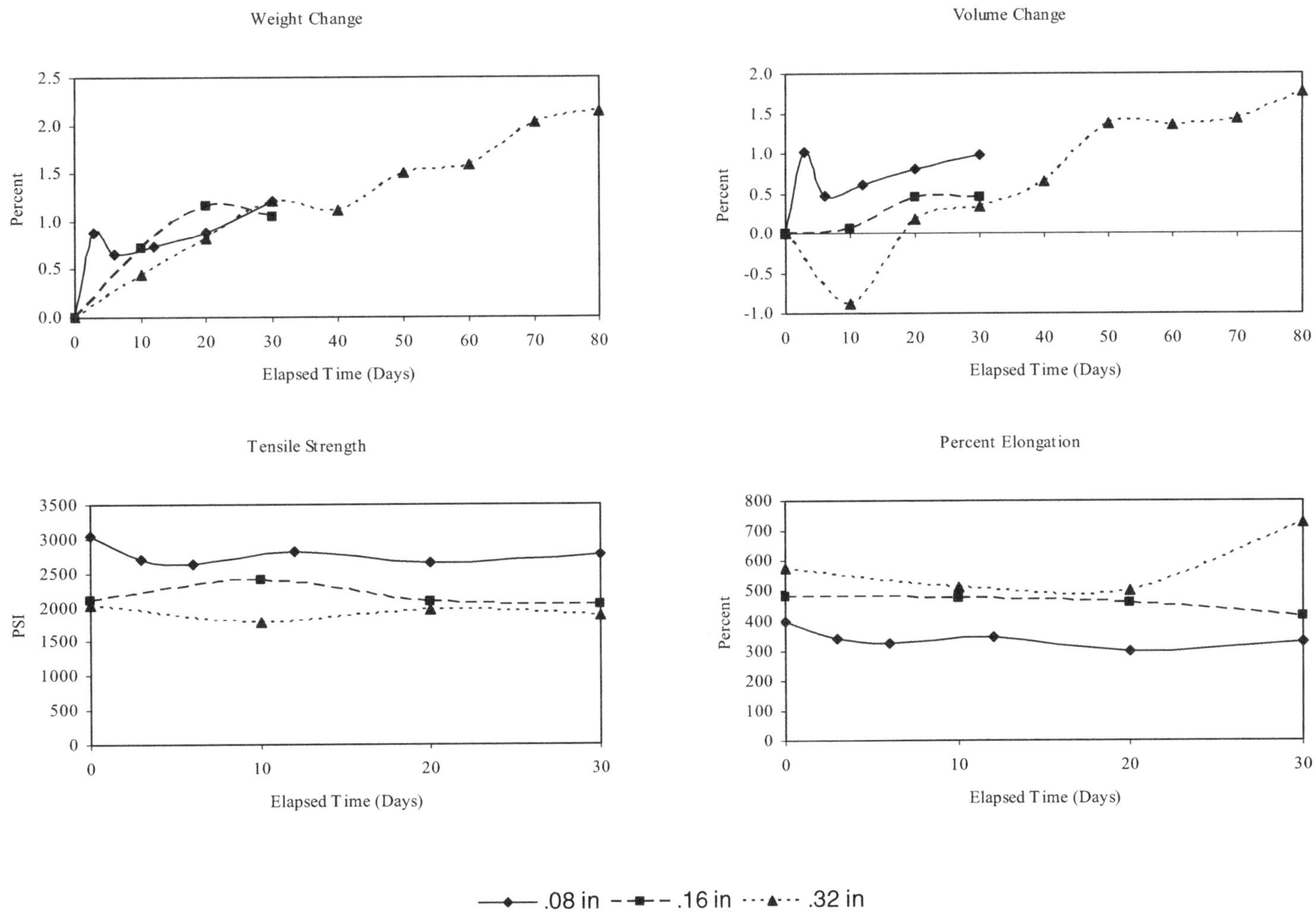

Figure 5.1 Effect of thickness on elastomeric properties at 1 ppm chloramine, 25°C, and pH 8.3 of natural rubber

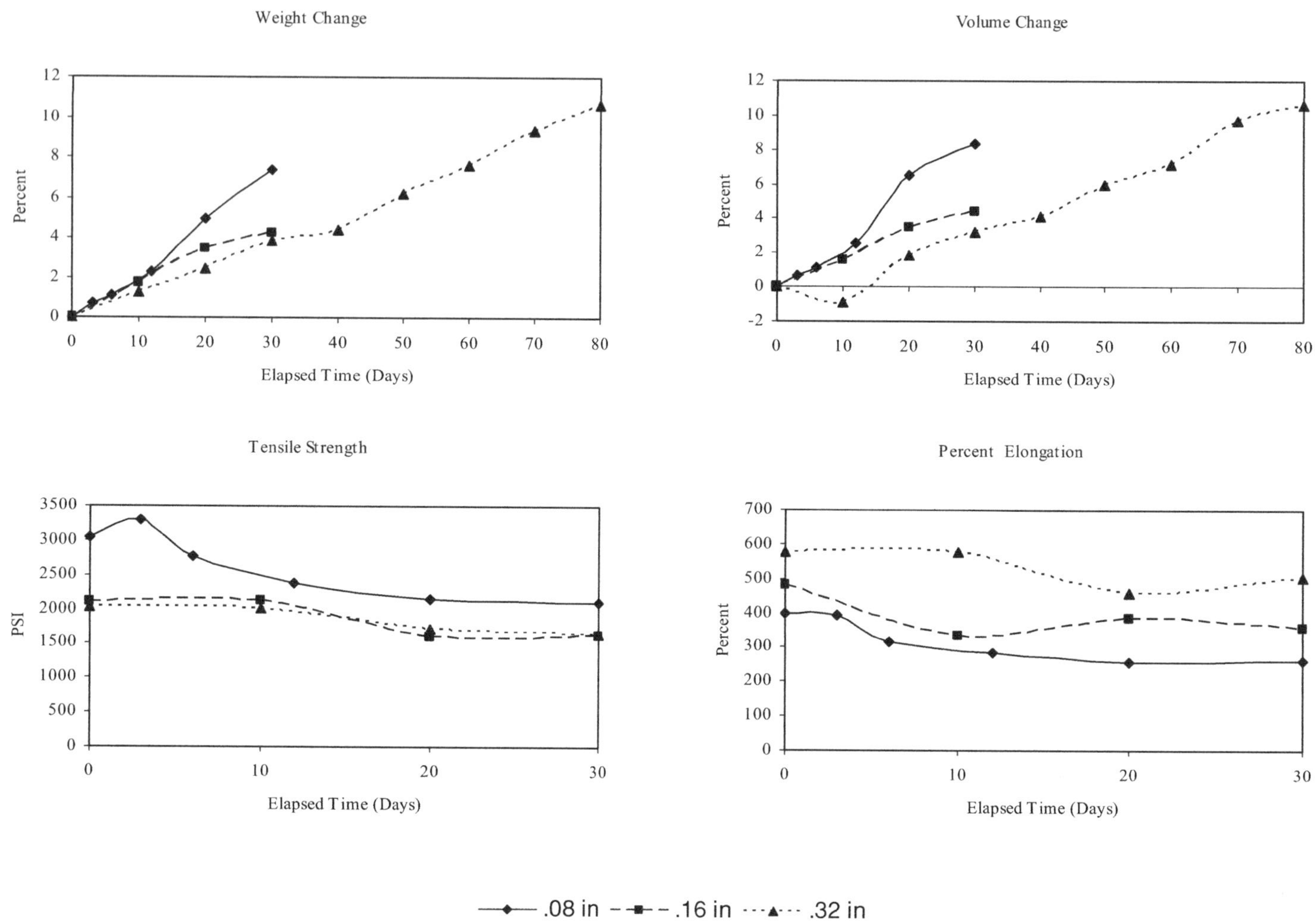

Figure 5.2 Effect of thickness on elastomeric properties at 30 ppm chloramine, 25°C, and pH 8.3 of natural rubber

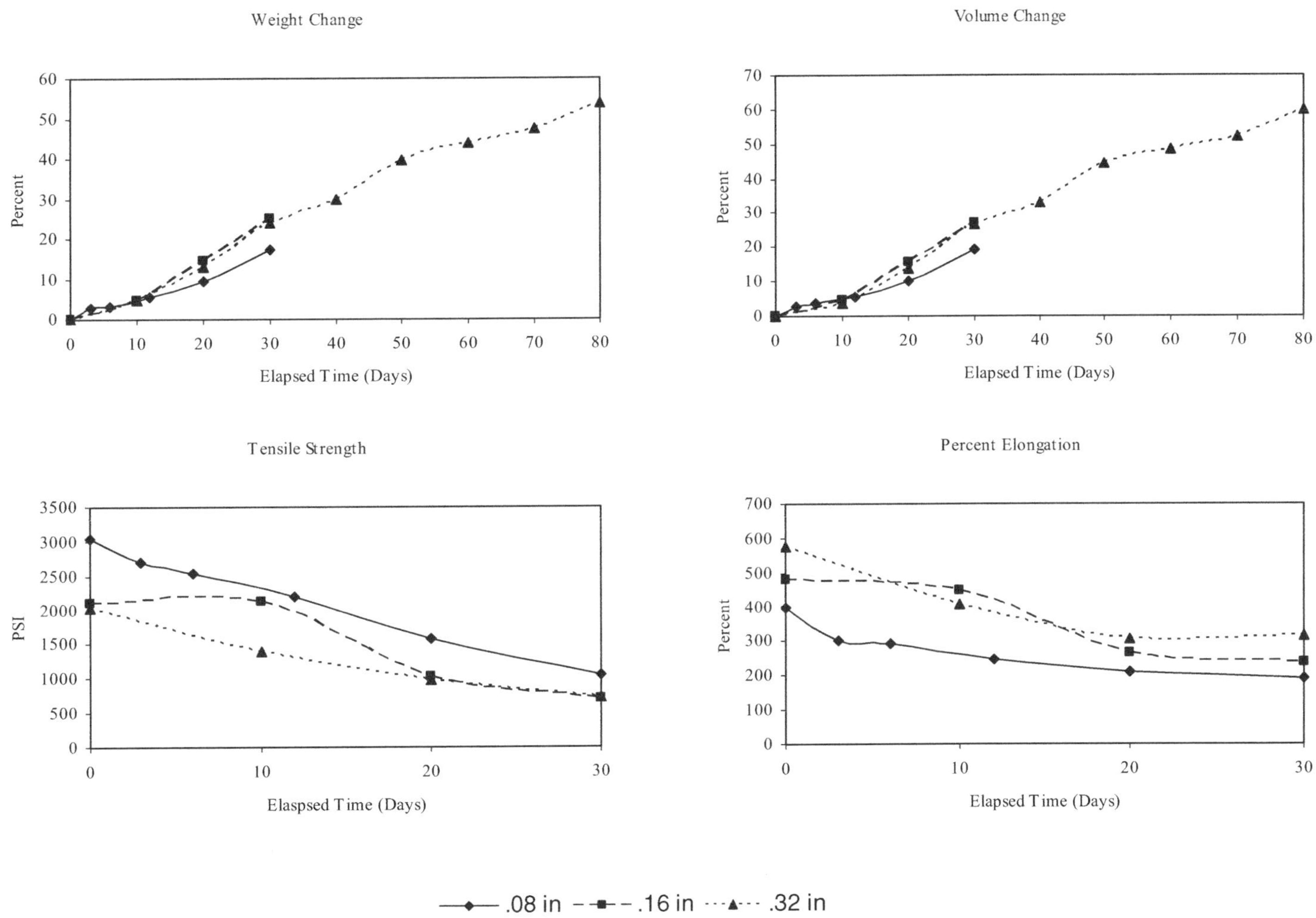

Figure 5.3 Effect of thickness on elastomeric properties at 1 ppm chloramine, 70°C, and pH 8.3 of natural rubber

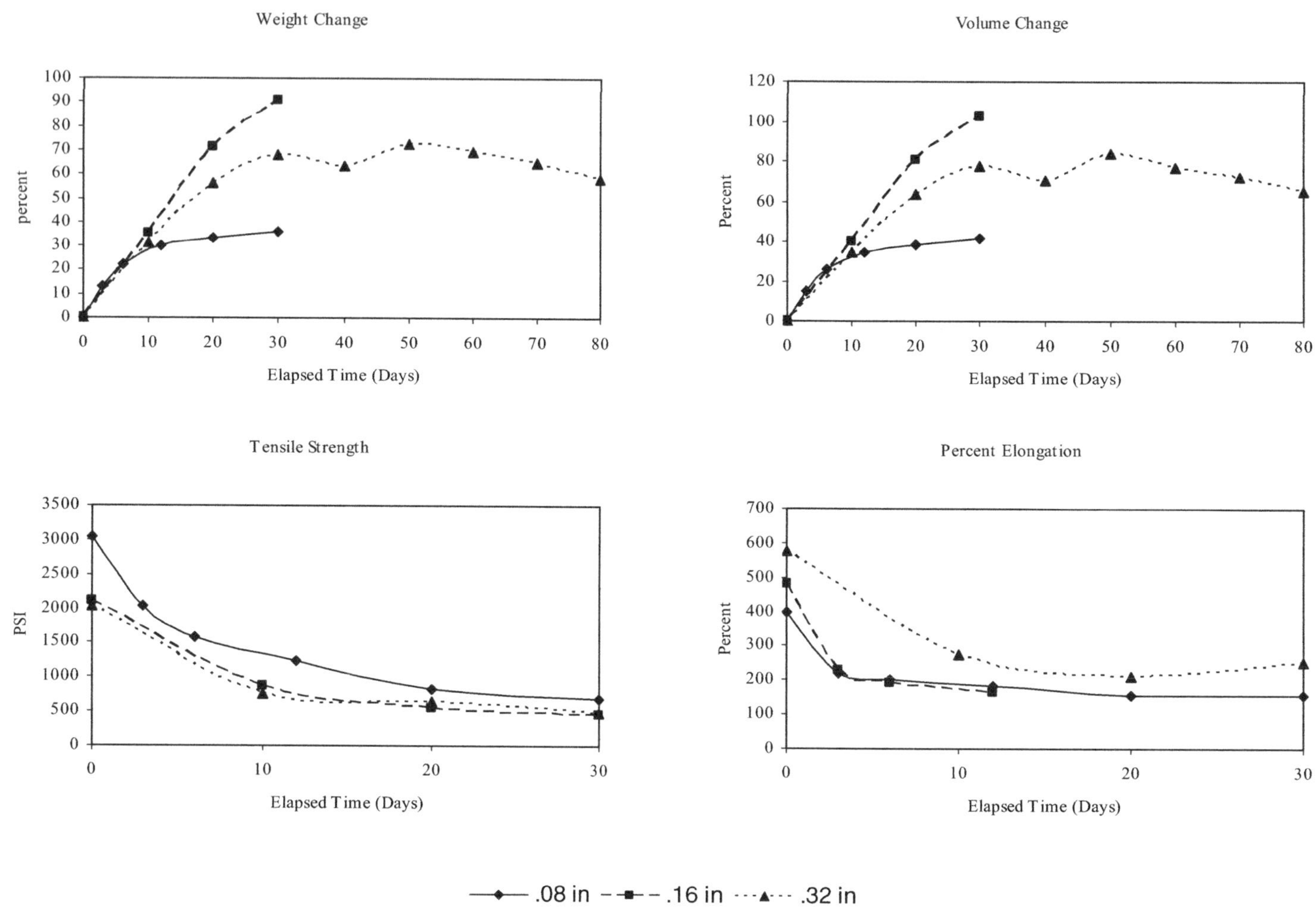

Figure 5.4 Effect of thickness on elastomeric properties at 30 ppm chloramine, 70°C, and pH 8.3 of natural rubber

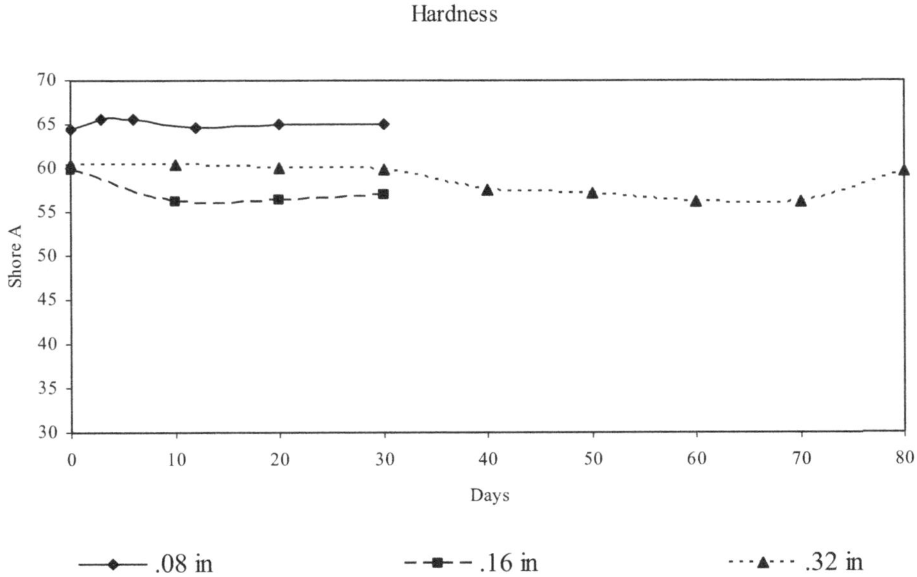

Figure 5.5 Effect of thickness on hardness at 1 ppm chloramine, 25°C, and pH 8.3 of natural rubber

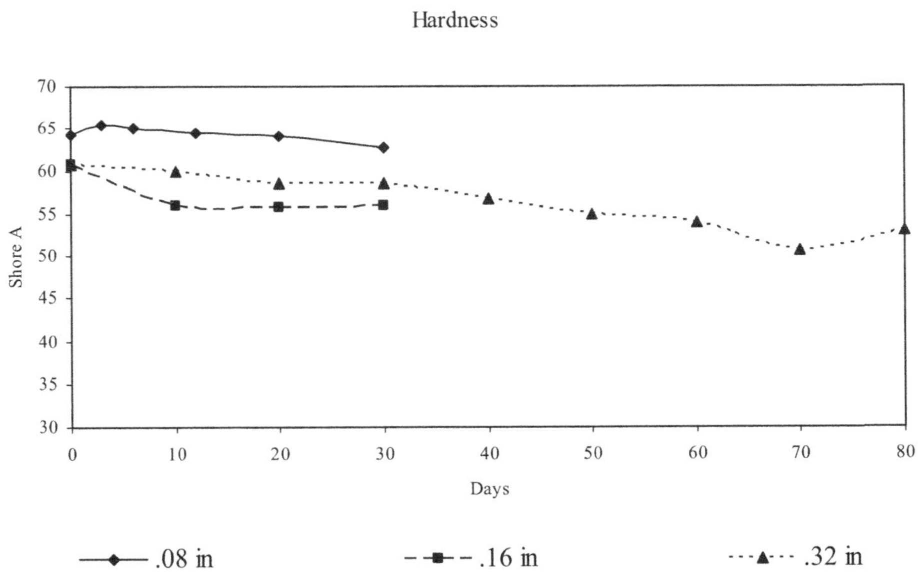

Figure 5.6 Effect of thickness on hardness at 30 ppm chloramine, 25°C, and pH 8.3 of natural rubber

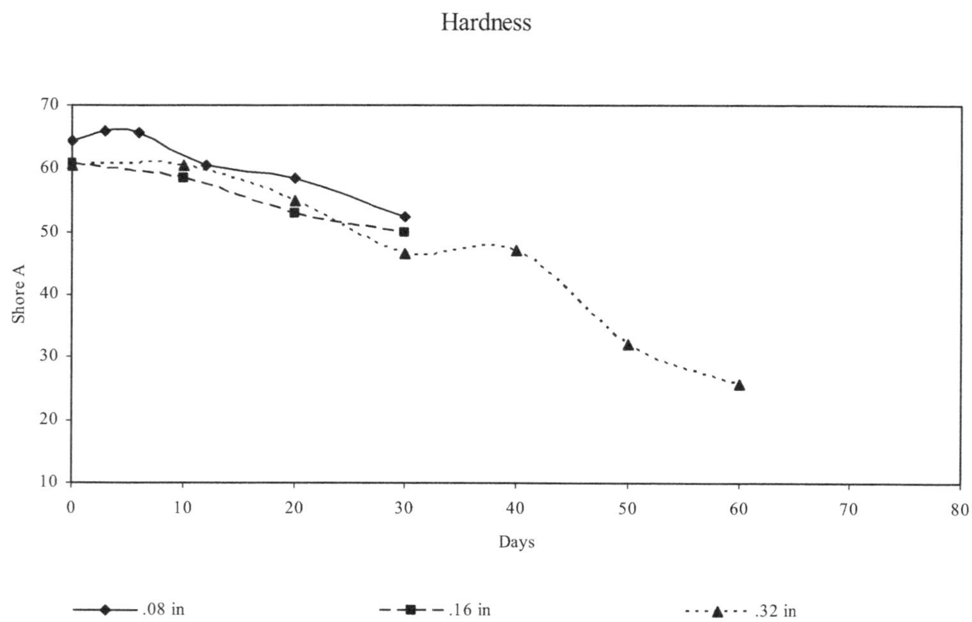

Figure 5.7 Effect of thickness on hardness at 1 ppm chloramine, 70°C, and pH 8.3 of natural rubber

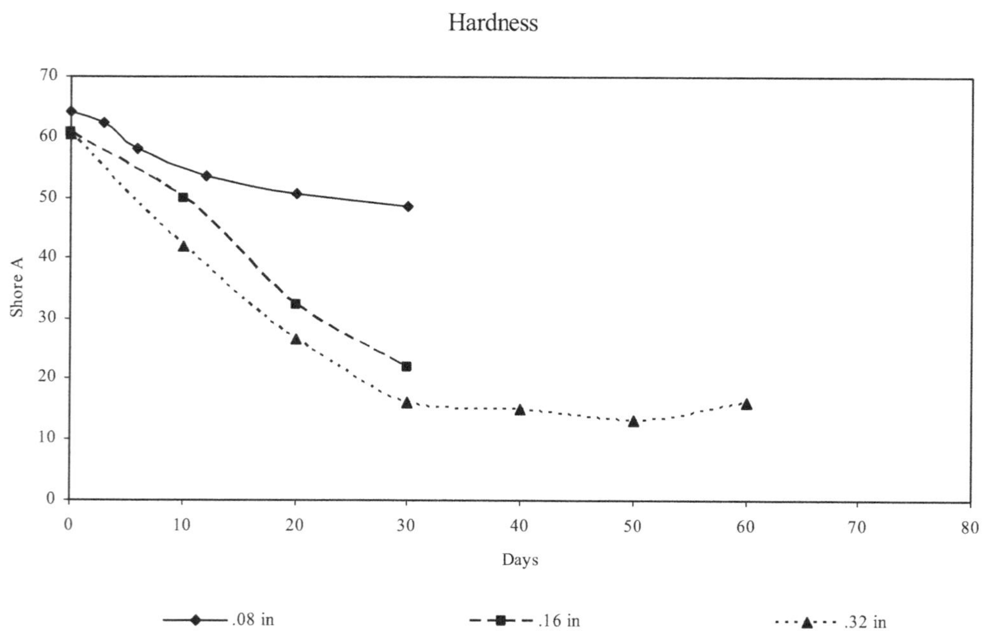

Figure 5.8 Effect of thickness on hardness at 30 ppm chloramine, 70°C, and pH 8.3 of natural rubber

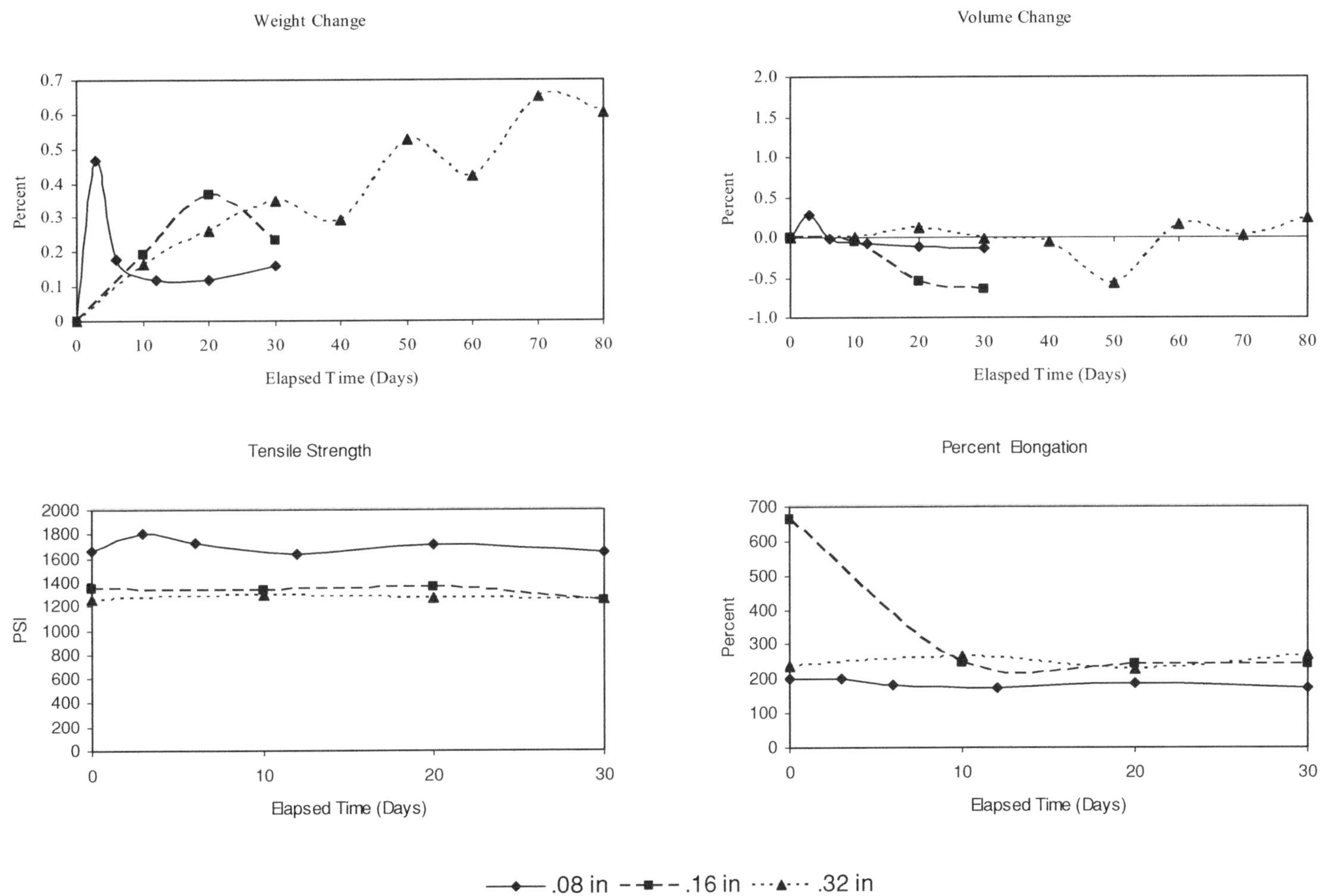

Figure 5.9 Effect of thickness on elastomeric properties at 1 ppm chloramine, 25°C, and pH 8.3 of EPDM-S

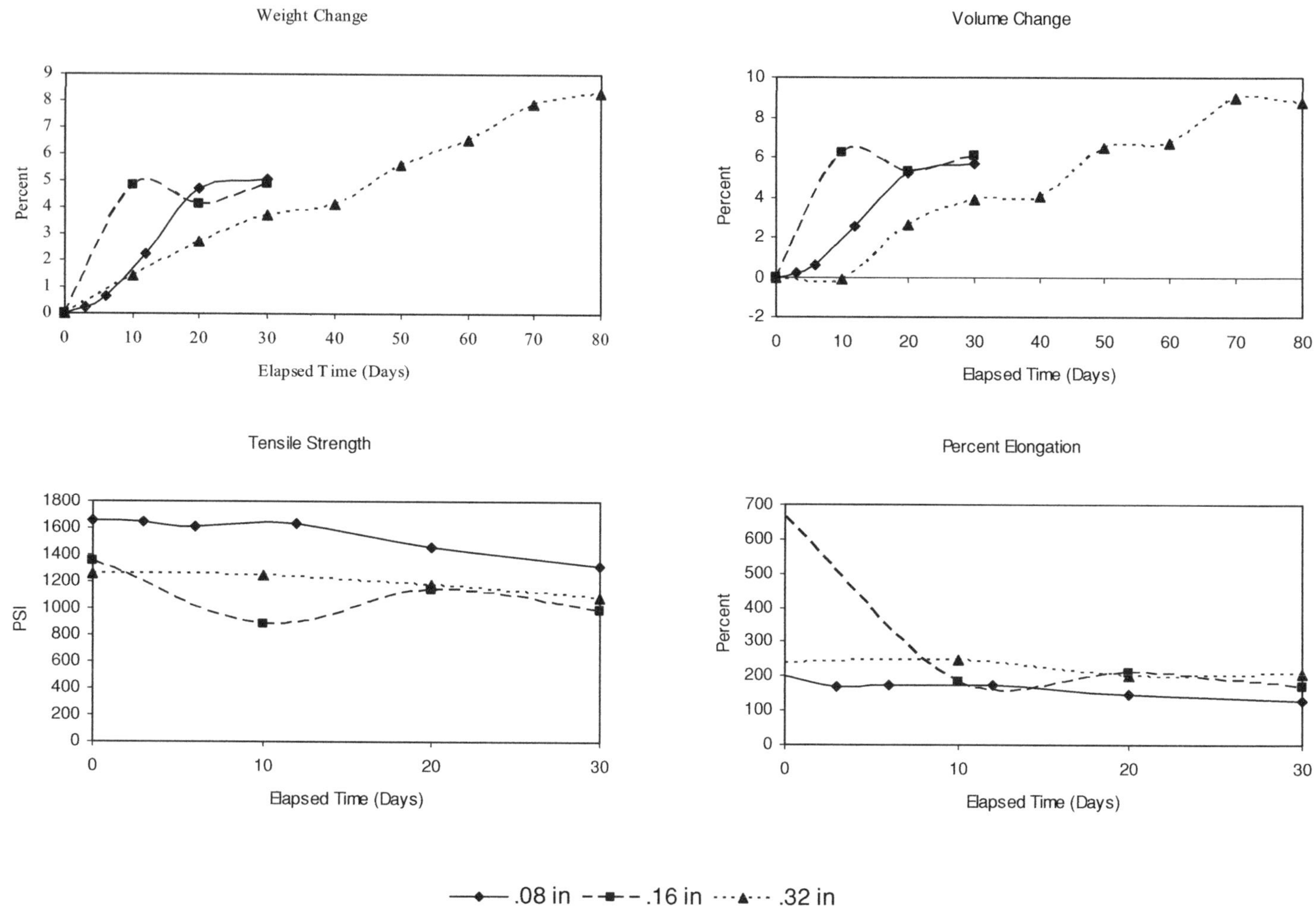

Figure 5.10 Effect of thickness on elastomeric properties at 30 ppm chloramine, 25°C, and pH 8.3 of EPDM-S

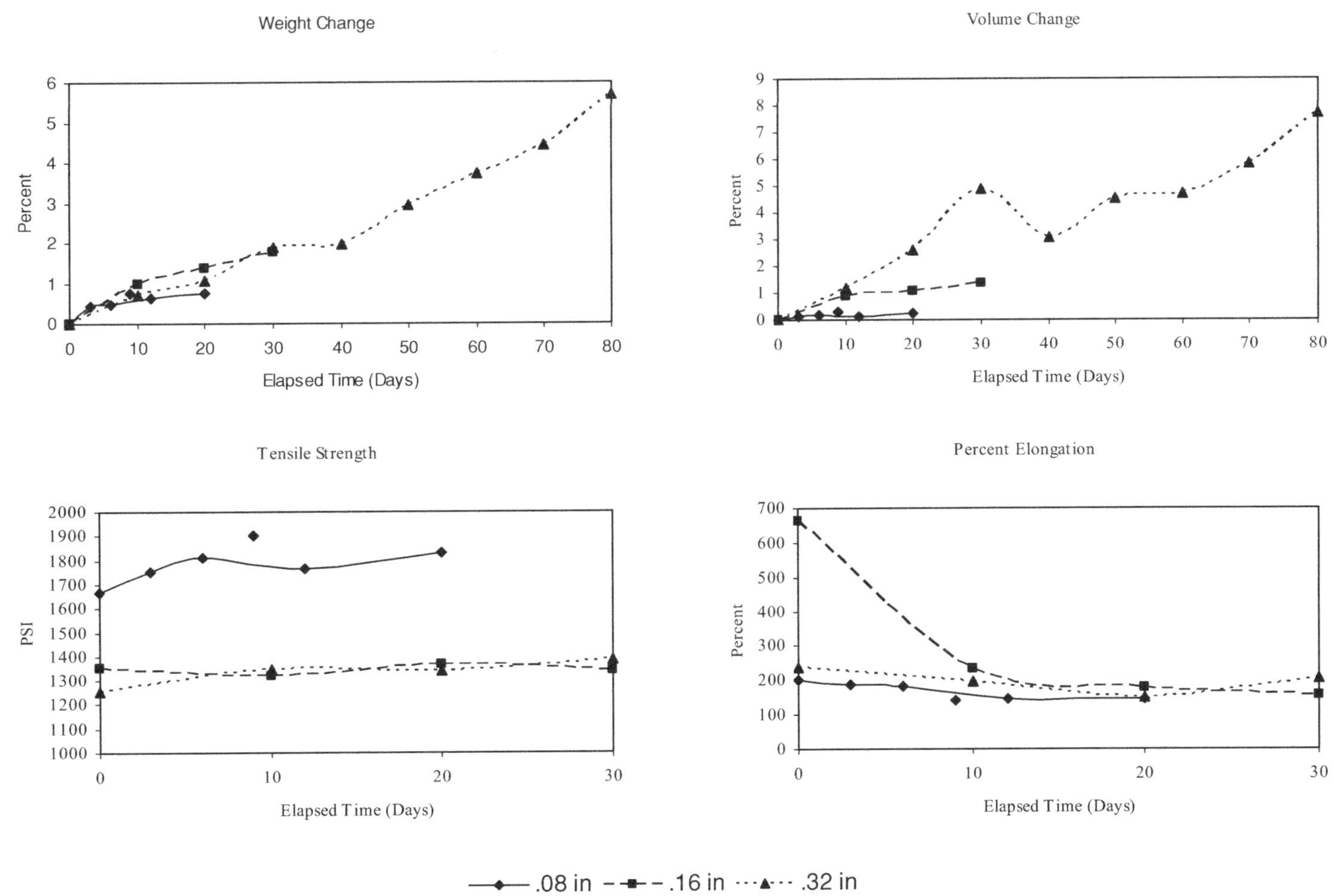

Figure 5.11 Effect of thickness on elastomeric properties at 1 ppm chloramine, 70°C, and pH 8.3 of EPDM-S

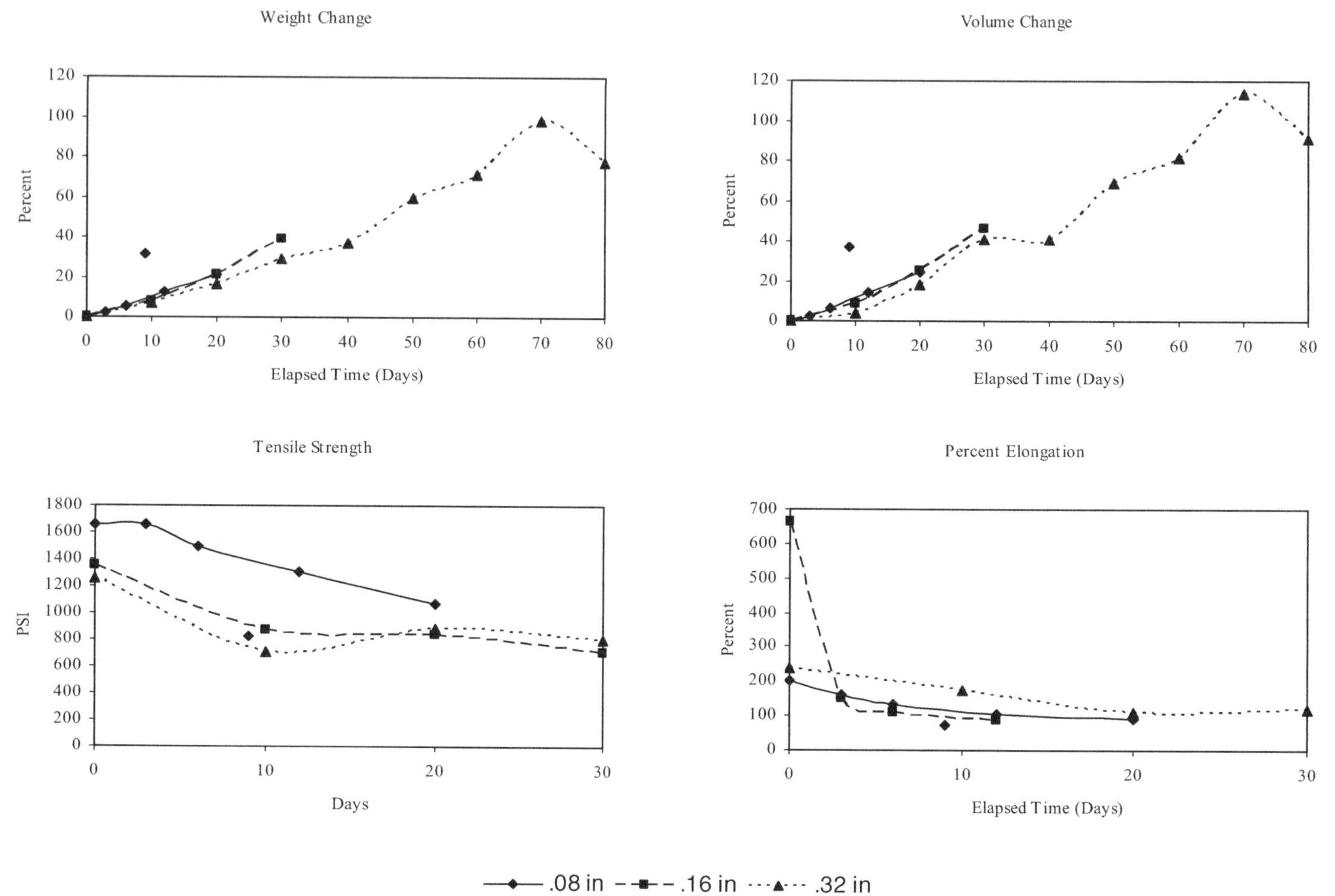

Figure 5.12 Effect of thickness on elastomeric properties at 30 ppm chloramine, 70°C, and pH 8.3 of EPDM-S

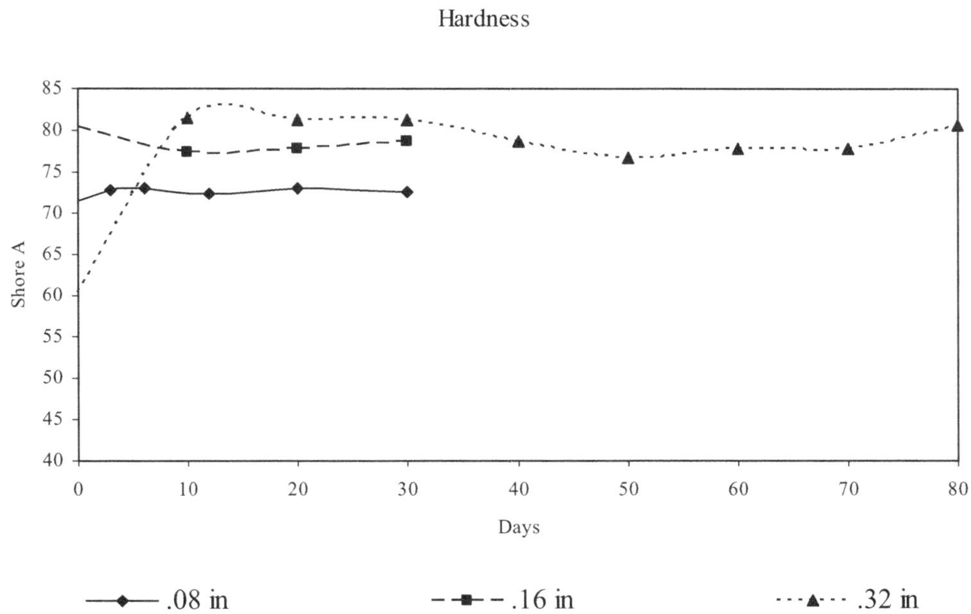

Figure 5.13 Effect of thickness on hardness at 1 ppm chloramine, 25°C, and pH 8.3 of EPDM-S

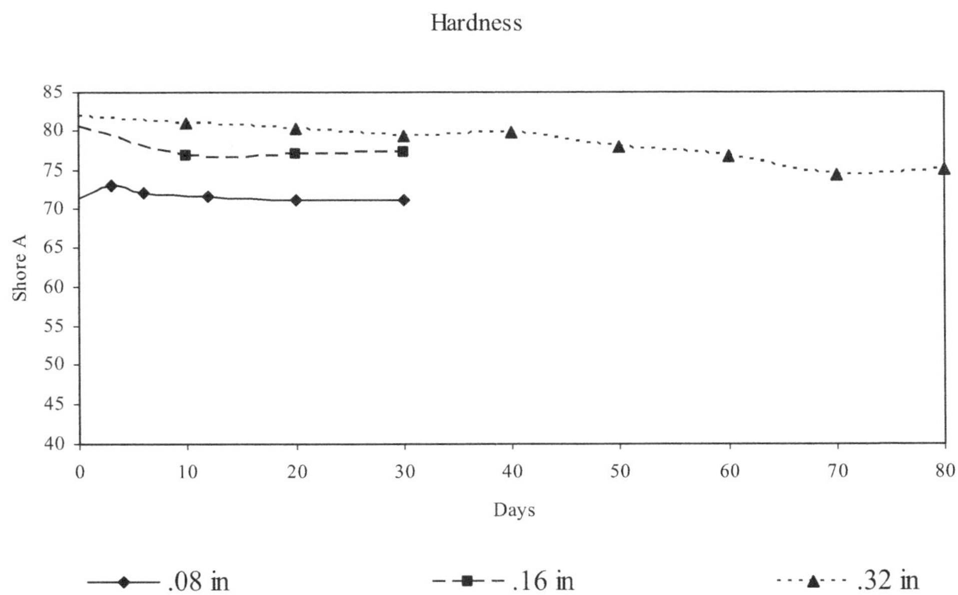

Figure 5.14 Effect of thickness on hardness at 30 ppm chloramine, 25°C, and pH 8.3 of EPDM-S

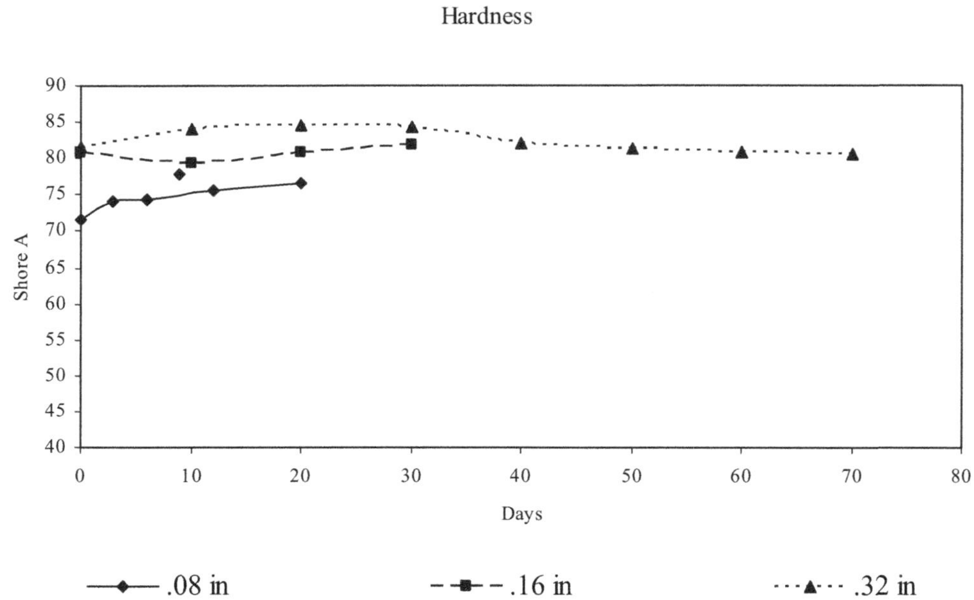

Figure 5.15 Effect of thickness on hardness at 1 ppm chloramine, 70°C, and pH 8.3 of EPDM-S

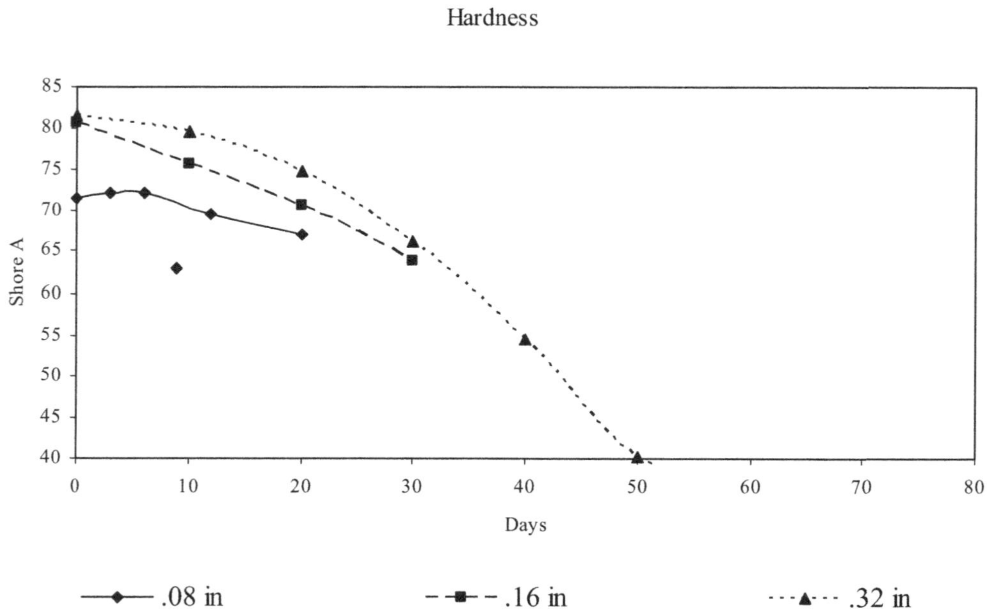

Figure 5.16 Effect of thickness on hardness at 30 ppm chloramine, 70°C, and pH 8.3 of EPDM-S

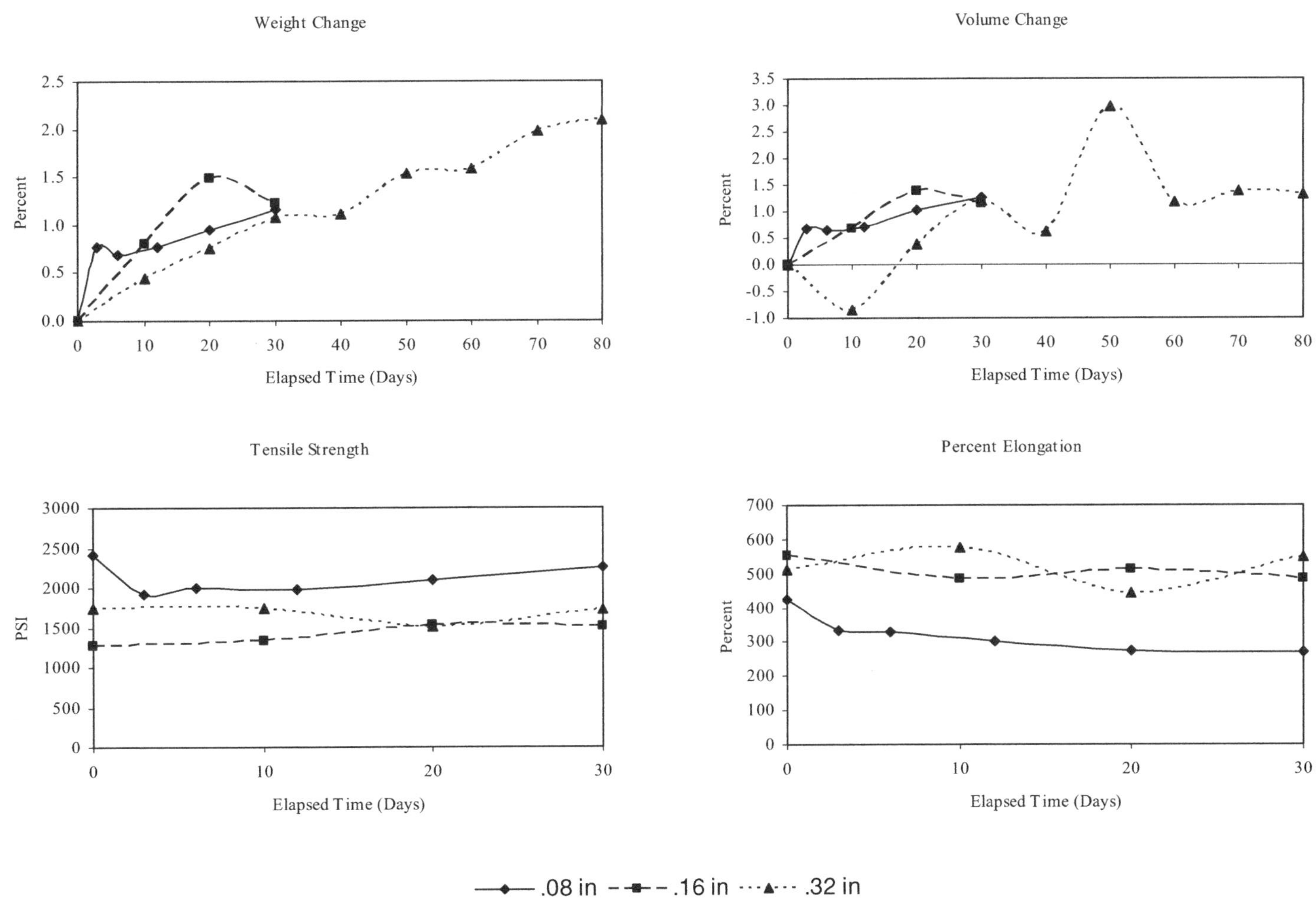

Figure 5.17 Effect of thickness on elastomeric properties at 1 ppm chloramine, 25°C, and pH 8.3 of SBR

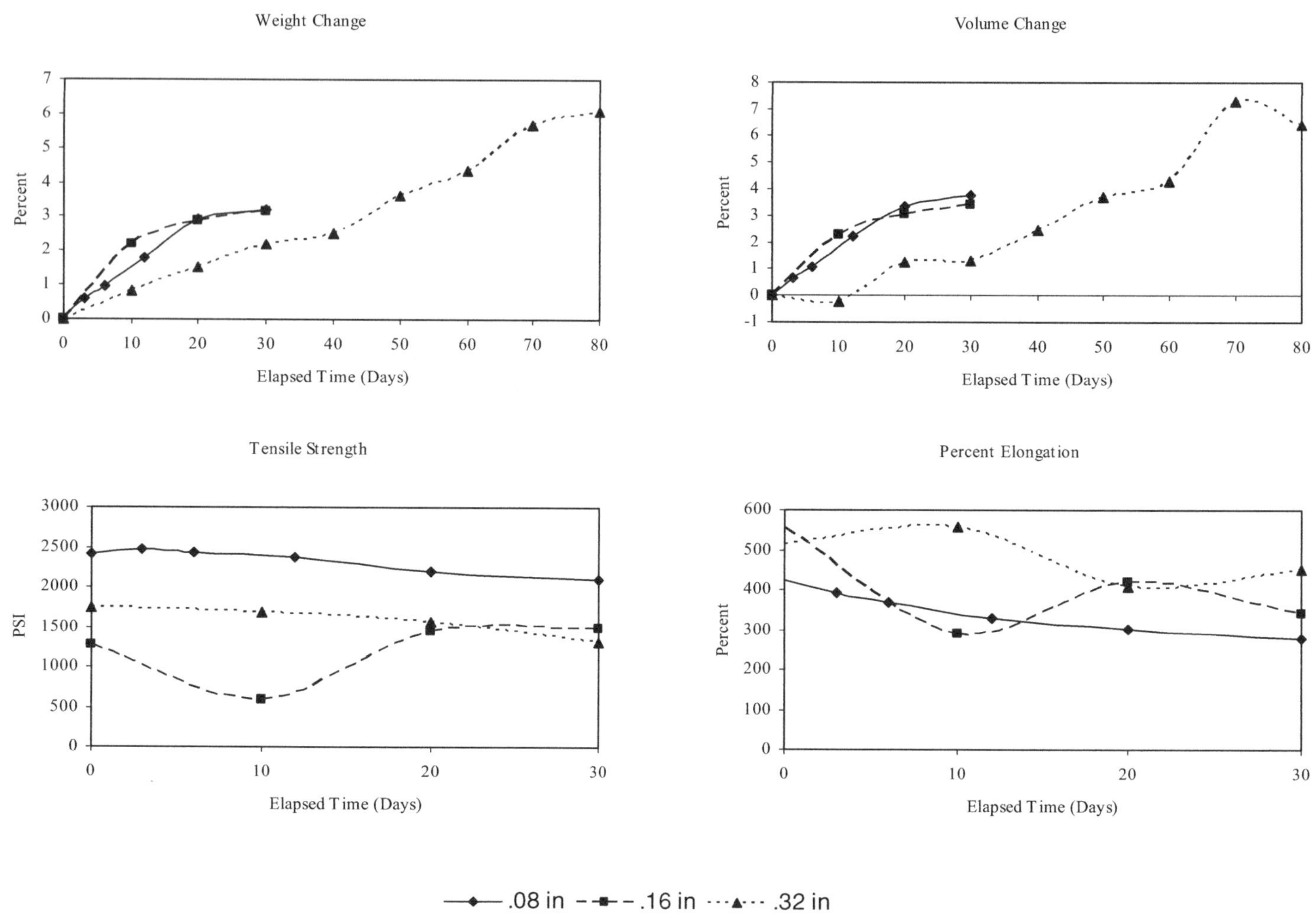

Figure 5.18 Effect of thickness on elastomeric properties at 30 ppm chloramine, 25°C, and pH 8.3 of SBR

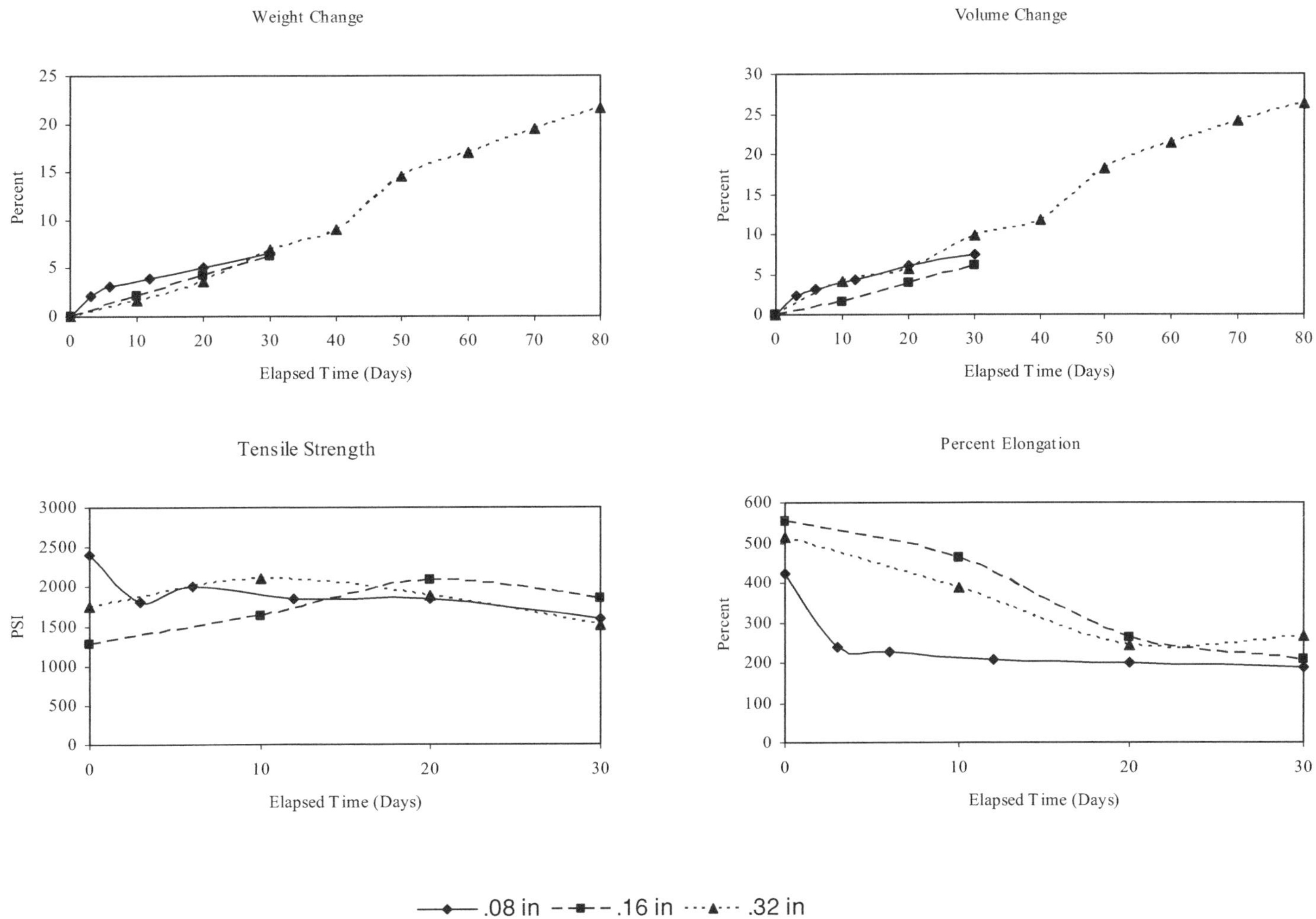

Figure 5.19 Effect of thickness on elastomeric properties at 1 ppm chloramine, 70°C, and pH 8.3 of SBR

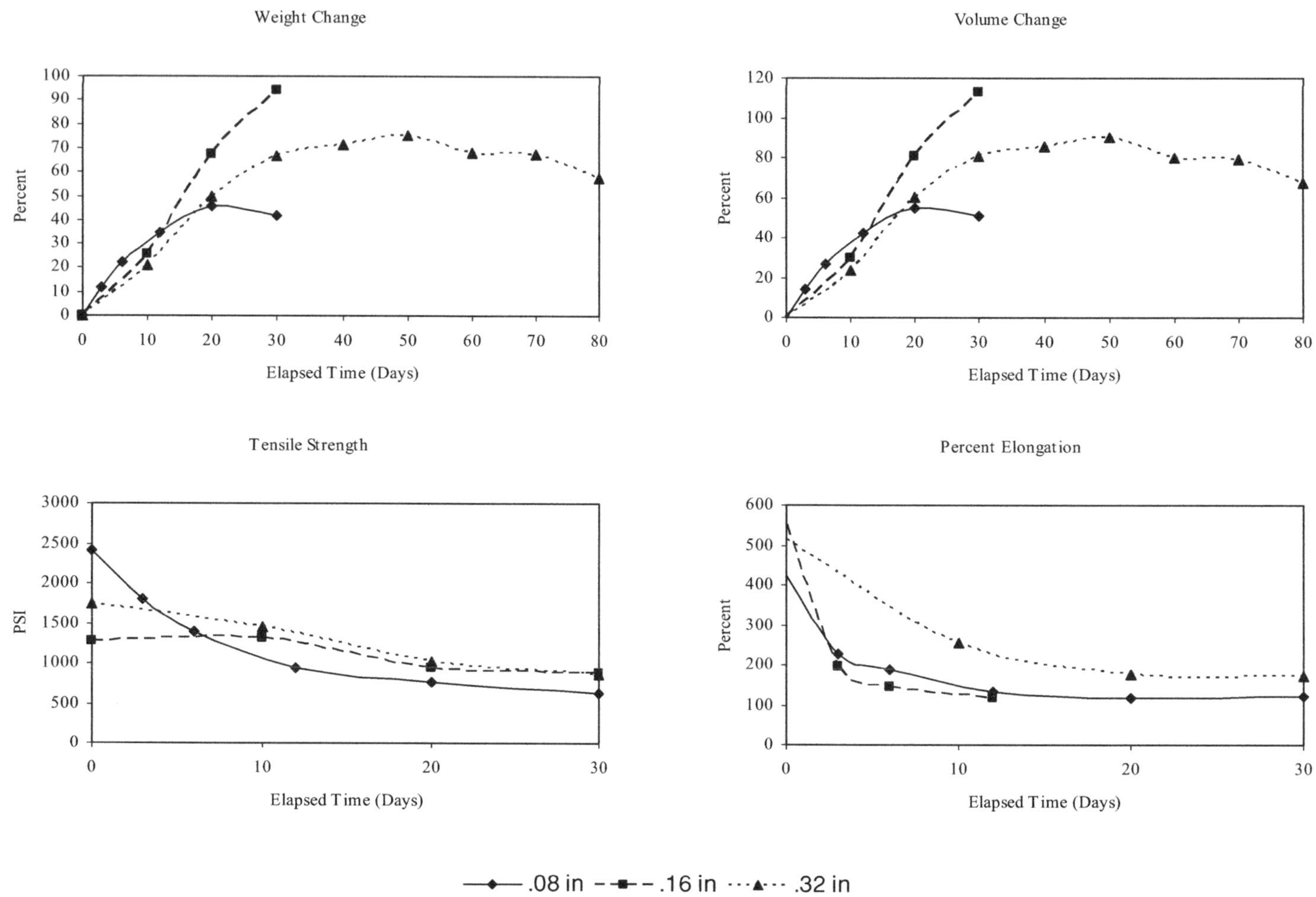

Figure 5.20 Effect of thickness on elastomeric properties at 30 ppm chloramine, 70°C, and pH 8.3 of SBR

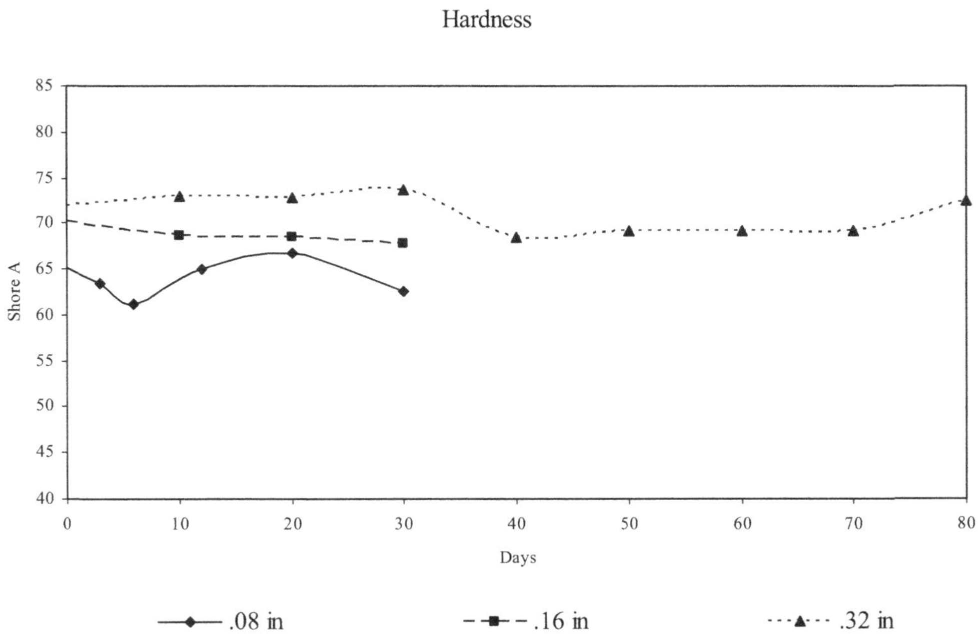

Figure 5.21 Effect of thickness on hardness at 1 ppm chloramine, 25°C, and pH 8.3 of SBR

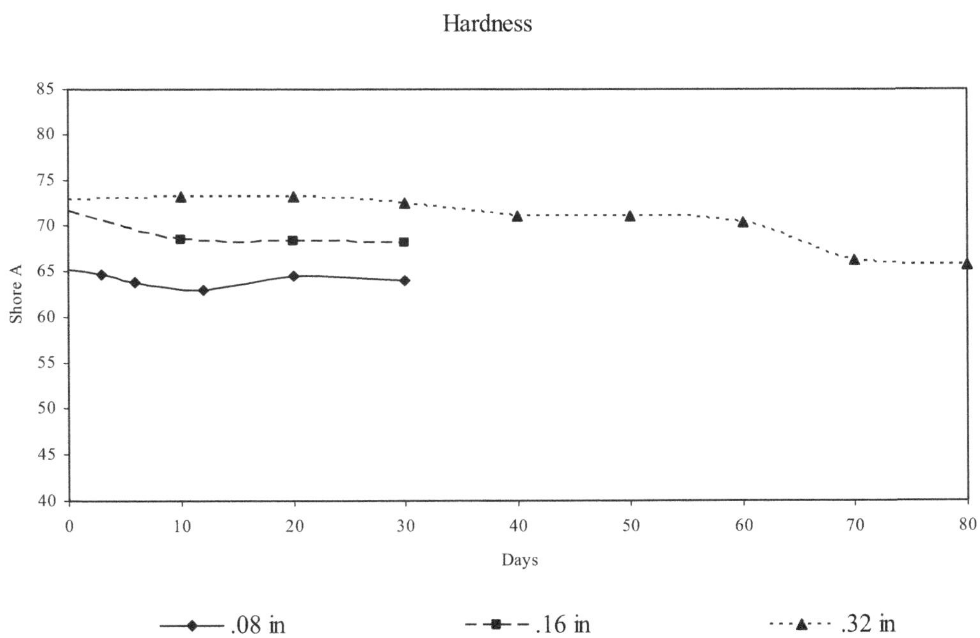

Figure 5.22 Effect of thickness on hardness at 30 ppm chloramine, 25°C, and pH 8.3 of SBR

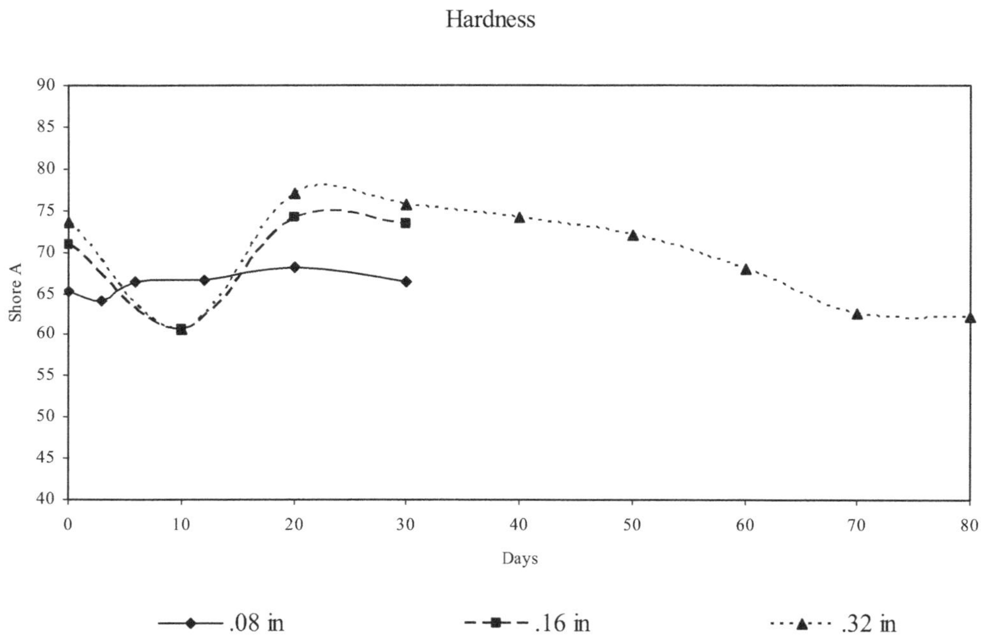

Figure 5.23 Effect of thickness on hardness at 1 ppm chloramine, 70°C, and pH 8.3 of SBR

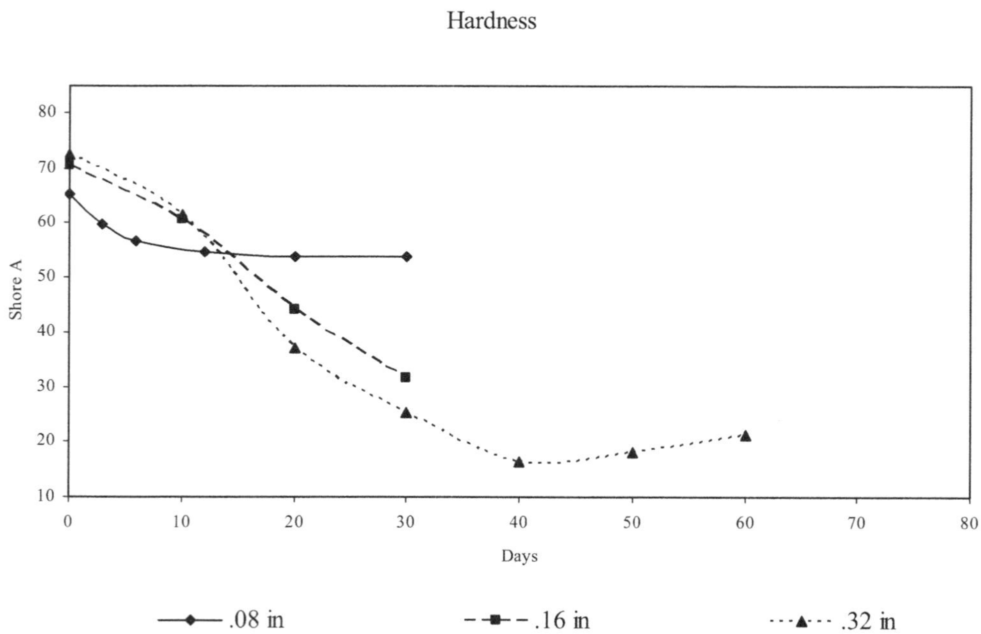

Figure 5.24 Effect of thickness on hardness at 30 ppm chloramine, 70°C, and pH 8.3 of SBR

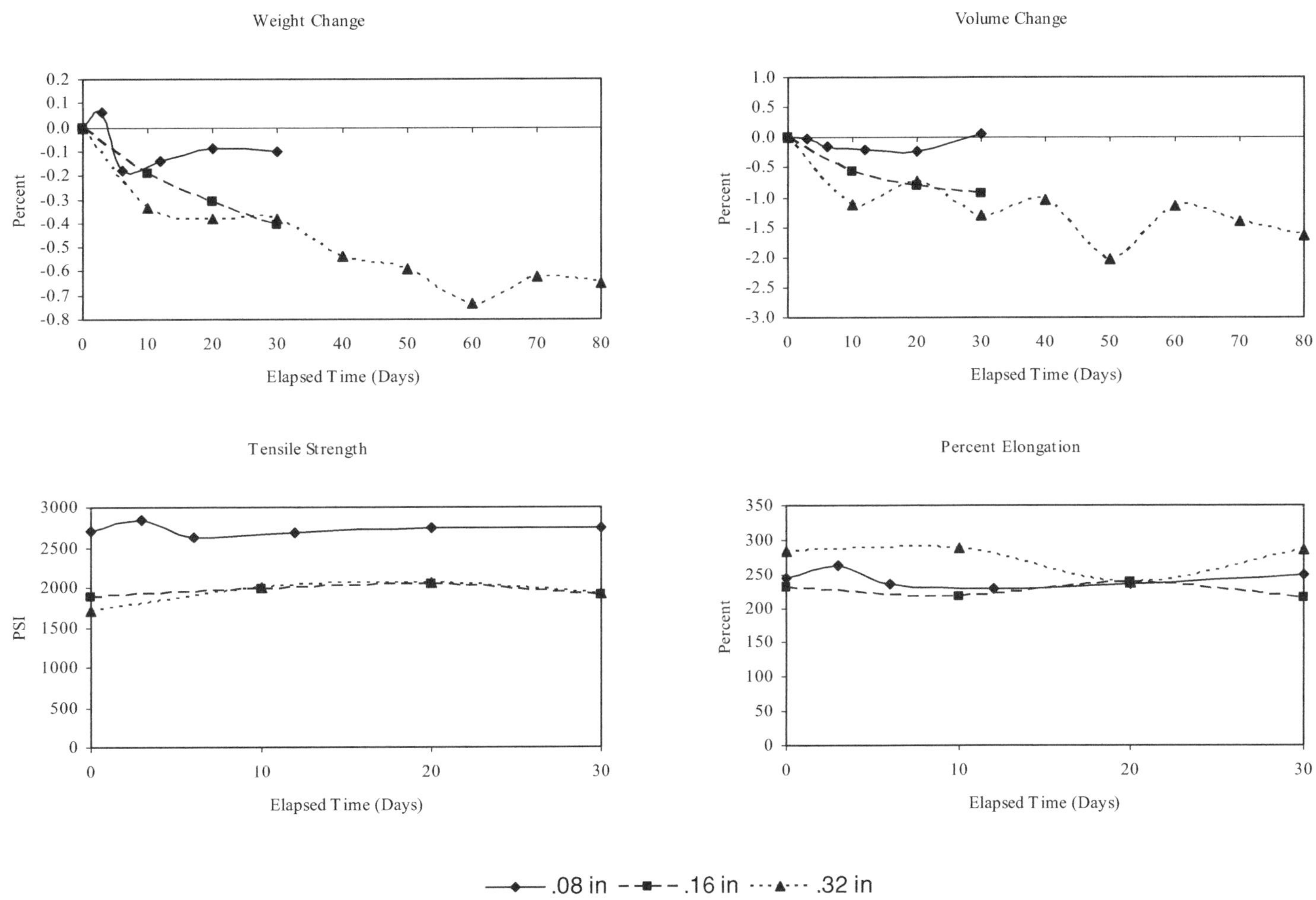

Figure 5.25 Effect of thickness on elastomeric properties at 1 ppm chloramine, 25°C, and pH 8.3 of EPDM-P

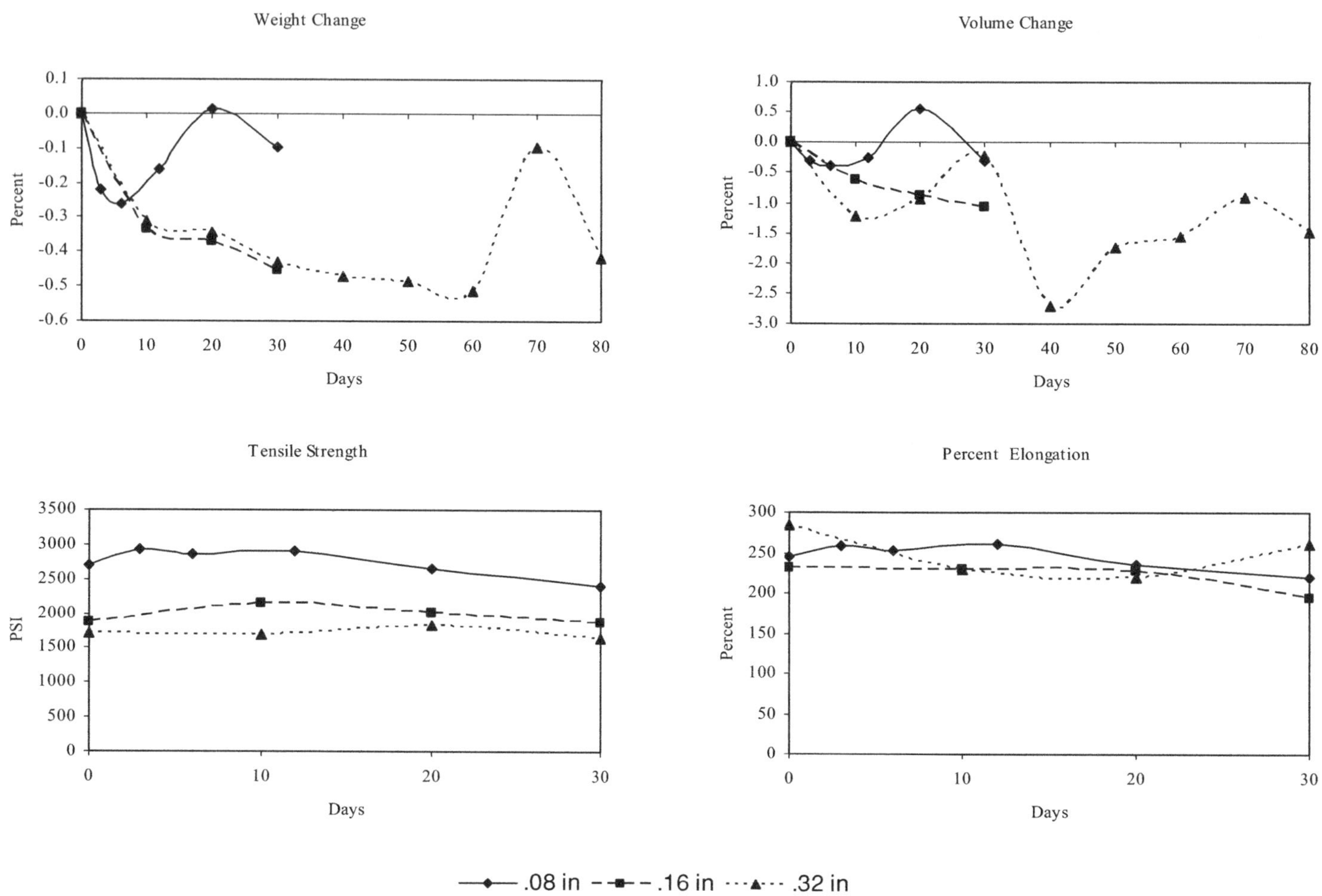

Figure 5.26 Effect of thickness on elastomeric properties at 30 ppm chloramine, 25°C, and pH 8.3 of EPDM-P

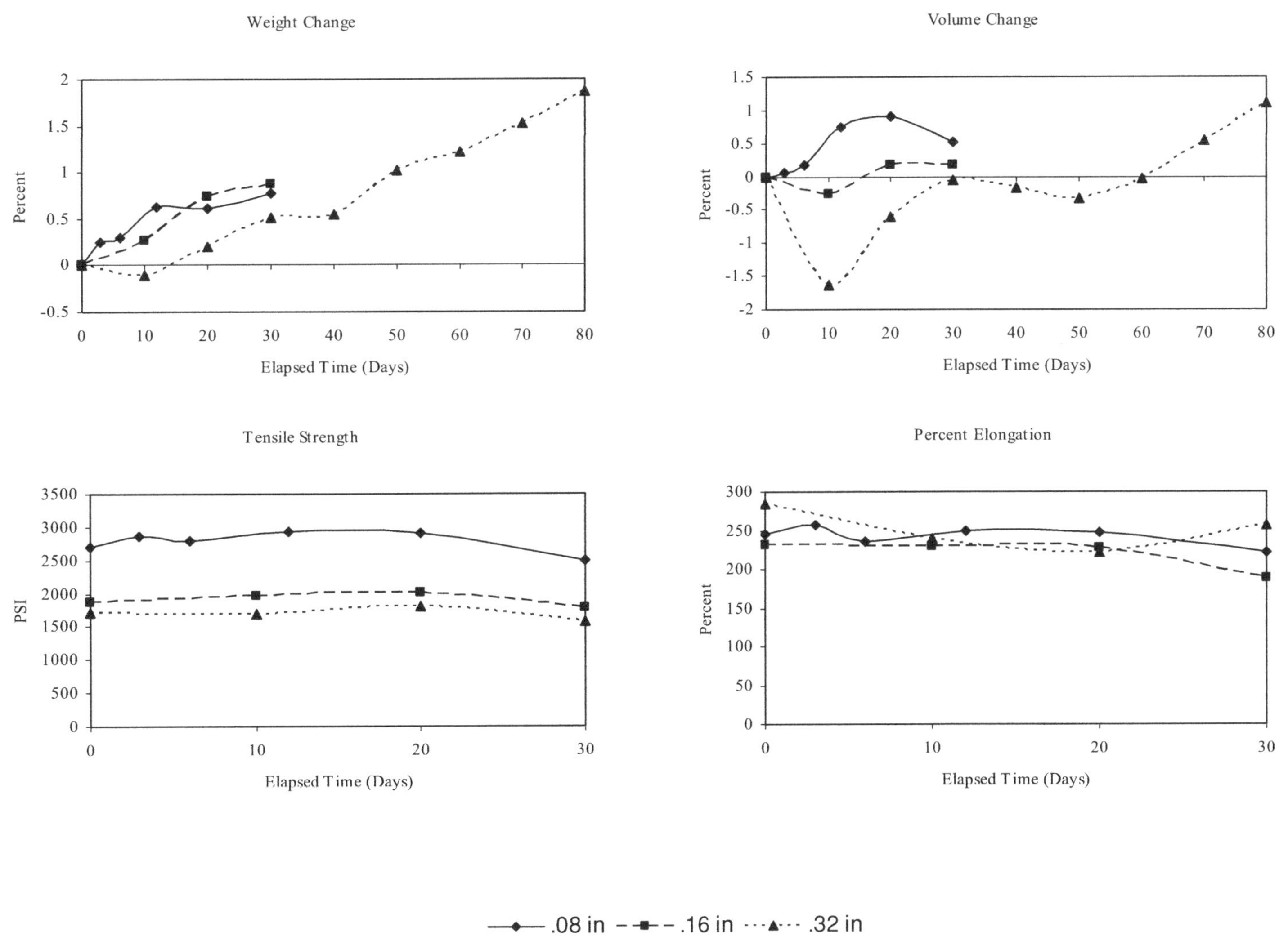

Figure 5.27 Effect of thickness on elastomeric properties at 1 ppm chloramine, 70°C, and pH 8.3 of EPDM-P

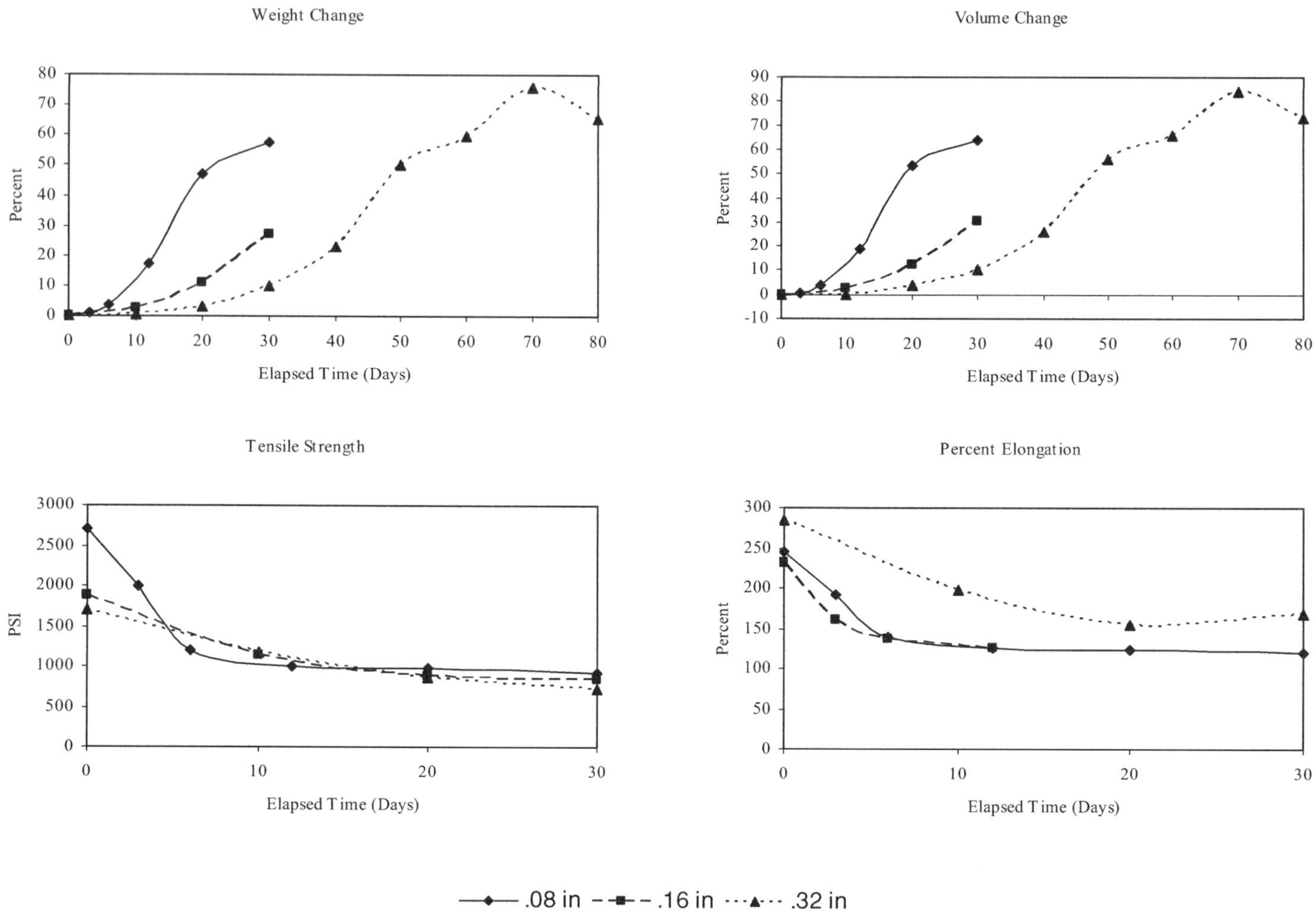

Figure 5.28 Effect of thickness on elastomeric properties at 30 ppm chloramine, 70°C, and pH 8.3 of EPDM-P

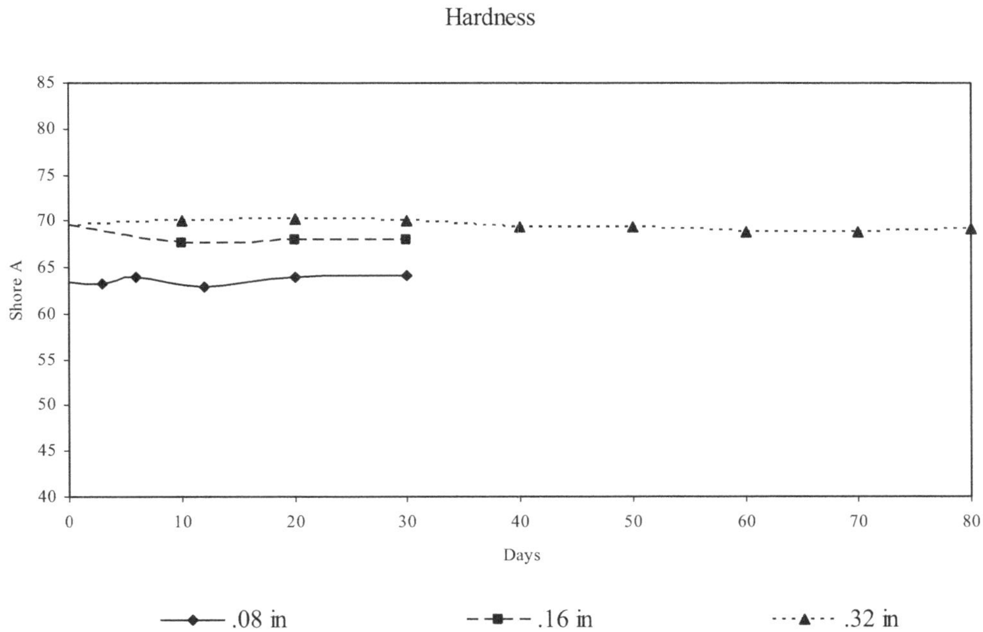

Figure 5.29 Effect of thickness on hardness at 1 ppm chloramine, 25°C, and pH 8.3 of EPDM-P

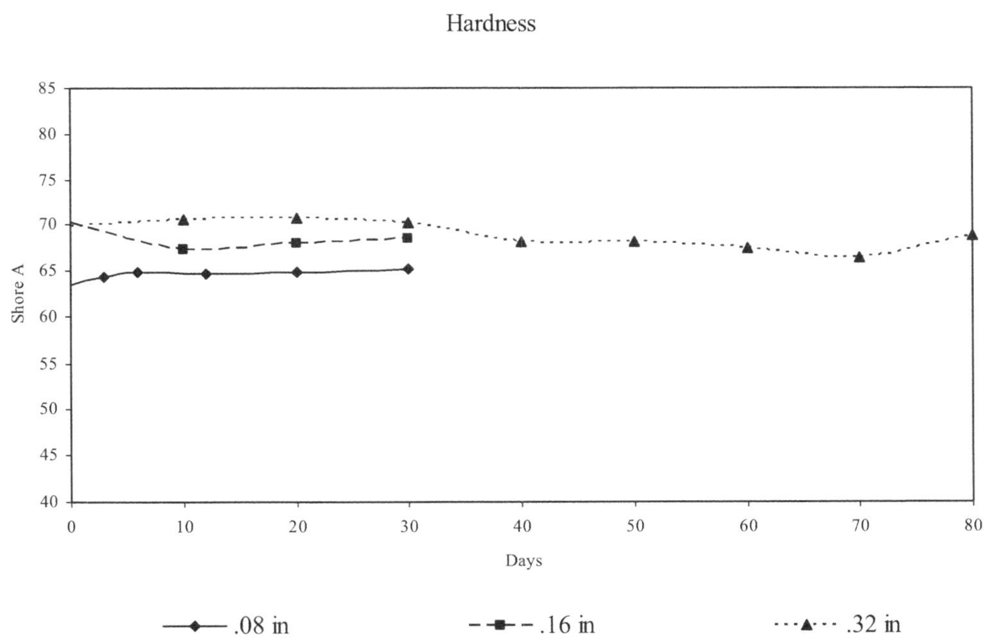

Figure 5.30 Effect of thickness on hardness at 30 ppm chloramine, 25°C, and pH 8.3 of EPDM-P

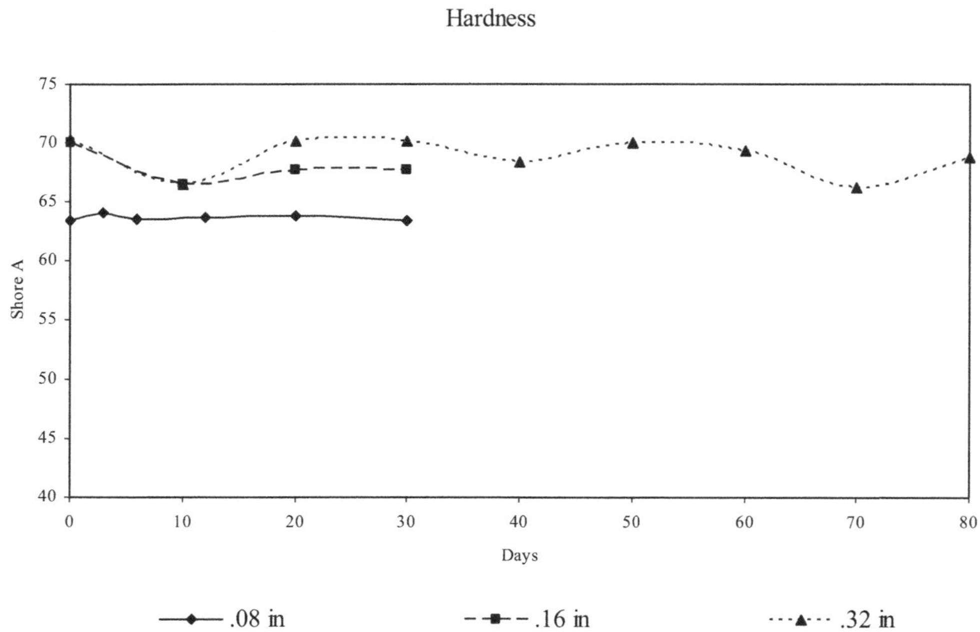

Figure 5.31 Effect of thickness on hardness at 1 ppm chloramine, 70°C, and pH 8.3 of EPDM-P

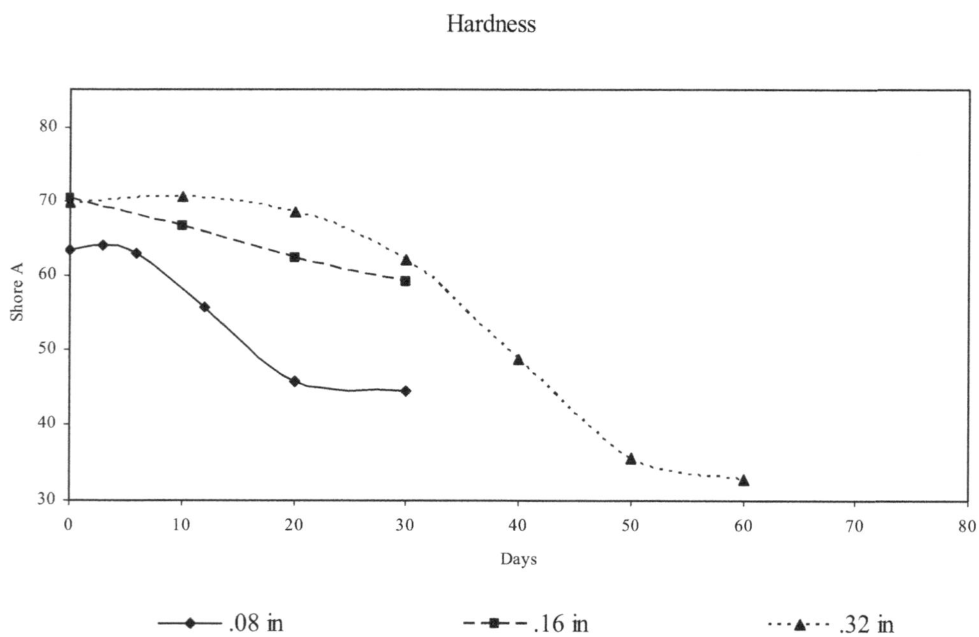

Figure 5.32 Effect of thickness on hardness at 30 ppm chloramine, 70°C, and pH 8.3 of EPDM-P

CHAPTER 6
ELASTOMER DEGRADATION OBSERVED VIA CARBON BLACK LOSS

OVERVIEW

Chapter 4 presented long-term degradation trends for six elastomers where both temperature and concentration were used to accelerate the testing window. While the effects of temperature are reasonably well understood and effectively modeled using the WLF equation (Equation 4.1), the accelerating effects due to the concentration component are less precise. For the results presented in Chapter 4, the concentration acceleration component was incorporated into the long-term life predictions using graphical procedures. However, as presented in Equation 4.2, a more rigorous analytical approach can be used if the diffusion rate of the chlorine or chloramines into the elastomer can be determined.

Assessing the diffusion rate of chlorine and chloramine disinfectants into an elastomer sample is a complex task. Not only does the diffusion rate depend upon time and temperature, but also on the elastomer matrix, cross linkages, filler materials, and carbon black content. Additionally, penetration into the sample is not expected to be uniform in area. It is a widely accepted hypothesis that loss of carbon black from the matrix allows propagation of cracks extending inward from the surface. In the presence of a relatively hard carbon black particle, the crack propagation may be stopped. (This is due to the hard physical nature of the particle and the energy-dispersing effect on the leading edge of the fissure due to molecular forces of attraction at the interface of the carbon black surface and the surrounding polymer matrix.) However, if the carbon black can be removed from the matrix, the resulting voids that have a diameter on the order of tens of nanometers have little energy-dissipating effects and enable propagation (Donnet and Voet 1976). Thus, the depth to which carbon black has been removed from the elastomer matrix should correlate to the depth of penetration of the chlorine or chloramine disinfectant into the elastomer. If a time component can be added to this, then an effective diffusion rate can be determined.

This chapter describes a series of experimental techniques, results, and conclusions from investigations into the effects that chloramines and chlorine have on the degradation of carbon black and the nature of their effect. Specifically, the experiment showed empirically that the carbon black particles are reduced in size as well as number as concentration within the elastomer sample is reduced and showed how deep within the sample these effects can be seen over time. A diffusion coefficient based on Fick's second law was also estimated using the optical micrograph data. Natural rubber, SBR, and EPDM-S elastomers were included in this study because of their common use in the water industry.

EXPERIMENTAL EQUIPMENT

The loss of carbon black material within the elastomer matrix can be used as an indicator of the depth of penetration of chlorine or chloramines into the elastomer. The analysis of the carbon black particles was performed mainly by employing techniques of Atomic Force Microscopy (AFM), invented by Gerd Binnig, Calvin Quate, and Christoph Gerber in 1986 (Binnig et al. 1986). The AFM scans over a surface by using a fine Si/SiO_2 tip. The tip is located at the end of a cantilever with a laser beam positioned to reflect off the end onto a photodiode. As the tip traverses the surface, the cantilever changes angle, and the position changes are recorded on the photodiode (Figure 6.1). The photodiode is composed of two major parts which separately detect normal and lateral motions. A plot of the laser deflection versus tip position in the x and y directions provide two- or three-dimensional surface images.

The three major modes of operation are contact, non-contact, and tapping. In contact mode the tip is placed in physical contact with the sample and drawn across the surface. The deflection of the cantilever is generated by variations of the surface topography. In the case of polymers whose soft surfaces can be easily damaged, contact mode is typically not recommended due to tip-induced sample damage.

Advancements in AFM technology now allow the user to produce images without contacting the surface. In non-contact mode, first introduced by Martin and coworkers (Martin et al. 1987), measurements of forces acting between the tip and a sample surface are used to produce the images. A tip-to-sample distance of 30 to 150 Angstroms is maintained, and long-range molecular forces such as Van der Waals forces induce deflections in the cantilever. As with the direct contact mode, cantilever deflections of amplitude, frequency, and phase can be used to develop topographic images (Martin et al. 1987).

The third AFM technique is tapping mode, sometimes referred to as intermittent-contact or more generally as Dynamic Force Mode (DFM). In tapping mode the cantilever is oscillated at its resonance frequency and is only in contact with the surface for a fraction of its oscillation period. Phase images are produced by measuring the difference between the oscillations of the cantilever driving piezo and the oscillations detected by the photodiode (Chernoff 1995). It is thought that image contrast is derived from the surface properties such as stiffness and viscoelasticity.

In order to produce images of the carbon black particles and evaluate their degradation during accelerated aging, the various modes of atomic force microscopy were investigated and evaluated for effectiveness. In tapping mode, phase imaging allowed for two-dimensional phase images to be produced based on variations of hard and soft areas across the sample surface. Tapping mode allowed some contact to be made and was thus determined to be the preferred approach in this experiment in order to take advantage of the sharp contrast in hardness between the carbon black and the surrounding matrix. Images and data were evaluated using the analysis tools in the Nanoscope Software (Nanoscope version 5.3, Copyright© 2002 Digital Instruments) and ImageJ analysis software. ImageJ is a freeware analysis tool, which converts the image data into a binary form, then evaluates the image for particle size and area fraction.

CARBON BLACK WITHIN AN ELASTOMER MATRIX

As the AFM analysis is based on inferred information, it is important to determine that the carbon black identified by the AFM method is actually carbon black. To correlate the AFM data a scanning electron microscope (SEM) was used to characterize the typical carbon black distribution within the natural rubber, SBR, and EPDM-S elastomer samples to be used in this experiment. While the SEM can provide a very detailed picture of an elastomer surface, the image quantities necessary for this application make it too expensive and time consuming to be practical for this large scale use.

The SEM was first used to assess the spatial distribution of the carbon black on the surface of the elastomers. Based on the SEM testing, carbon blacks within the elastomer matrix were of the expected 60 to 100 nm size range and rarely appeared as single spherical particles, but usually as aggregates of coalesced, fused elementary particles (Figure 6.2). The carbon black aggregates have a tendency to agglomerate into larger units, which may be re-dispersed by shear forces, but re-agglomerate upon standing (Donnet and Voet 1976).

Elastomer formulations obtained from the manufacturer were used to convert the theoretical volume fraction of carbon black into two-dimensional area fractions. Evaluation of day zero samples indicated a reasonable agreement between the theoretical area fraction determined by formula estimation and analysis using SEM, AFM, and ImageJ software. Based on the testing, natural rubber was estimated to have an area fraction of about 25 percent carbon black; SBR and EPDM-S elastomers were estimated to have an area fraction of about 30 percent. These values were in the anticipated range for the amount of carbon black expected within these materials.

The SEM also was used to assess the carbon black distribution of each elastomer throughout the depth of the elastomer materials. Physically cutting the sample for cross-sectioning was determined to be unacceptable because of striations left across the surface by the cutting implement, which made determination of the nanoscale particles very difficult. Cutting the samples also was determined to be too invasive because artificial tearing would have removed the carbon black particles from the sample surface. Thus, cross-sectioning was accomplished using liquid nitrogen to freeze the coupon and then breaking it by force, either by hand or with the aid of a flat surface with a sharp edge. The freezing and breaking technique resulted in a surface virtually free of mechanical abrasion.

After creating the cross-section edge, the samples were coated with a thin layer of gold by chemical deposition to allow the electron beam to non-destructively image the sample surface. SEM technology uses an electron beam to scan the sample surface, thereby causing charging to occur, which can damage softer surfaces such as polymers.

Based on the SEM testing, particles in the expected size range for carbon black appeared to occur regularly over the exposed cross-section surface. Consistent dispersal of the carbon black throughout the polymer matrix appeared to have been achieved during manufacture of the elastomer. Moreover, analysis using ImageJ software showed that the area fraction of the two-dimensional surface was in the expected range for each type of rubber.

The micrographs and analyses generated by the SEM testing showed that it is reasonable to assume that differences in particle size and distribution in aged samples are due mainly to the effect of the aging process, and not due to improper mixing during the elastomer manufacture. Thus, the loss of carbon black that can be inferred from the AFM testing is a reasonable approach to assess penetration of the chlorine or chloramines.

AFM TESTING DATA

To assess the rate of degradation using the AFM, representative elastomer samples were first artificially aged using increased temperatures and concentrations. The test samples were exposed to 30 ppm chloramines, or chlorine, at 45°C for 30 days. During the 30-day test window, sacrificial samples were periodically removed from the test solution and tested for hardness, mass, and volume changes and cross-sectioned for AFM testing. It was anticipated that samples with a longer exposure time would show a carbon black loss occurring to a greater depth. Sample images showing a decrease in area fraction are shown in Figure 6.3 as an example of the changes seen within an image as the sample is aged. The performance variables, i.e. mass and volume change and hardness, for the samples obtained during the period were consistent with the effects seen in previous testing done at similar conditions.

AFM images of the 0.08 inch thick cross-section were taken at regular intervals from one side of the sample to the other. Figures 6.4, 6.5, and 6.6 graphically represent the ratio of carbon black to the area surface (area fraction) from the center of the rubber coupon to the outside edge. As the samples and exposure conditions were symmetrical, the data presented was obtained from both sides of the elastomer midpoint. For the original, zero day, samples, and for those samples in which very little change in volume had occurred, the X-axis extends from 0 to 0.04 inches. The 0 point represents the middle of the sample and extends out to the edge of the sample. For samples where a significant amount of expansion occurred, the X-axis has been represented appropriately, such as in natural rubber and SBR at 30 days where the X-axis is shown to extend out to 0.06 inches.

In addition to the area fraction calculations, three-dimensional surface images of the sample were made at the same time that the AFM phase images were produced. As the AFM "colors" the sample based on size, both a carbon atom and a missing carbon atom could both appear as a similarly colored spot in the expected size range on the sample image. The purpose of rendering the 3-D surface plots was to distinguish between a void space left by a carbon black particle and that of a protruding carbon black particle. In order to distinguish between the two, height images were analyzed in conjunction with the phase images. Samples of typical phase and height images are shown in Figure 6.7. Similar to the format used for the area fraction, the average particle size (diameter) from the midpoint of the sample to the edge are displayed in Figures 6.8, 6.9, and 6.10.

AFM TESTING ANALYSIS

The AFM was used to assess the depth to which carbon black was extracted from natural rubber, SBR, and EPDM-S elastomer samples. It was assumed that this loss of carbon black could be closely associated with the depth of chlorine or chloramine attack. Additionally, differences were expected to be observable between chlorine and chloramine exposure.

The data in Figures 6.5 to 6.10 show the fractional amount of carbon black versus cross-sectional depth after the specified chloramines exposure period. Based on the SEM test results, it was expected that the carbon black distribution would be uniform throughout the entire sample thickness at zero days exposure. It was expected that as the chloramines penetrated the sample, carbon black would be systematically removed along the diffusion front. Thus, tests at longer exposure periods should show a greater carbon black loss at the surface, and the subsequent carbon black loss should be observed deeper into the sample. The data shows time elapsed graphical representation of the area fraction covered by carbon black, as well as particle diameters, for each of the elastomer types tested. Trends in the average area fractions were as expected and showed a decrease in the area covered by the carbon black. With respect to the three elastomers tested, the following observations were noted:

- Natural rubber showed the least resistance to attack as a substantial loss of carbon black is indicated due to chloramine exposure. A decrease in the area fraction at the edges of the 30-day chloramine sample is clearly evident.
- Data from the AFM scans indicated only a small amount of carbon black loss in the EPDM-S materials when exposed to chloramine solutions. Similar to the natural rubber, EPDM-S showed some advanced degradation near the edges in the 30-day sample.

- The SBR samples also displayed a relatively large degradation effect due to chloramine exposures. It is noted that data from days 10 and 20 in the chloramine system follow trends in the loss of the average area fraction covered by carbon black.

To assess the differences between chlorine- and chloramine-induced degradation, elastomer samples were exposed to similar concentrations and temperatures for chlorine and then AFM techniques were used to assess sample degradation. As shown in Figures 6.11, 6.12, and 6.13, it is evident that chlorine affected the sample in a similar manner as chloramines. A major difference between the two disinfectants was the amount of expansion of the elastomer volume in the mixture with chloramines. The vertical line denoting the original sample size is hardly exceeded in the case of the chlorine samples even after 30 days. This may therefore be taken to be the relative edge of the sample. The sharply curved trend lines of the chlorine samples also indicates that erosion of the particles is almost entirely at the outer edge of the samples. This appears to coincide with the extent of diffusion into the samples interior by the penetrating fluid. In most cases, the amount of erosion of the carbon black particles was greater in the solutions containing chloramines than in those containing chlorine. Error bars have been included in the data pertaining to chlorine but has not been repeated for the chloramines; chloramines error bars appear in the previous set of data.

DIFFUSION RATE CALCULATIONS

Ultimately, the purpose of the AFM testing was to assess the depth of carbon black removal (which is closely related to chlorine/chloramines penetration) as a function of time. With this information it was possible to determine a diffusion coefficient for the materials.

As the sample aged, the fluid penetrated the sample, and a zone of expansion formed around the perimeter, and degradation proceeded to the interior. The effective rate of fluid penetration (diffusion) can be calculated using a simplified form of Fick's second law. The simplified form of the equation assumes a semi-infinite distance, such that chloramines moving in from the surfaces do not meet in the center of the coupon. The liquid diffusing in from the sides of the rubber coupon did not effectively reach the elastomer center after 30 days exposure. Therefore, the supposition of a semi-infinite distance was upheld. The effective penetration distance is defined as the point at which the concentration of the diffusing species has an equal value to the average of the concentration at the surface and the initial concentration within the sample (Schaffer et al. 1999).

Micrographs illustrating the diffusion of water mixtures under conditions of accelerated aging have been provided in Figure 6.14. Each of the micrographs depicts a cross-sectioned sample of the 0 day material to the left, 30 days accelerated aging in water containing chlorine in the center, and after 30 days aging with chloramines to the right. The zone of expansion at the edges of the specimens can clearly be seen in samples aged with chloramines. Zones of expansion appear as raised or bulbous ridges at the sides of the sample. In samples where accelerated aging was accomplished in a mixture containing chlorine, the ridges are hardly noticeable and do not extend very far into the sample. The depth of the ridges coincide well with the AFM data shown above. The loss of area fraction and decrease in particle diameter extending from the edge into the interior of the samples appear similar to the depth of the ridges viewed under an optical microscope.

In this analysis the effective penetration distance was obtained by the clearly discernable zone of expansion at the edges of the sample measured with digital calipers accurate to 0.01 millimeters.

Table 6.1
Diffusion coefficients determined after accelerated aging for 30 days

System	Natural rubber (cm^2/sec)	EPDM-S (cm^2/sec)	SBR (cm^2/sec)
Chloramines	1.42E–9 ±1.2E–10	3.70E–10 ±3.0E–11	1.26E–9 ±1.8E–10
Chlorine	3.19E–11 ±1.3E–11	1.03E–12 ±8.3E–13	2.40E–11 ±1.1E–11

Several distance measurements were obtained for each sample to assess mechanical accuracy of the calipers and to account for uneven expansion within the sample. The multiple distance measurements were combined to provide an error estimate in the resulting diffusion coefficients. The formula used for calculating the effective diffusion rate *D*, is given as x_{eff} ? $(Dt)^{0.5}$ (Schaffer et al., 1999). The calculated diffusion coefficients for each sample are as shown in Table 6.1.

AFM CONCLUSIONS

Data procured during this experiment is representative of typical results seen in earlier experiments concerning performance of natural rubber, SBR and EPDM-S under conditions of accelerated aging. In earlier testing of tensile strength it was found that, similar to the other elastomeric properties, the maximum stress and strain values decreased much more rapidly when exposed to mixtures with chloramines than with chlorine. Also, EPDM-S retained its integrity much better than natural rubber and SBR.

The diffusion of the penetrating fluid into the sample was likely responsible for the loss of carbon black, which in turn contributed to the deterioration of the bulk properties. AFM imaging of the samples exposed to conditions of accelerated aging showed an overall loss of carbon black material and a decrease in the average particle size as fluids penetrated the sample. Degradation of the carbon black coincided with deterioration of bulk properties and occurred in an increasing region around the perimeter of the sample. These results, along with earlier observations based on the turbidity of the extracted fluids, strongly suggest that the chloramine-induced erosion of the carbon black particles is a strong factor in the decrease in elastomer performance over time.

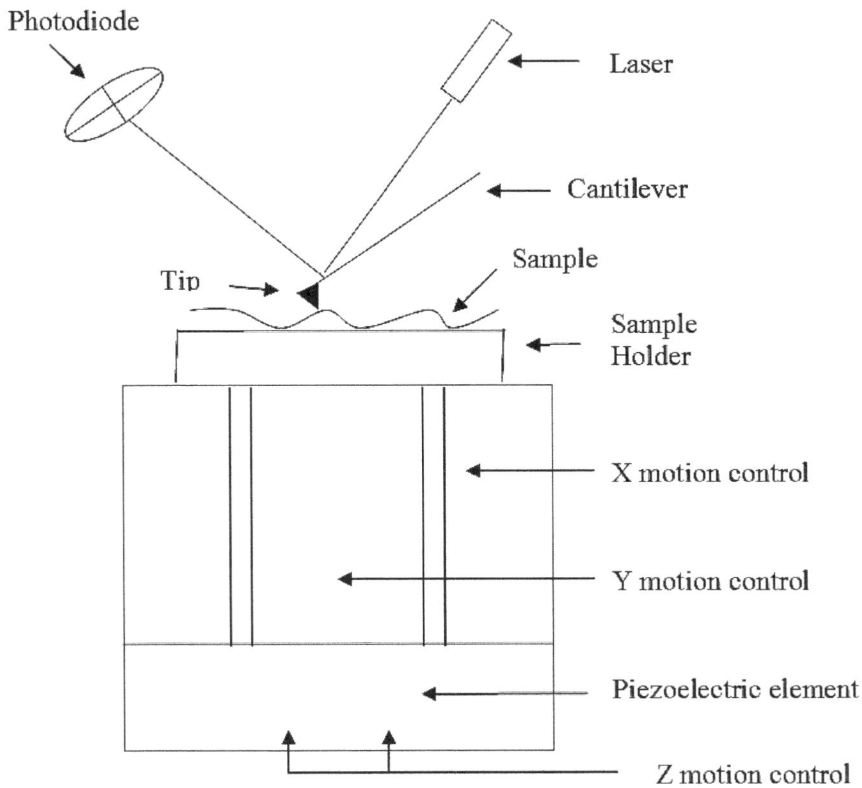

Figure 6.1 Concept and major components of AFM

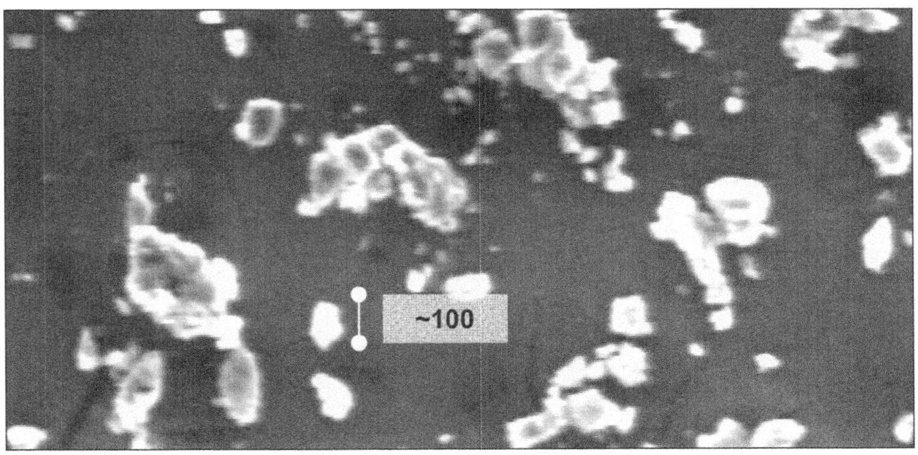

Figure 6.2 SEM image of carbon black powder (SBR) coated with gold

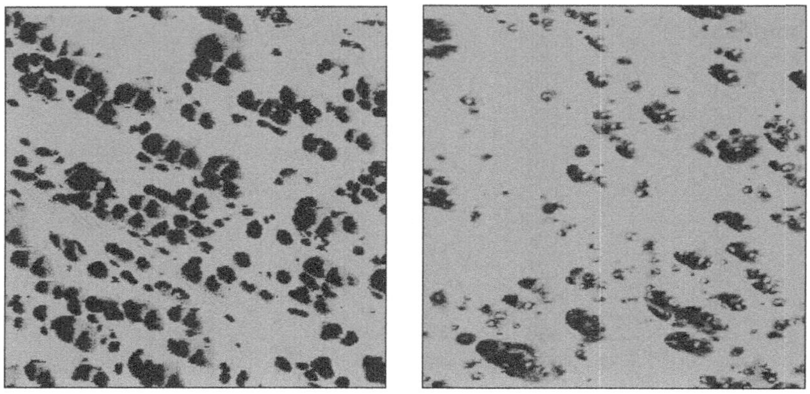

Figure 6.3 AFM images depicting near-edge concentrations and size of carbon black particles at 0 days on the left and after 30 days aging with chloramines on the right

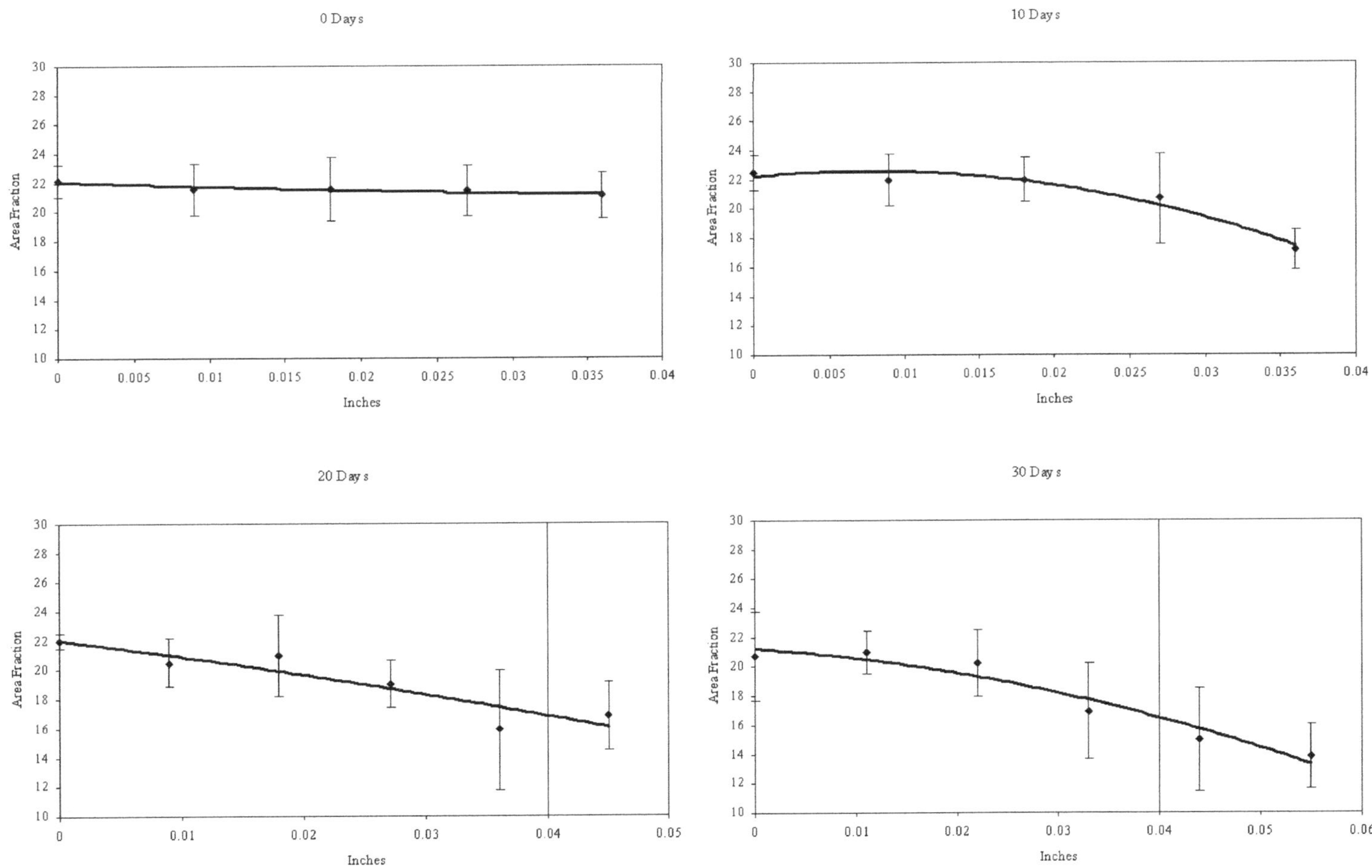

Figure 6.4 Area fractions of natural rubber cross-section aged in water mixture with chloramines

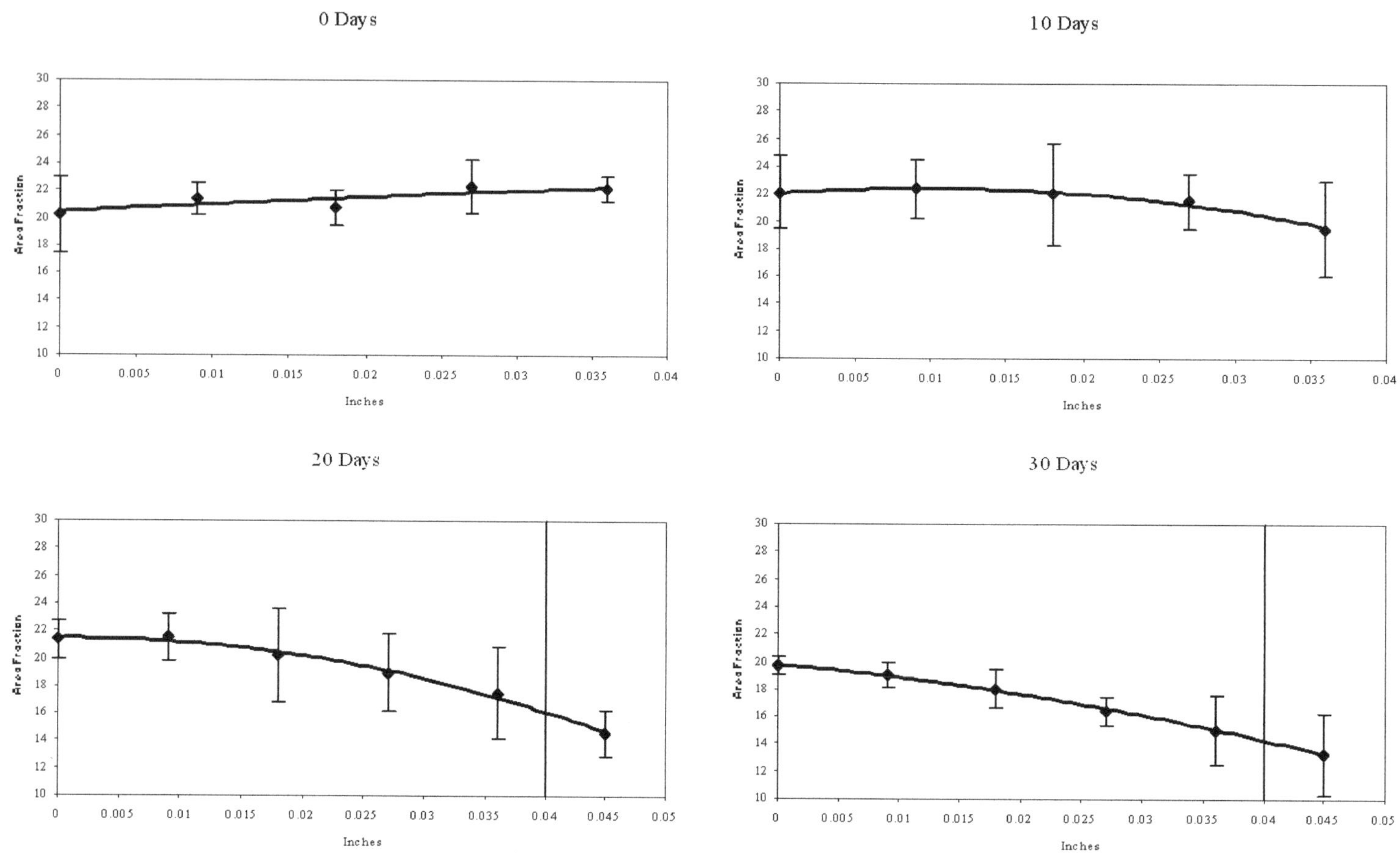

Figure 6.5 Area fractions of EPDM-S cross-section aged in water mixture with chloramines

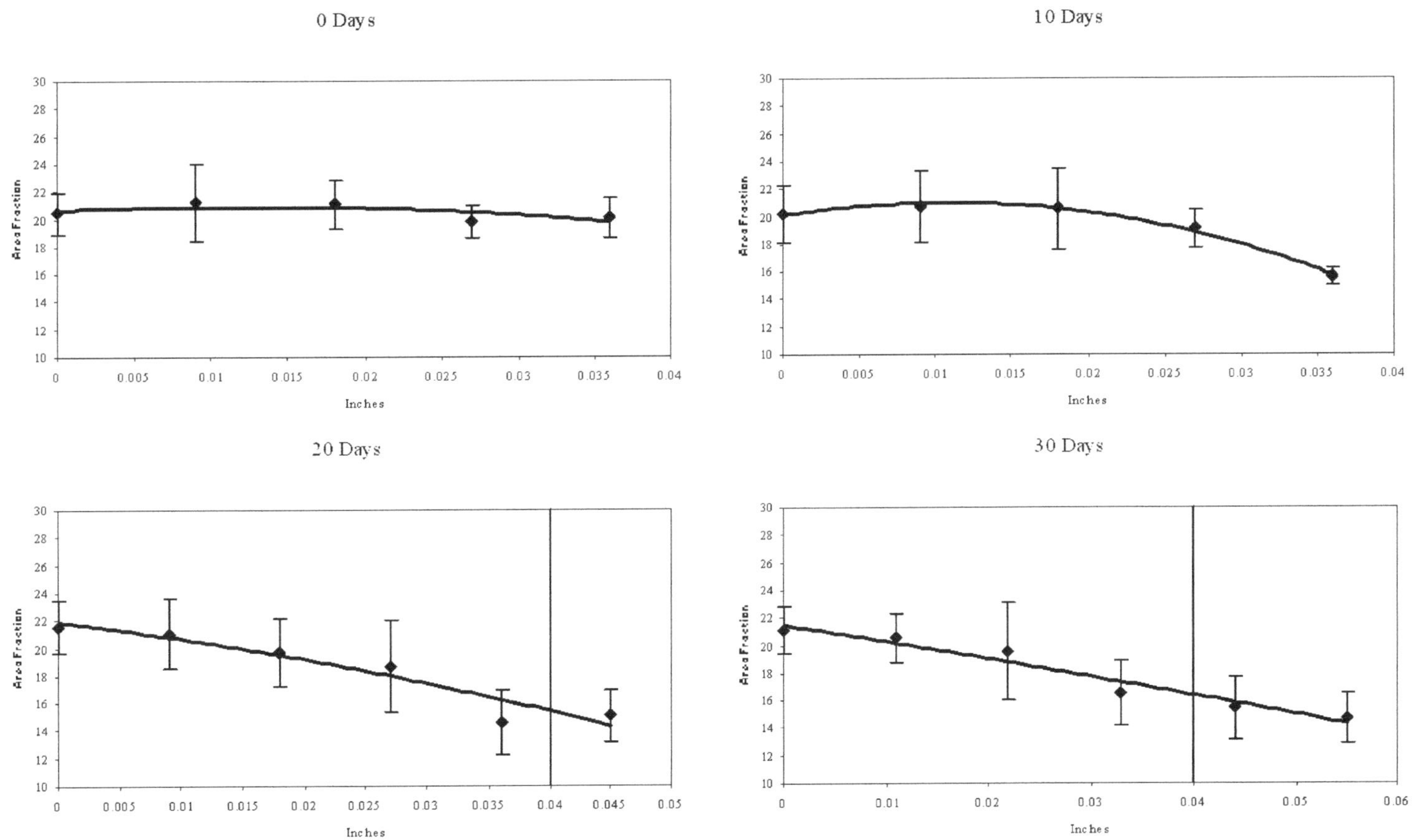

Figure 6.6 Area fractions of SBR cross-section aged in water mixture with chloramines

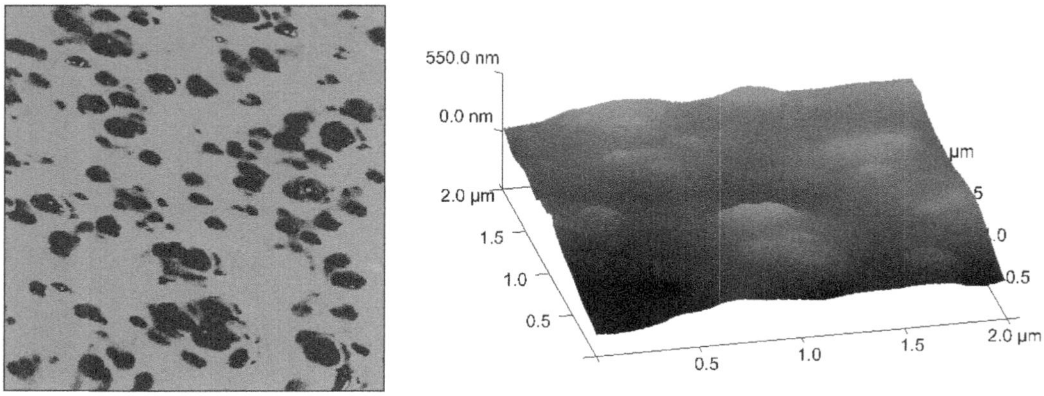

Figure 6.7 Phase image on left and 3-D surface plot on right of EPDM-S sample after 10 days accelerated aging in chloramines mixed with water

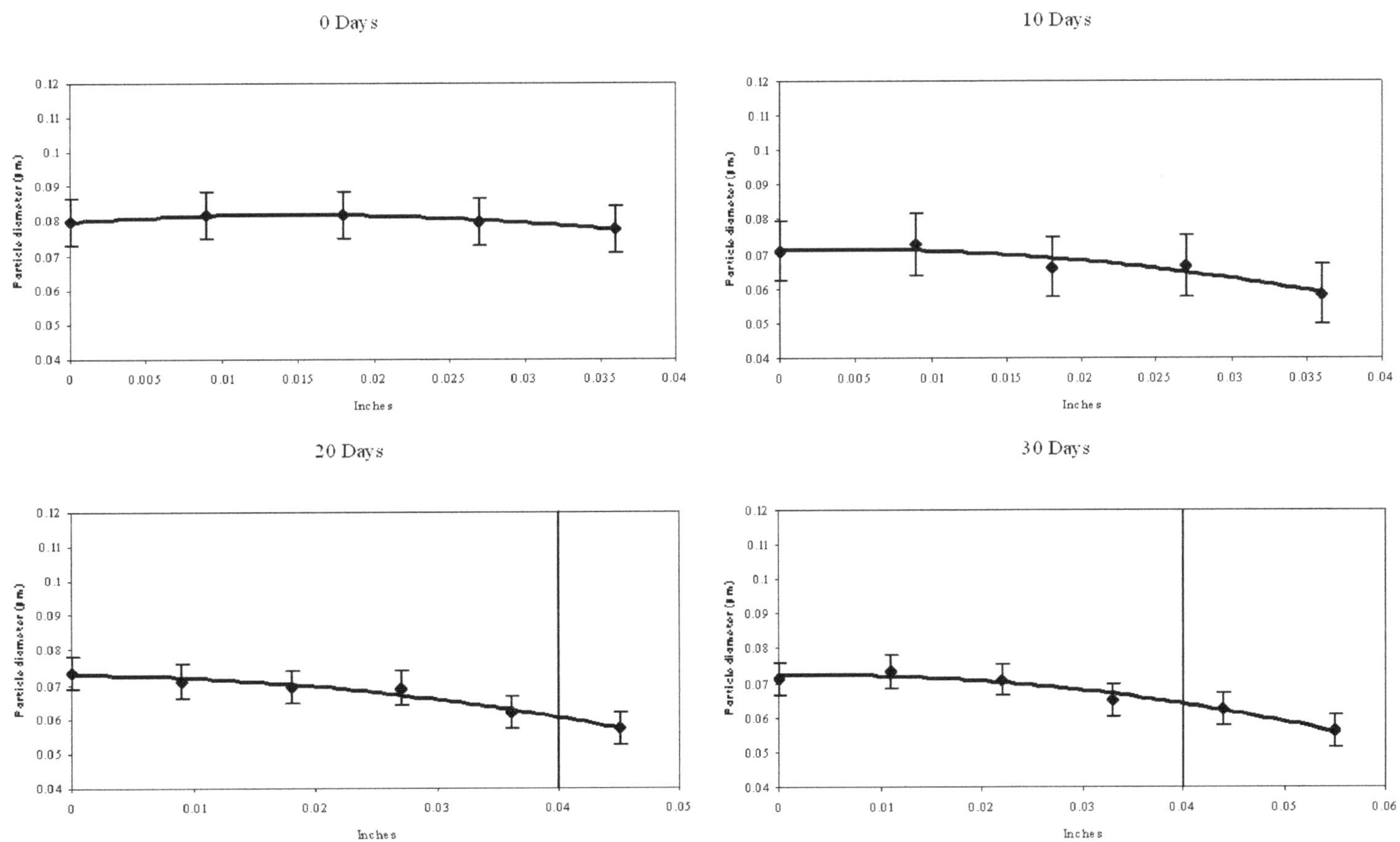

Figure 6.8 Particle diameter of natural rubber cross-section aged in water mixture with chloramines

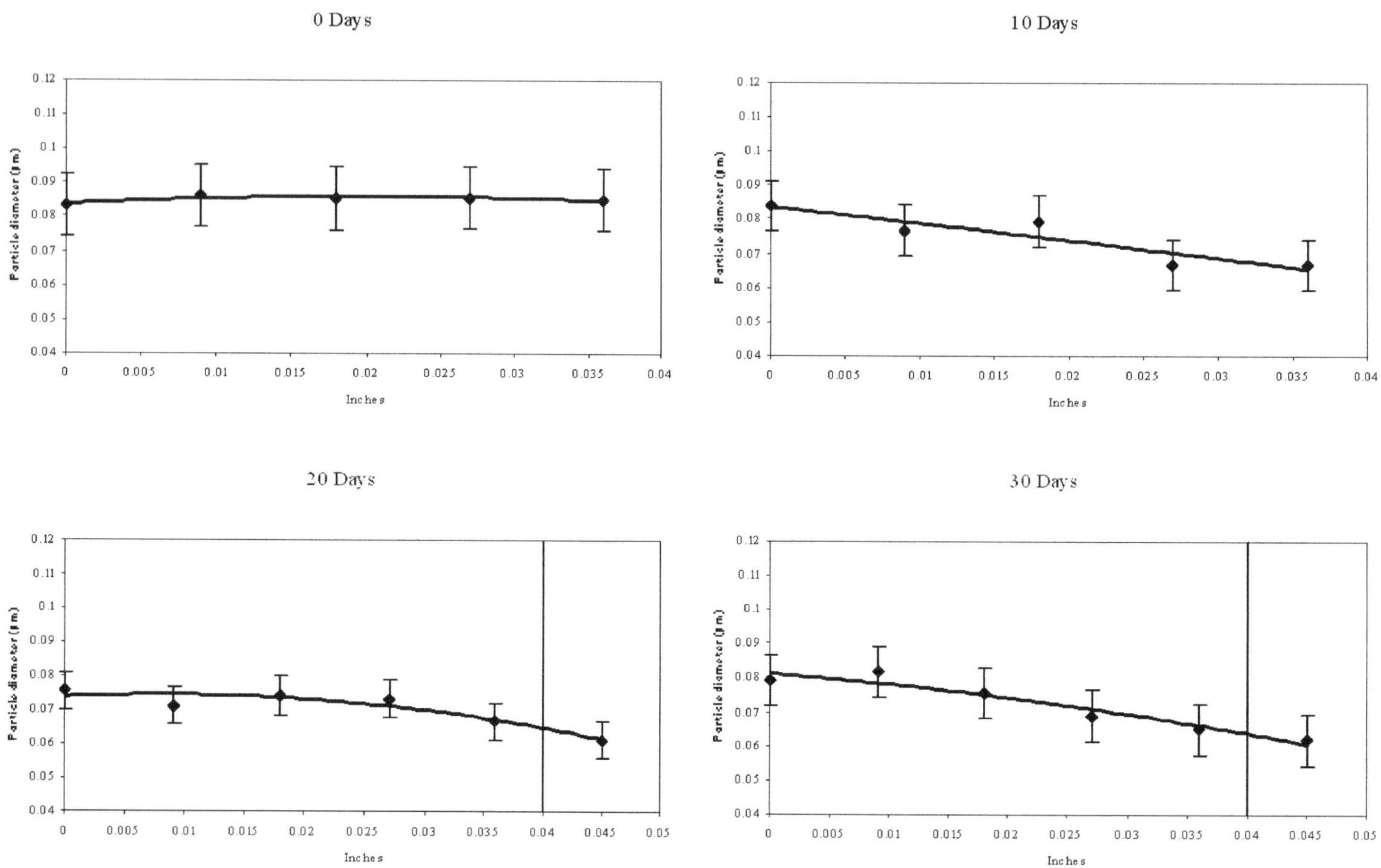

Figure 6.9 Particle diameter of EPDM-S cross-section aged in water mixture with chloramines

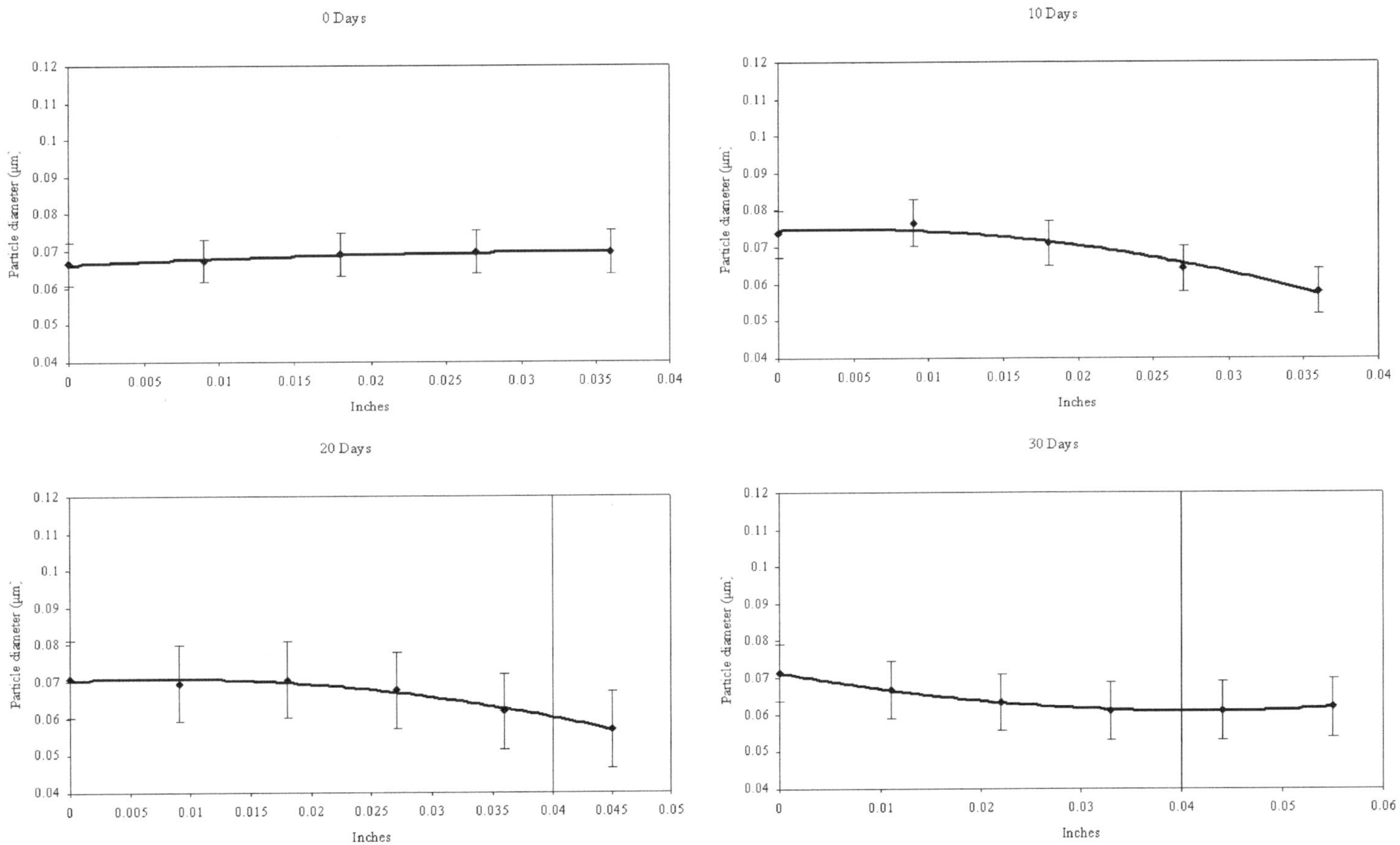

Figure 6.10 Particle diameter of SBR cross-section aged in water mixture with chloramines

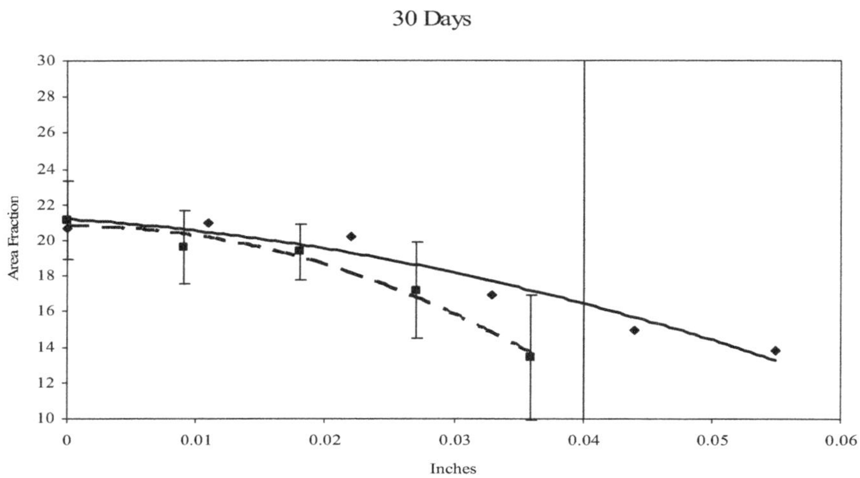

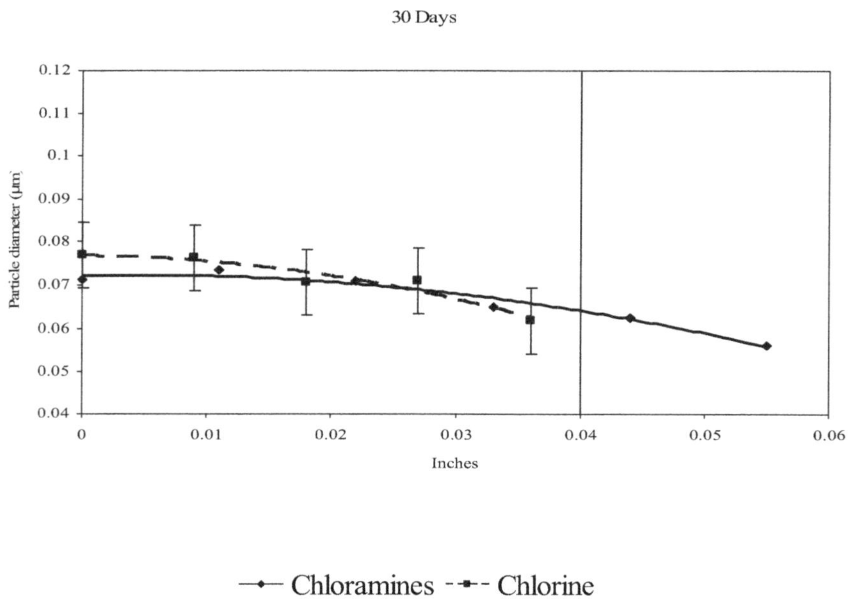

Figure 6.11 Area fraction (top) and particle diameter (bottom) in natural rubber sample

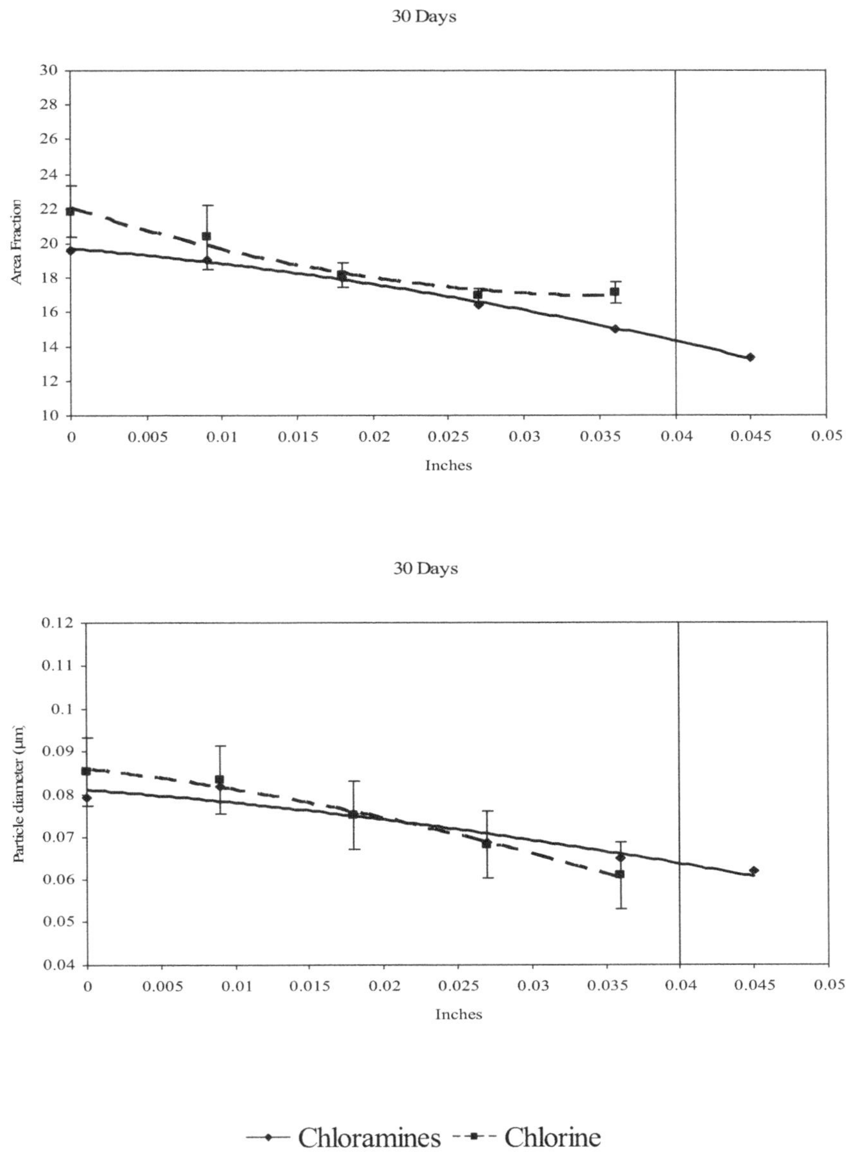

Figure 6.12 Area fraction (top) and particle diameter (bottom) in EPDM-S sample

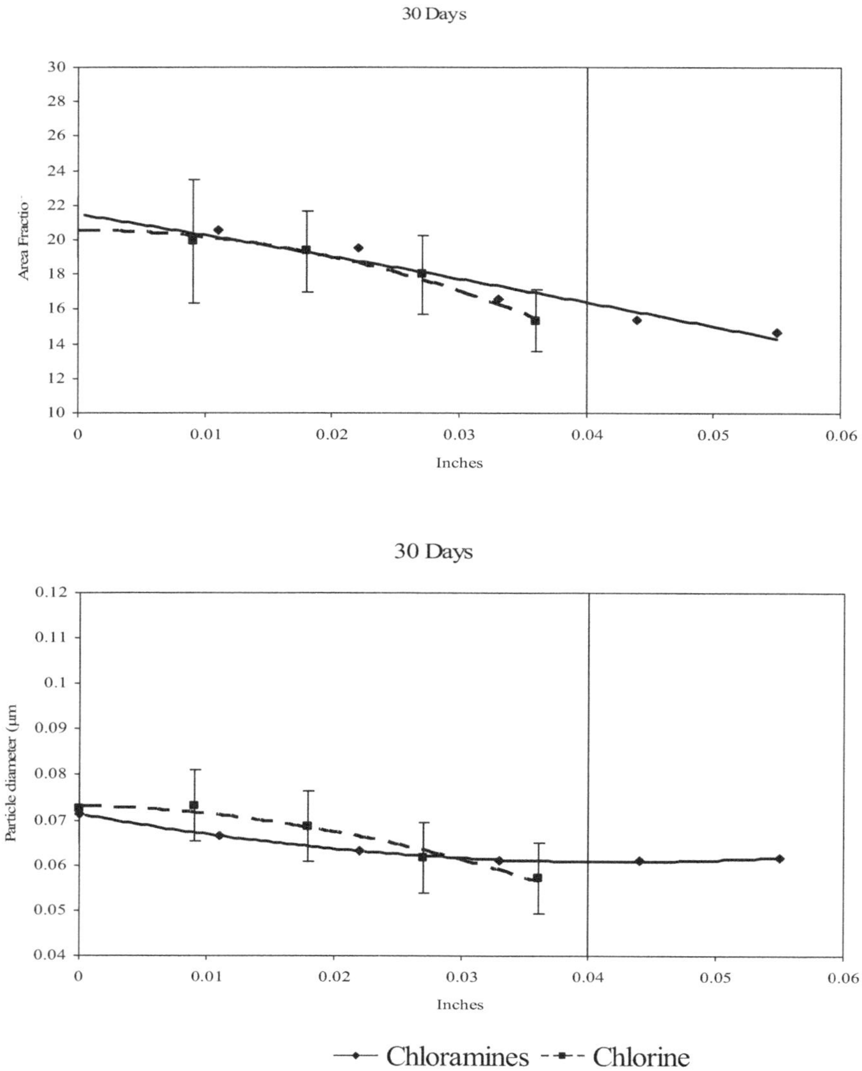

Figure 6.13 Area fraction (top) and particle diameter (bottom) in SBR sample

Figure 6.14 Sample cross-sections of natural rubber (left), EPDM-S (center), and SBR (right). There are three samples for each elastomer, the first taken at 0 days, the second taken at 30 days in a chlorine mixture, and the third taken at 30 days in a chloramines mixture.

CHAPTER 7
CONCLUSIONS

ELASTOMER DEGRADATION BEHAVIOR

This report has provided data and discussion driven by three primary objectives:

- To establish a quantitative method for predicting the life expectancy of in-service elastomer components when exposed to potable water
- To identify elastomeric materials with high performance when subjected to chloramines
- To provide the water industry better tools for managing the risks and financial costs associated with elastomer failure

The research presented here is intended to provide design and maintenance engineers with the information that they will need in order to determine which elastomer meets their needs based on the required performance criteria. While many properties and behaviors were observed that may ultimately influence material decisions, there can be some general comments made about the performance of each of the materials examined.

Natural Rubber

Being the first elastomeric material used in the potable water infrastructure, many of the researchers' observations refer to the performance of natural rubber. While the material is being phased out of use, a significant number of natural rubber components likely are still in use within water systems. Natural rubber degradation is influenced by both the concentration and temperature of the solution, though more so by temperature. Consistently, natural rubber was last in nearly every performance category at every temperature and concentration. The one area where it does outperform other elastomers is in breaking strain, where it maintained is plasticity for far longer, essentially resisting embrittlement. While not the best choice of the materials currently available, natural rubber components already in service can perform for long periods of time if they are left undisturbed.

EPDM-S

This class of elastomeric components is a relative newcomer for use in the water works industry. Judging from the results reported here, it can be used very effectively and over a long term. EPDM-S performed at or near the top of every category measured, showing that it was affected far more by temperature than chloramine concentration. Given the thermoplastic nature of EPDM, this is not surprising. High chloramine concentrations do influence the rubber more so than at lower concentrations, which leads to a drop off in performance. At standard operation conditions, however, EPDM-S remains dimensionally stable and retains its physical properties.

NBR

While not the worst performing elastomer studied, NBR is also not the best performer in any category. Degradation of NBR seems to be equally influenced by temperature and solution concentration. By all appearances, NBR performs well initially, but this performance begins to drop off rapidly as aging progresses. This suggests that NBR may be suitable for some short-term applications, but the performance will drop off more severely with time. However, much like natural rubber, it seems to retain its plasticity for longer periods of time.

SBR

SBR is the most abundant elastomer currently utilized in the water works industry. While it is not the best performer in any category, it maintains its mid-level performance across every parameter. Like EPDM-S, SBR seems to be more influenced by temperature than chloramine concentration. At operating temperature, SBR maintains a good level of dimensional stability, but it tends to be susceptible to embrittlement far more than the other elastomers studied. This suggests that movement or disruption of SBR components that have been in place for longer periods of time will cause them to fail.

Neoprene

By far the worse performer in this group, neoprene was at the bottom for almost every performance indicator evaluated in this study. Neoprene is by far the least dimensionally stable material and shows a tendency toward embrittlement. Neoprene as a material is formulated for resistance to organic-based solvents and, as such, is used in environments where oil and water are found. It is readily apparent through the data presented in this report that this resistance does not extend to attack by chloramines.

EPDM-P

EPDM-P was the top performer in each category examined in this study. In many cases, the performance variables changed so little that it became almost impossible to accurately predict the performance life of this elastomer. The change in curing method from sulfur to peroxide imparts a high degree of resistance to chloramine attack, as was discussed earlier. Like EPDM-S, EPDM-P is more susceptible to temperature effects due to its thermoplastic nature. Under standard operating conditions, however, this degradation mechanism would be less significant, so components should have a longer overall lifespan.

ELASTOMER SELECTION BASED ON PERFORMANCE INDICATORS

The master curves for the various elastomers provide a good deal of insight into the expected lifespan of rubber parts in a chloramine system (Chapter 4). However, how the elastomers perform in a real-world application is dependent on a variety of factors. Most importantly, the function of the part must be assessed in relation to the critical elastomer failure mode. For example, in some conditions, it may be entirely suitable to form a gasket out of a highly swelling material and have adequate long-term performance. However, it is unlikely that the same material

Table 7.1
Common failure modes for typical elastomer components

Component	Failure mode
Resilient seat gate valve	Surface degradation (hardness), swell
Butterfly valve	Swell
Gasket	Swell
Meter seal	Surface degradation, swell, elongation, and tensile strength

would perform as well on a gate valve where swelling would adversely affect the closure operation. Thus, it is important to consider elastomers as engineering materials and assess the suitability of each formulation based on anticipated critical failure modes. Common failure modes for typical elastomer components within the water industry are shown in Table 7.1.

Elastomer Swell Performance

Figure 7.1 displays the relative elastomer performance for change in mass over time when exposed to a 1 ppm chloramine solution. The change in mass is an important parameter to consider when selecting elastomers for a specific application, as it correlates directly to swell and breakdown of the internal elastomer structure. While the surface-area-to-mass ratio does not influence the rate of swelling, it is important to note that components that have a higher mass will swell to greater extents. In applications where long-term elastomer size is critical because of clearance or other issues, it is important to select an elastomer that will have limited swell. However, in water utility applications where swell is not a critical factor, such as gaskets, high swell elastomers may be suitable, provided they meet other critical performance requirements.

Table 7.2 compares the 50 percent swell degradation values for the elastomers evaluated in this study then exposed to a 1 ppm chloramine solution. It is noted that EPDM-P does not have a discrete value associated with the time to reach 50 percent weight gain. The EPDM-P material was observed to be very resistant to the accelerated degradation and thus long-term performance could not be estimated with the test conditions. Table 7.2 should be used as a general guide to selecting elastomers with respect to swelling performance. It is noted that formulators can modify elastomer compounds to meet specific performance criteria, and specific materials may perform significantly differently from the generic formulations used in this study.

Elastomer Stress/Strain Performance

Figure 7.2 displays the relative elastomer performance for change in breaking strain over time when exposed to a 1 ppm chloramine solution. The change in breaking strain often is considered in elastomer selection because it correlates to the integrity of the elastomer molecular backbone structure. As the mass of a particular component increases in relation to its surface area, the breaking strain will increase due to the additional material. However, but the rate of decrease in the breaking strain over time will remain unchanged. In applications where long-term elastomer elongation or strength is critical, such as diaphragms, it is important to select an elastomer that will have limited loss of strain over time. However, in water utility applications where strength or

Table 7.2
The 50 percent swell degradation of common elastomers

Elastomer name	Number of years to reach 50% of weight change
EPSM-P	> 30
EPDM-S	21.9
Nitrile	3.9
SBR	2.3
Natural rubber	1.9
Neoprene	0.4

Table 7.3
The 50 percent strain degradation of common elastomers

Elastomer name	Number of years to reach 50% of initial breaking strain
EPDM-P	> 30
Natural rubber	17.4
Nitrile	13.3
SBR	3.4
Neoprene	1.8
EPDM-S	1.1

strain is not a critical factor, elastomers that exhibit a loss of strain with time may be suitable, provided they meet other critical performance requirements.

Table 7.3 compares the 50 percent strain degradation values for the elastomers evaluated in this study then exposed to a 1 ppm chloramine solution. EPDM-P does not have a discrete value associated with the time to reach 50 percent strain. The EPDM-P materials was observed to be very resistant to the accelerated degradation and thus long-term performance could not be estimated with the test conditions. Table 7.2 should be used as a general guide to selecting elastomers with respect to breaking strain performance. It is noted that formulators can modify elastomer compounds to meet specific performance criteria, and specific materials may perform significantly differently from the generic formulations used in this study.

Elastomer Hardness Performance

Figure 7.3 displays the relative elastomer performance for change in hardness over time when exposed to a 1 ppm chloramine solution. It is noted that accelerated-age predictions could only be completed for natural rubber and neoprene elastomers. In applications where long-term elastomer surface hardness is critical, it is important to select an elastomer that will have limited loss of hardness over time. Decreasing the surface-area-to-mass ratio of a part has an initial benefit of increasing the hardness of that component, but certain elastomers, such as SBR, tend to undergo large changes in hardness over time due to the mechanisms of carbon black loss. An application of this in the water utility industry would be when considering wear resistance of an elastomer part. However, in water utility applications where hardness is not a critical factor, elastomers that exhibit a loss of hardness with time may be suitable, provided they meet other critical performance requirements.

CONCLUSIONS

Based on their research and observations of elastomeric performance, the project team arrived at the following conclusions:

- Overall, the EPDM-P materials performed significantly better than any of the test elastomers exposed to chlorine and chloramine disinfectants. The EPDM-P material exhibited very little degradation in any of the measured parameters throughout the testing. The exception was at the most extreme testing condition (70°C temperature and 60 ppm average concentration), in which the EPDM-P materials degraded rapidly. This rapid degradation likely was due to higher order failure modes occurring within the sample and not anticipated to occur within normal water distribution systems.
- The presence of loop strain actually *reduced swelling* (less weight and volume change) and *improved tensile performance* for all of the elastomers tested. This behavior, combined with the small surface-to-mass ratio and limited exposure area, is the likely reason that pipe gaskets made of any elastomeric material have been observed to have adequate long-term performance.
- In general, 30-day testing of un-aged samples under the specified test conditions showed that at low pH, where dichloramine species is preferentially formed, performance degradation was higher than in the corresponding monochloramine condition. (It should be noted that the differences were not extreme, generally less than 10 percentage points, and may be within typical elastomer sampling variability.) Except for EPDM-S, all elastomer materials tended to exhibit higher swelling, loss of strength, and greater softening in the dichloramine environment. SBR and natural rubber tended to be more sensitive to pH than the other elastomers tested.
- The surface-area-to-mass ratio was observed to have little influence on the bulk mechanical property degradation of a given elastomer. The hardness values, which are highly surface specific, tended to follow the same trends regardless of surface-area-to-mass ratio. There were a few exceptions, namely, SBR and natural rubber at high temperature and concentration. These exceptions could be due to a change in the surface microstructure as the new samples were prepared differently from the original 0.080 inch samples, or they could be due to some other aging mechanism. Further studies could aid in determining the exact cause of these departures. While the surface-to-mass ratio does not influence the rate of swelling, it is important to note that components that have a higher mass will swell to greater extents.
- Elastomer materials exposed to monochloramine under the test conditions specified were observed to have a much greater swelling and loss in hardness than those exposed to chlorine at the same conditions. Tensile performance loss also was much worse in monochloramine than chlorine. The difference in performance between the two environments is large (10 to 50 percent) for all of the elastomers with the

exception natural rubber, which showed a change in performance in the range of 1 to 10 percent.

- The materials exposed to dichloramine also degraded substantially more than those in chlorine under the same conditions.
- The Atomic Force Microscope testing indicated that loss of carbon black from within the elastomer matrix appeared to be closely related to the degradation of the measured performance parameters.

This research confirmed that elastomeric components must be considered an "engineering material." As such, the selection of the appropriate elastomeric material for a specific application must take into consideration the critical failure mode for the component. An elastomeric material choice should be based on the anticipated product application and associated failure condition.

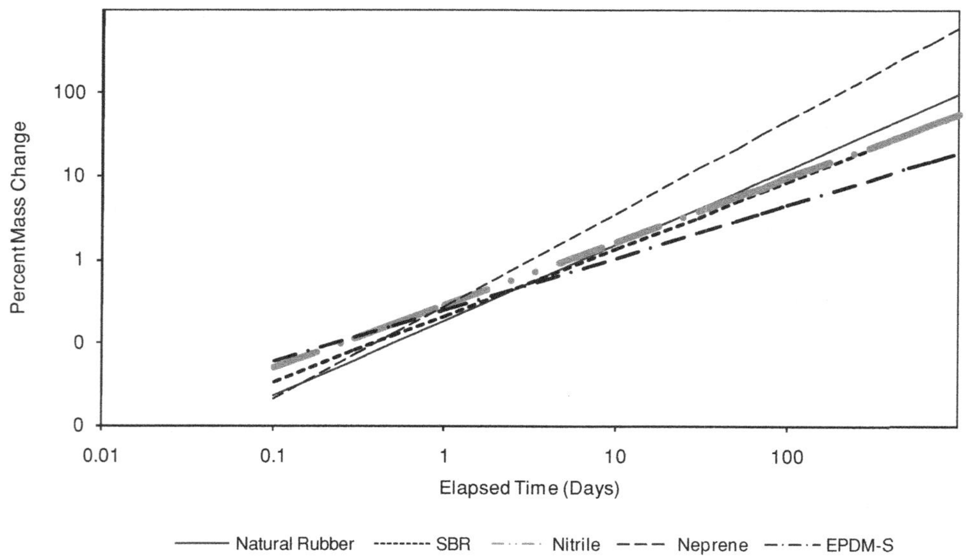

Figure 7.1 Relative performance of percent mass change for selected elastomers

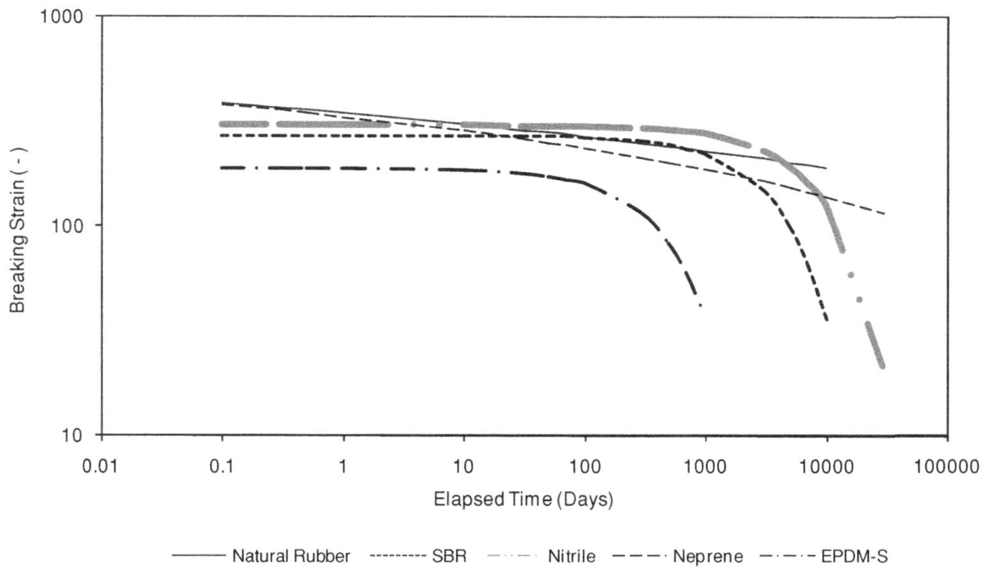

Figure 7.2 Relative performance of breaking strain for selected elastomers

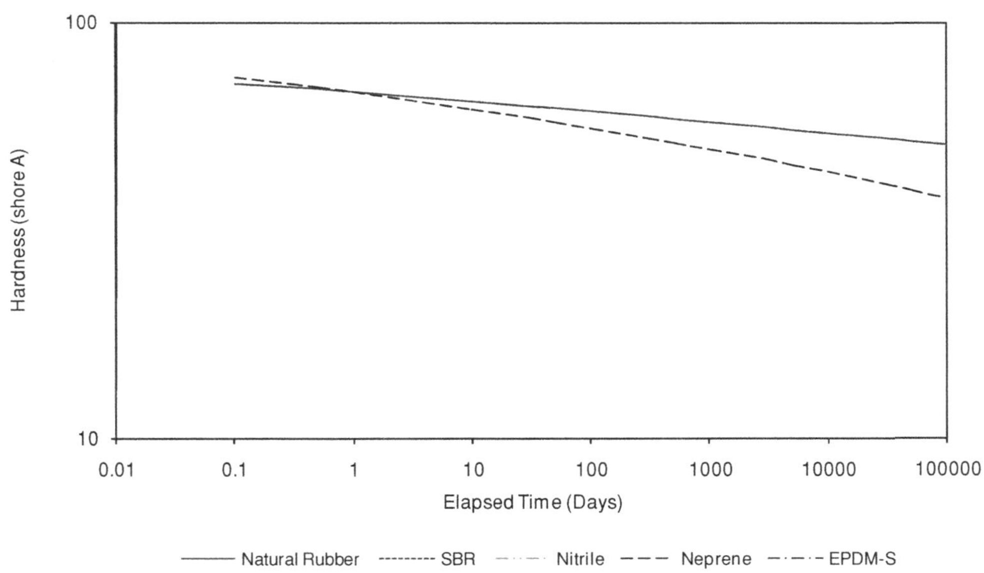

Figure 7.3 Relative performance of hardness for selected elastomers

APPENDIX A
ACCELERATED-AGING TEST DATA

TEMPERATURE EFFECTS

CONCENTRATION EFFECTS

PH EFFECTS

LOOP STRAIN EFFECTS

CHLORINE VERSUS CHLORAMINE EFFECTS

TEMPERATURE EFFECTS

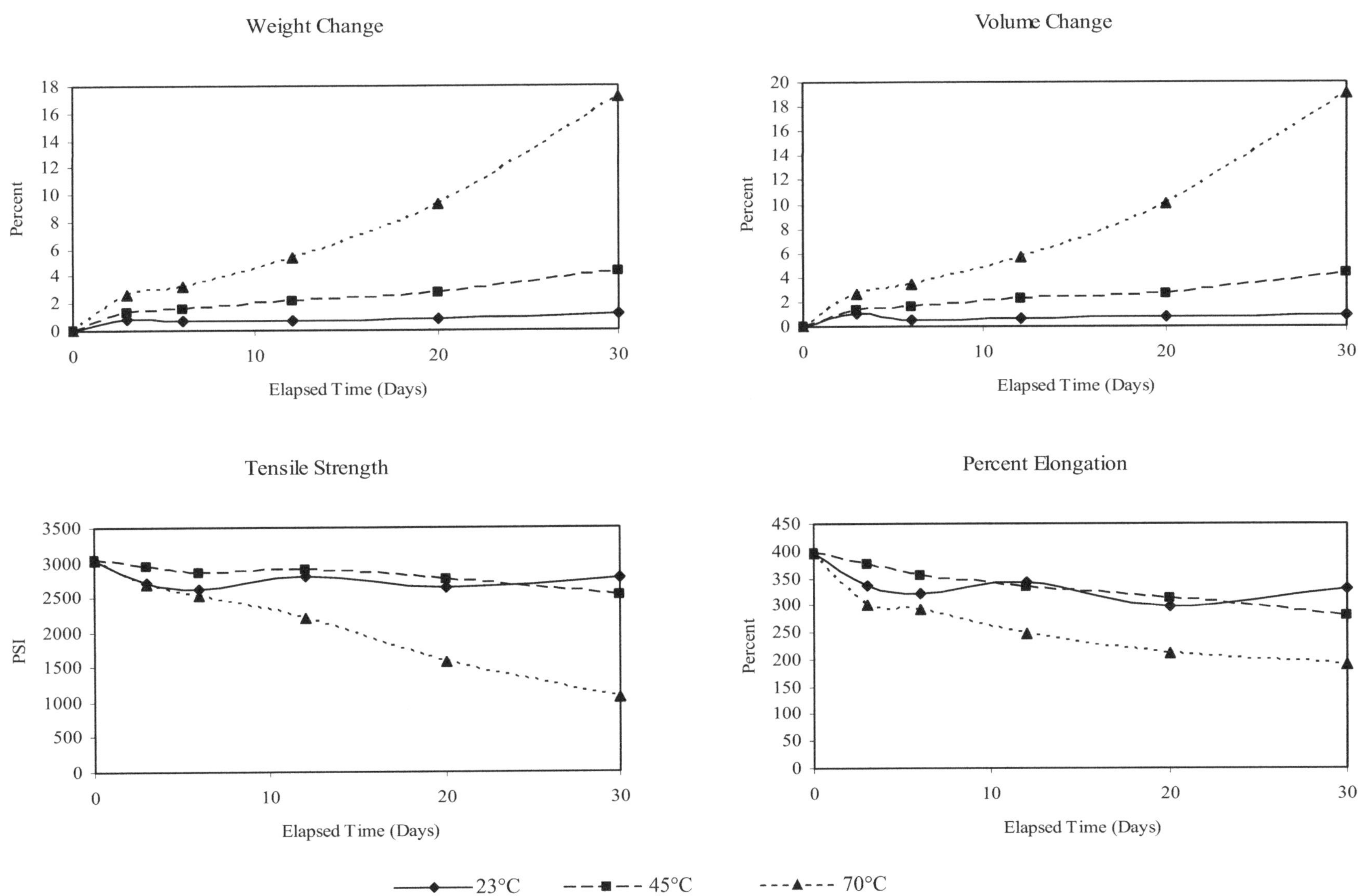

Figure A.1 Temperature effect on elastomeric properties at 1 ppm chloramine and pH 8.3 of natural rubber

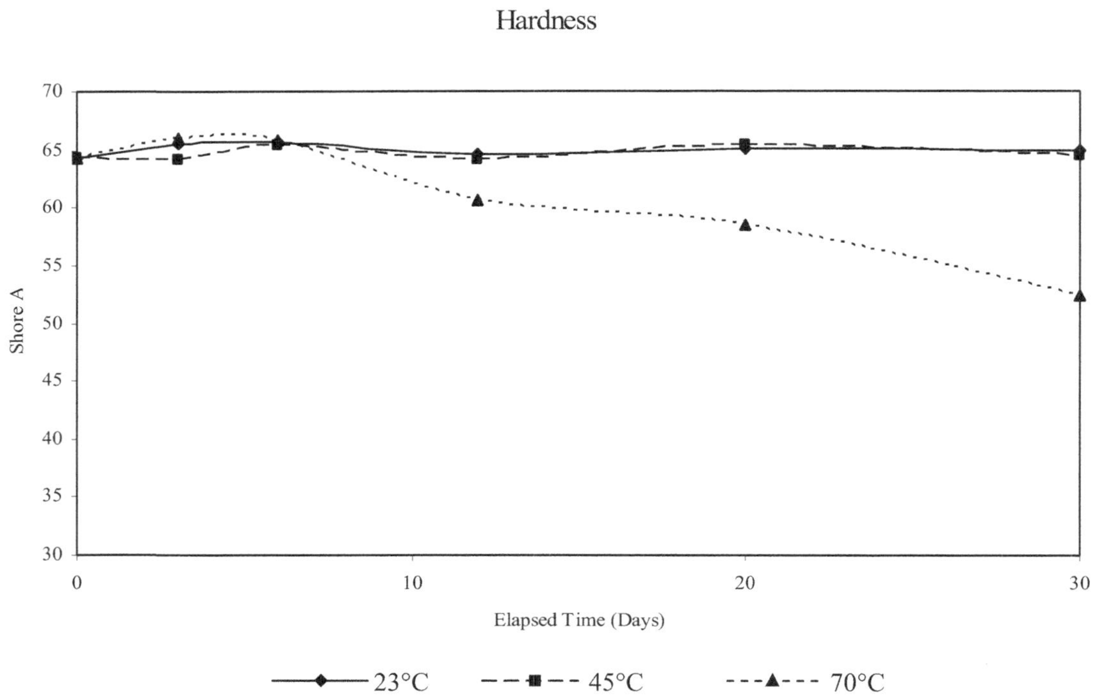

23°C at 30 days

45°C at 30 days

70°C at 30 days

Figure A.2 Temperature effect on hardness at 1 ppm chloramine and pH 8.3 of natural rubber with 40× magnified pictographs

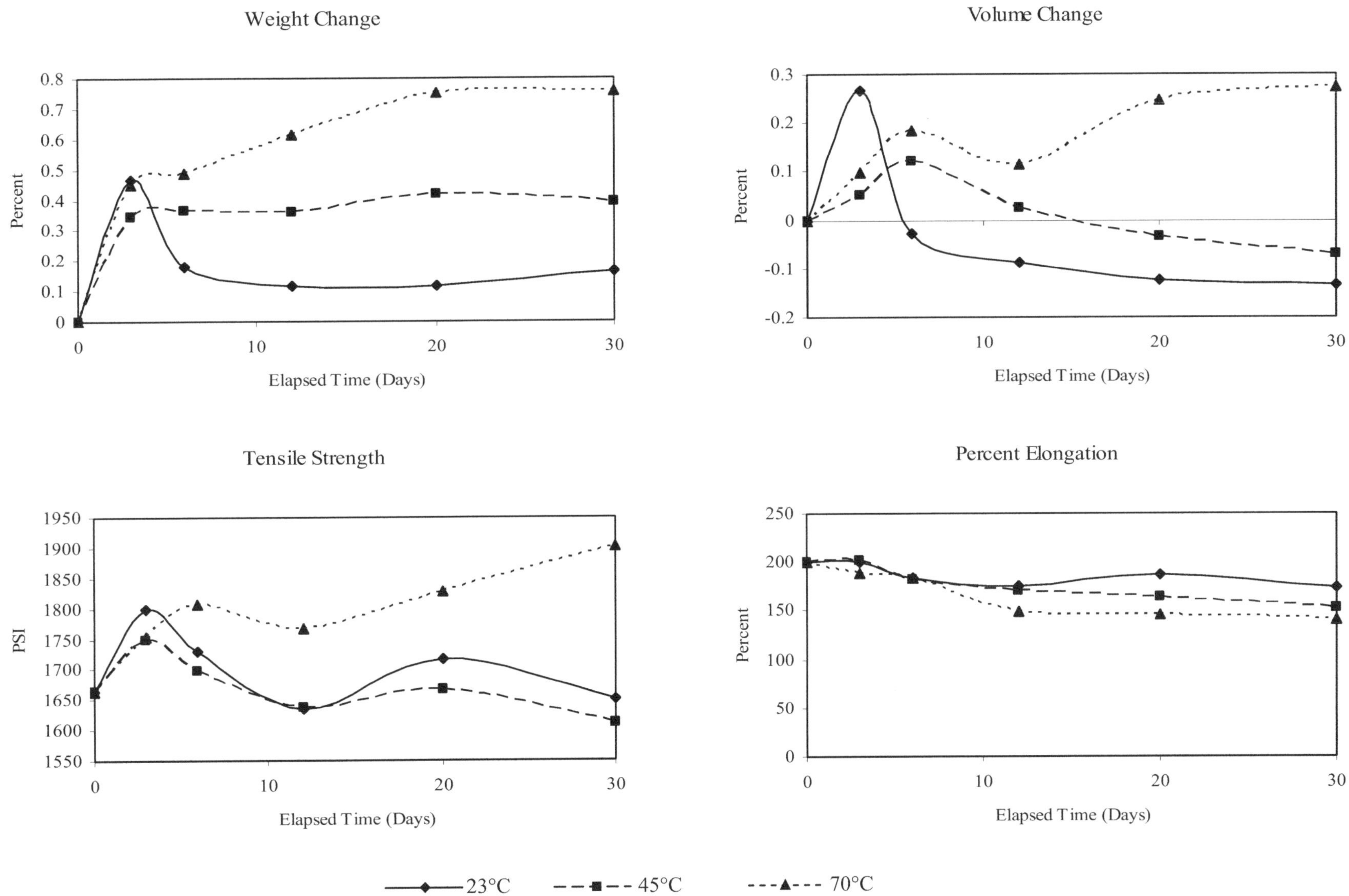

Figure A.3 Temperature effect on elastomeric properties at 1 ppm chloramine and pH 8.3 of EPDM-S

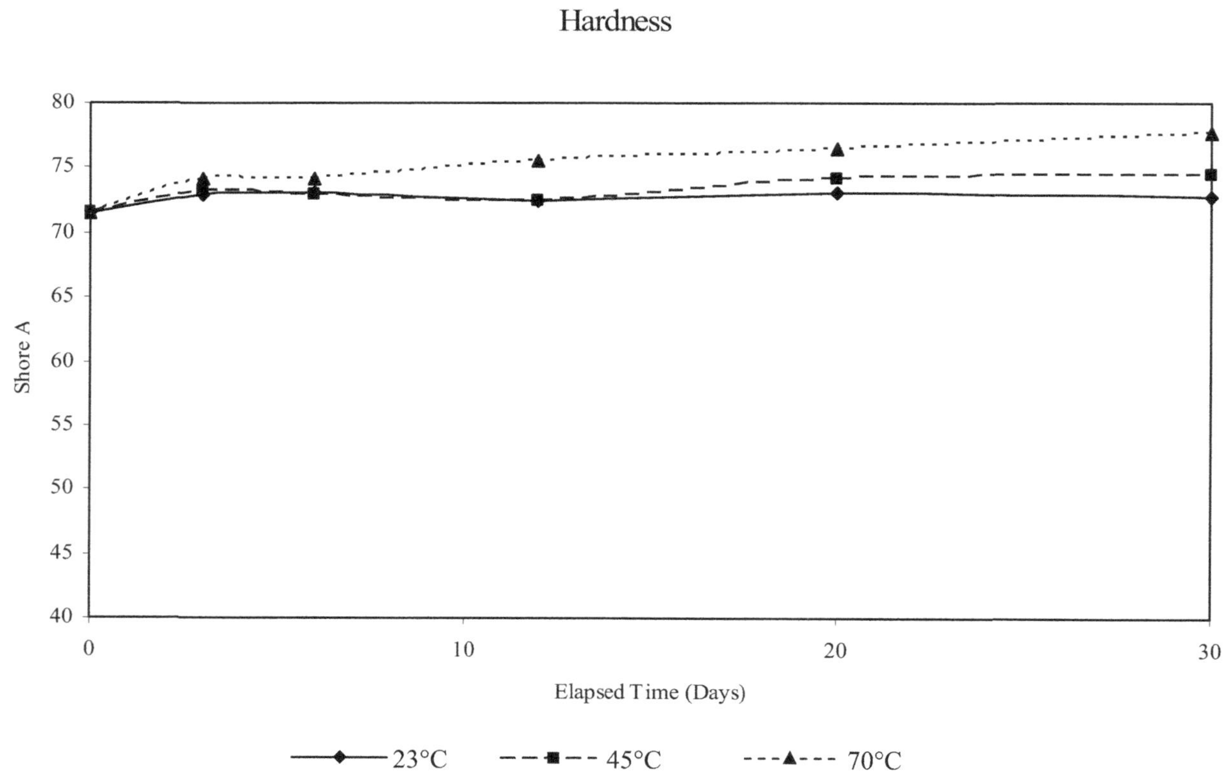

23°C at 30 days

45°C at 30 days

70°C at 30 days

Figure A.4 Temperature effect on hardness at 1 ppm chloramine and pH 8.3 of EPDM-S with 40× magnified pictographs

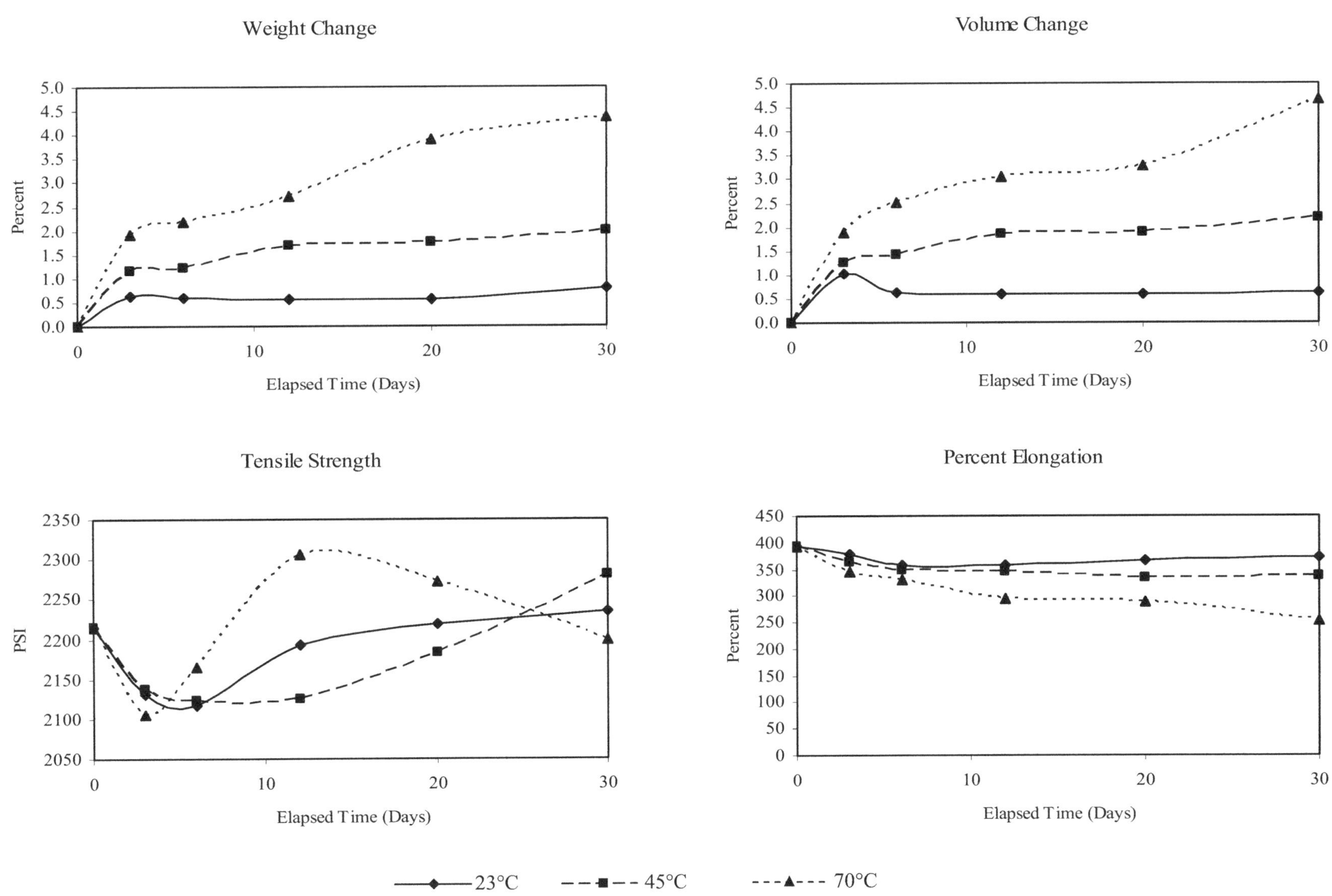

Figure A.5 Temperature effect on elastomeric properties at 1 ppm chloramine and pH 8.3 of nitrile (NBR)

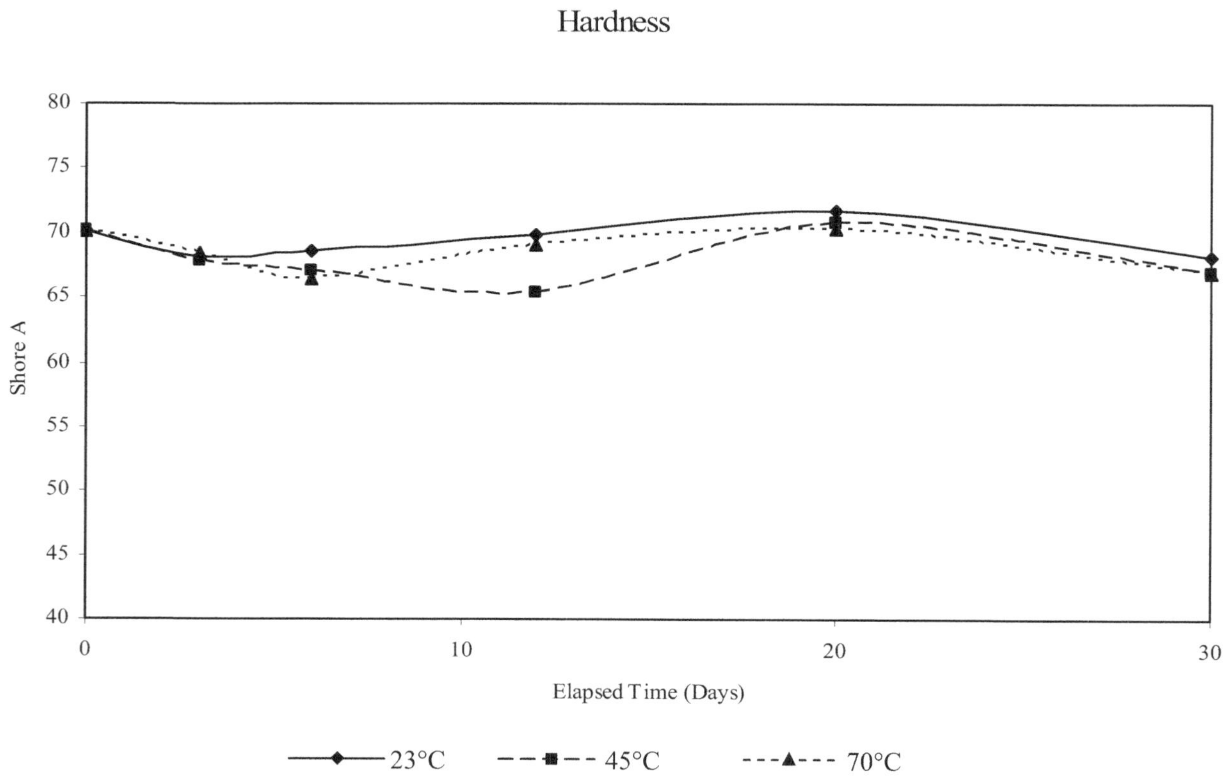

23°C at 30 days

45°C at 30 days

70°C at 30 days

Figure A.6 Temperature effect on hardness at 1 ppm chloramine and pH 8.3 of nitrile (NBR) with 40× magnified pictographs

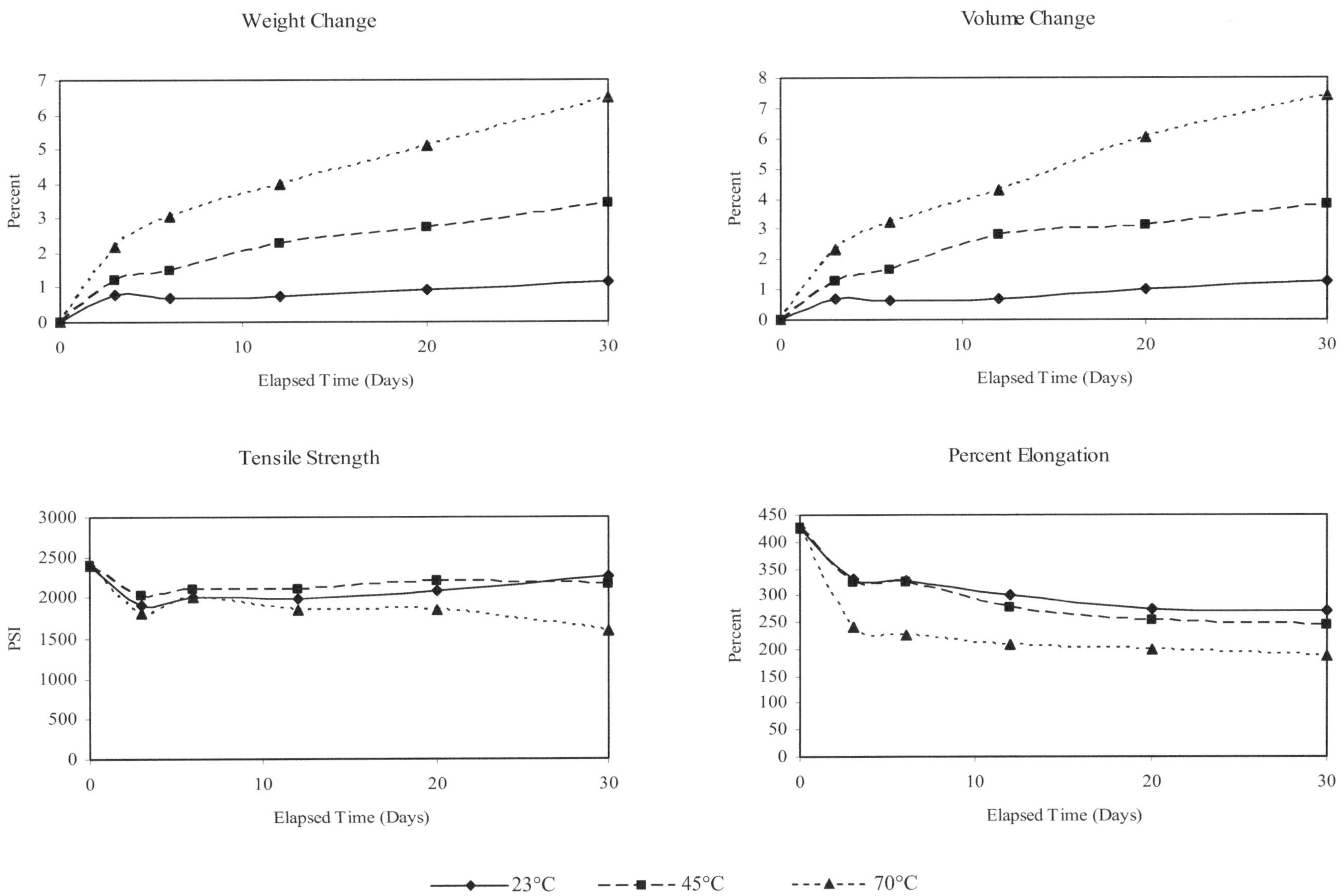

Figure A.7 Temperature effect on elastomeric properties at 1 ppm chloramine and pH 8.3 of SBR

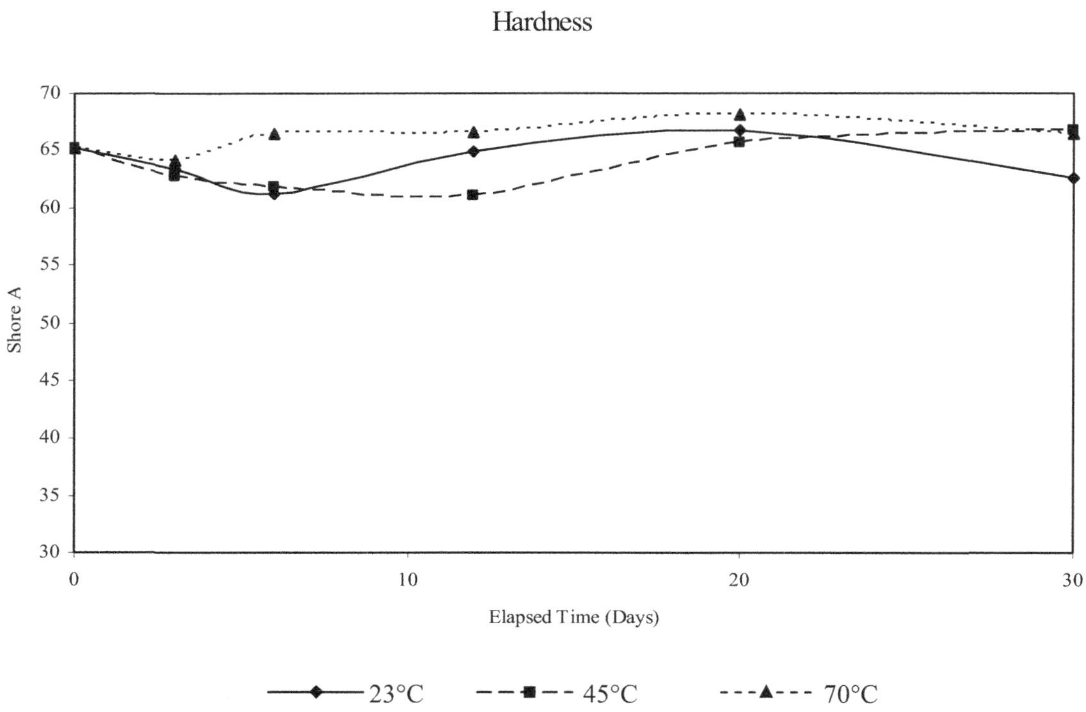

23°C at 30 days

45°C at 30 days

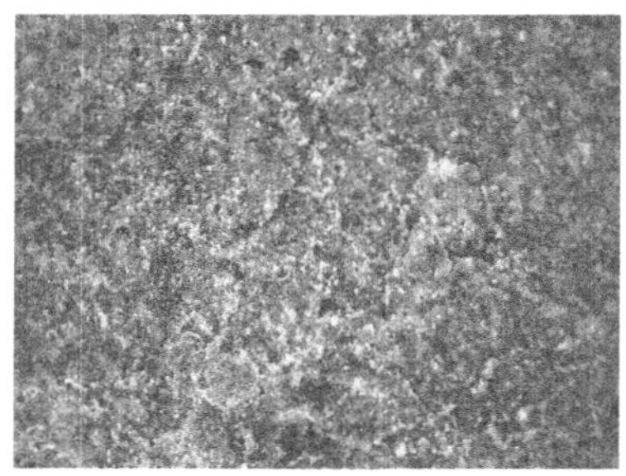

70°C at 30 days

Figure A.8 Temperature effect on hardness at 1 ppm chloramine and pH 8.3 of SBR with 40× magnified pictographs

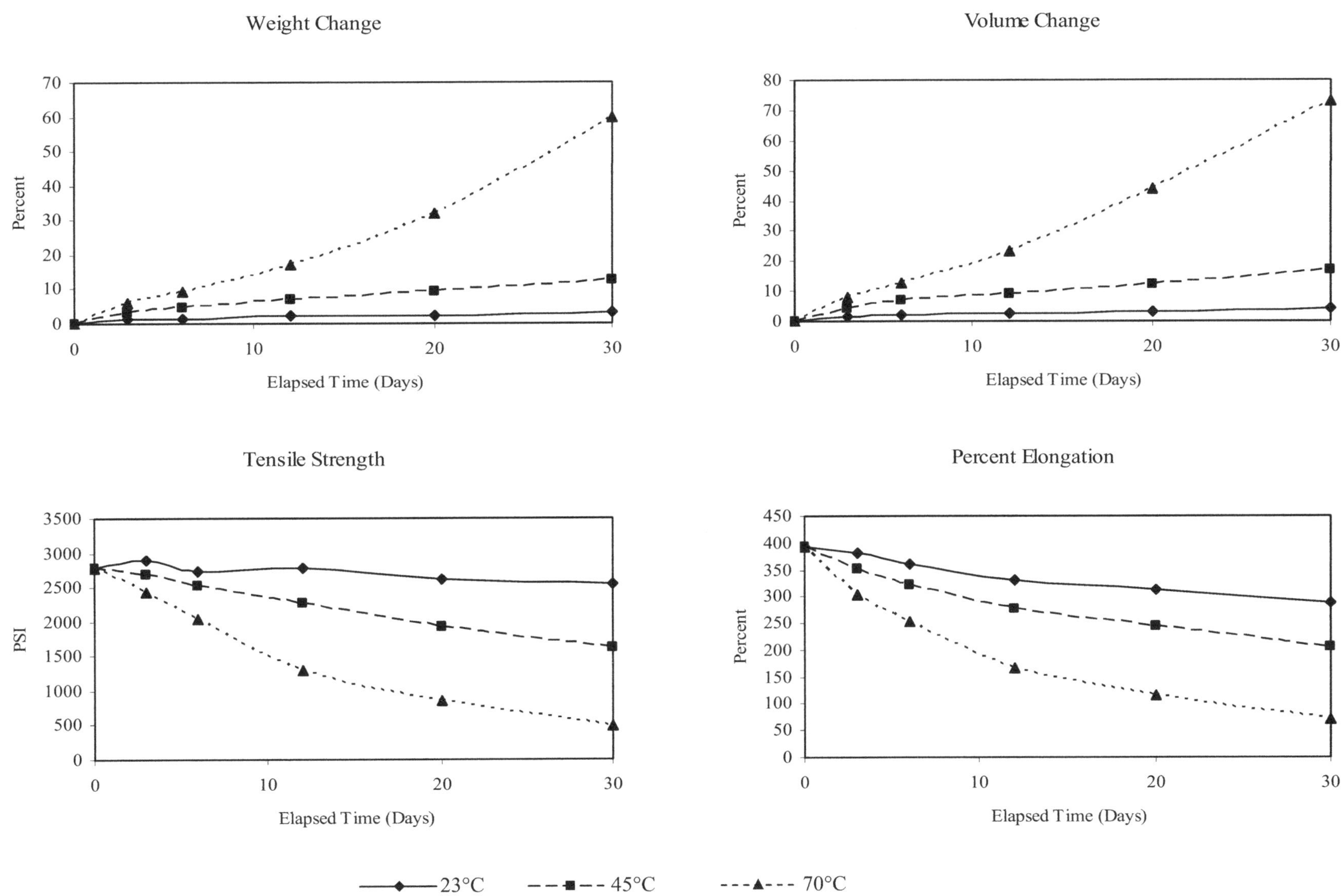

Figure A.9 Temperature effect on elastomeric properties at 1 ppm chloramine and pH 8.3 of neoprene

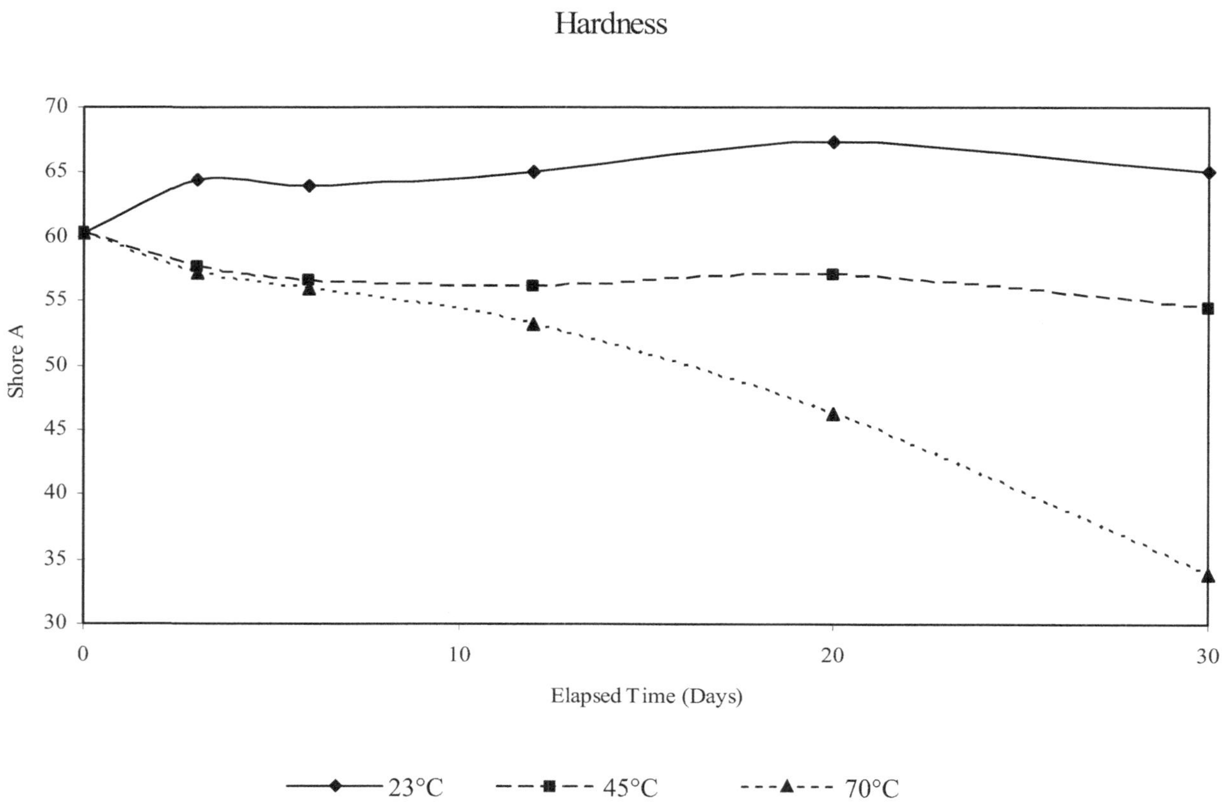

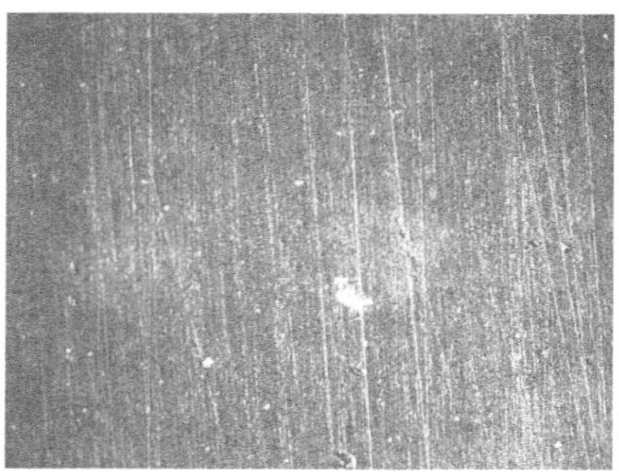

23°C at 30 days

45°C at 30 days

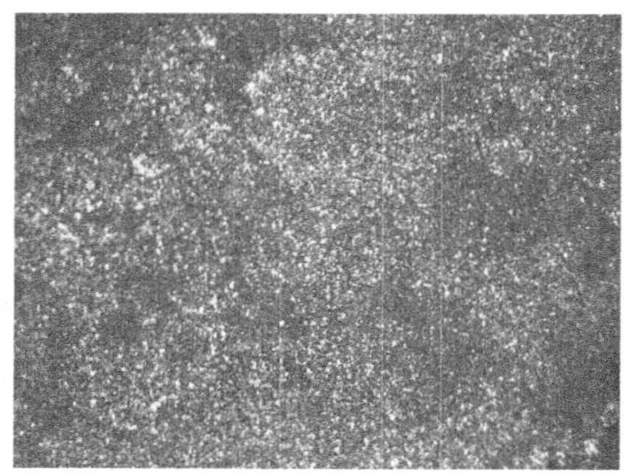

70°C at 30 days

Figure A.10 Temperature effect on hardness at 1 ppm chloramine and pH 8.3 of neoprene with 40× magnified pictographs

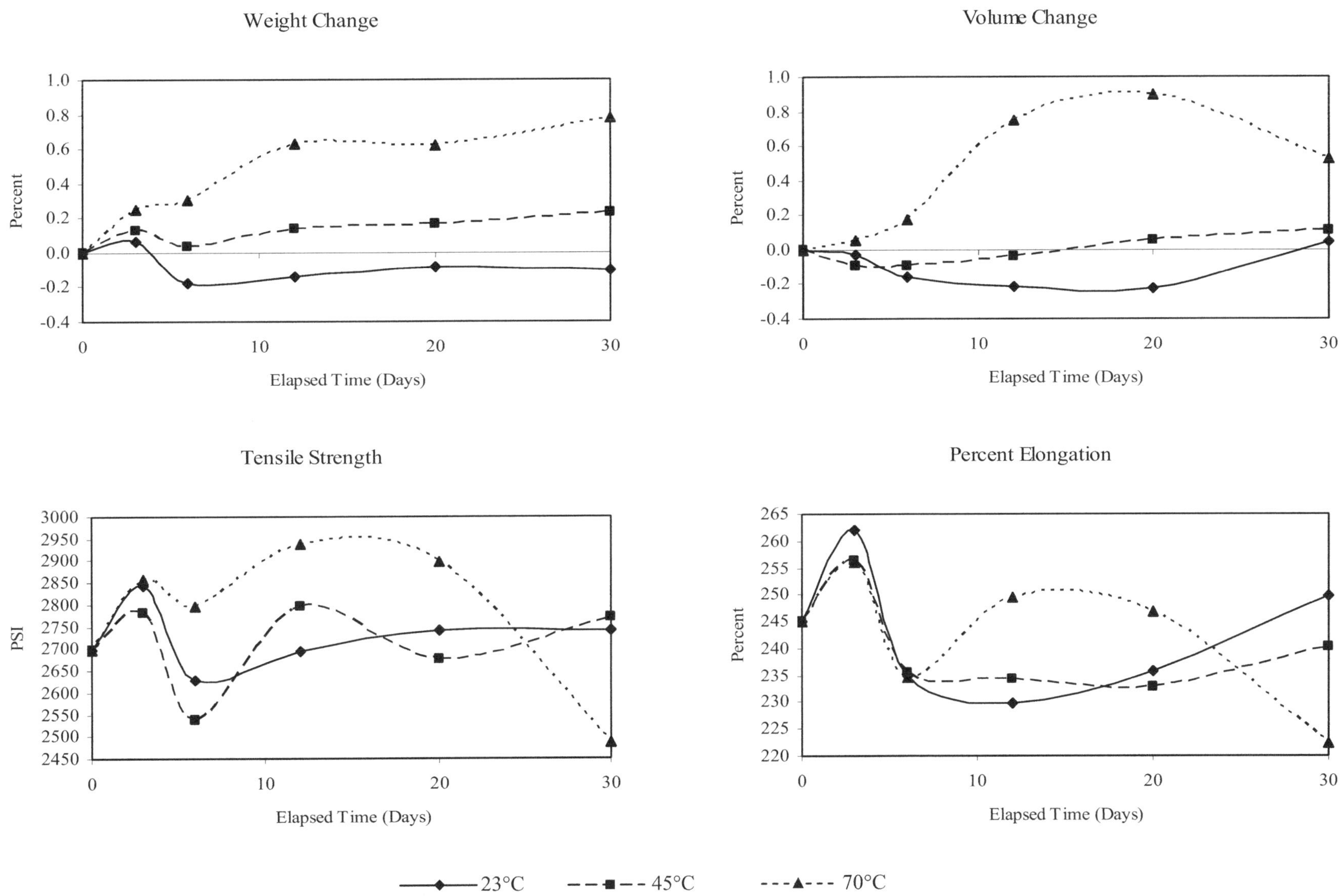

Figure A.11 Temperature effect on elastomeric properties at 1 ppm chloramine and pH 8.3 of EPDM-P

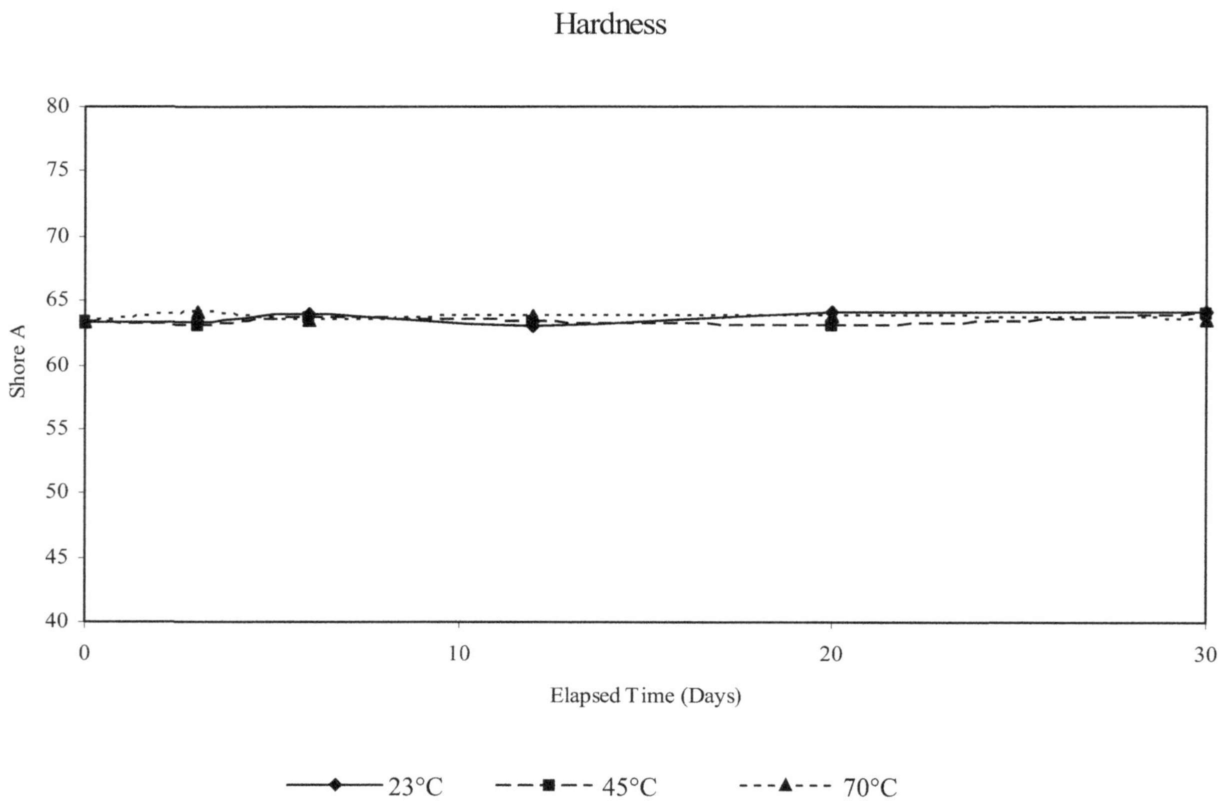

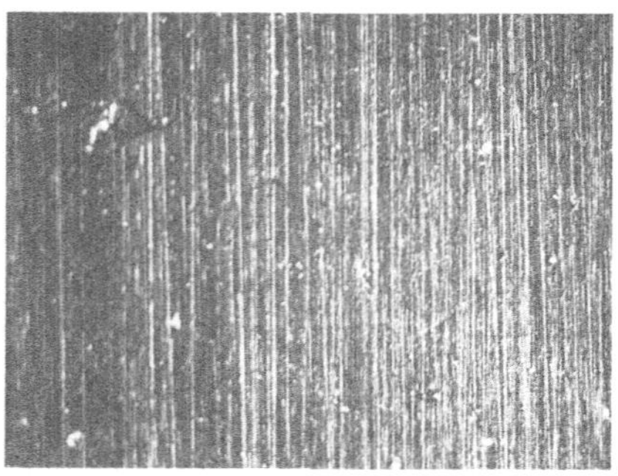

23°C at 30 days

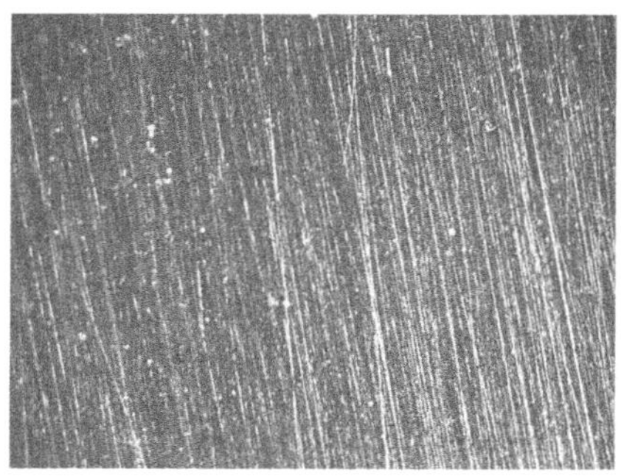

45°C at 30 days

70°C at 30 days

Figure A.12 Temperature effect on hardness at 1 ppm chloramine and pH 8.3 of EPDM-P with 40× magnified pictographs

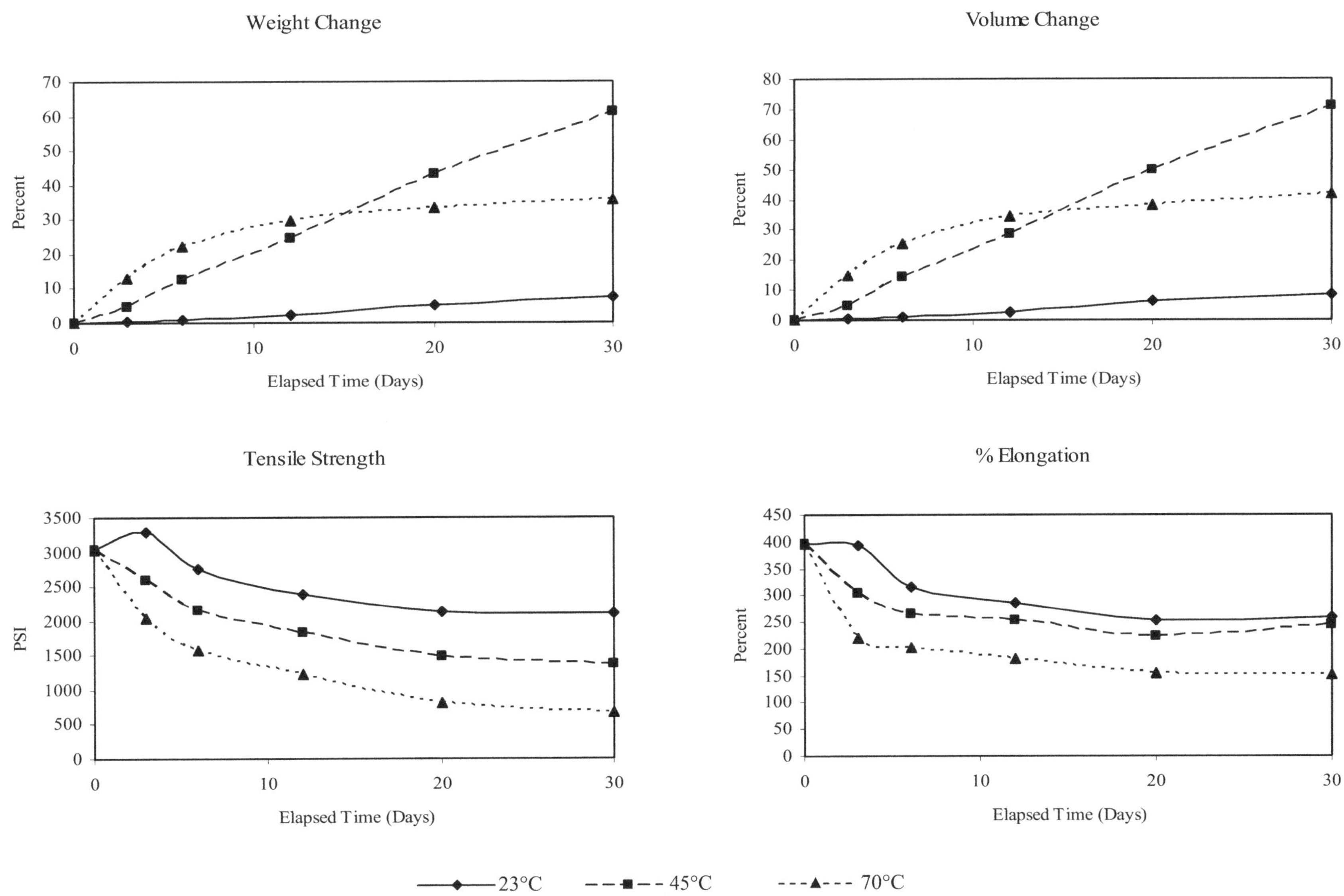

Figure A.13 Temperature effect on elastomeric properties at 30 ppm chloramine and pH 8.3 of natural rubber

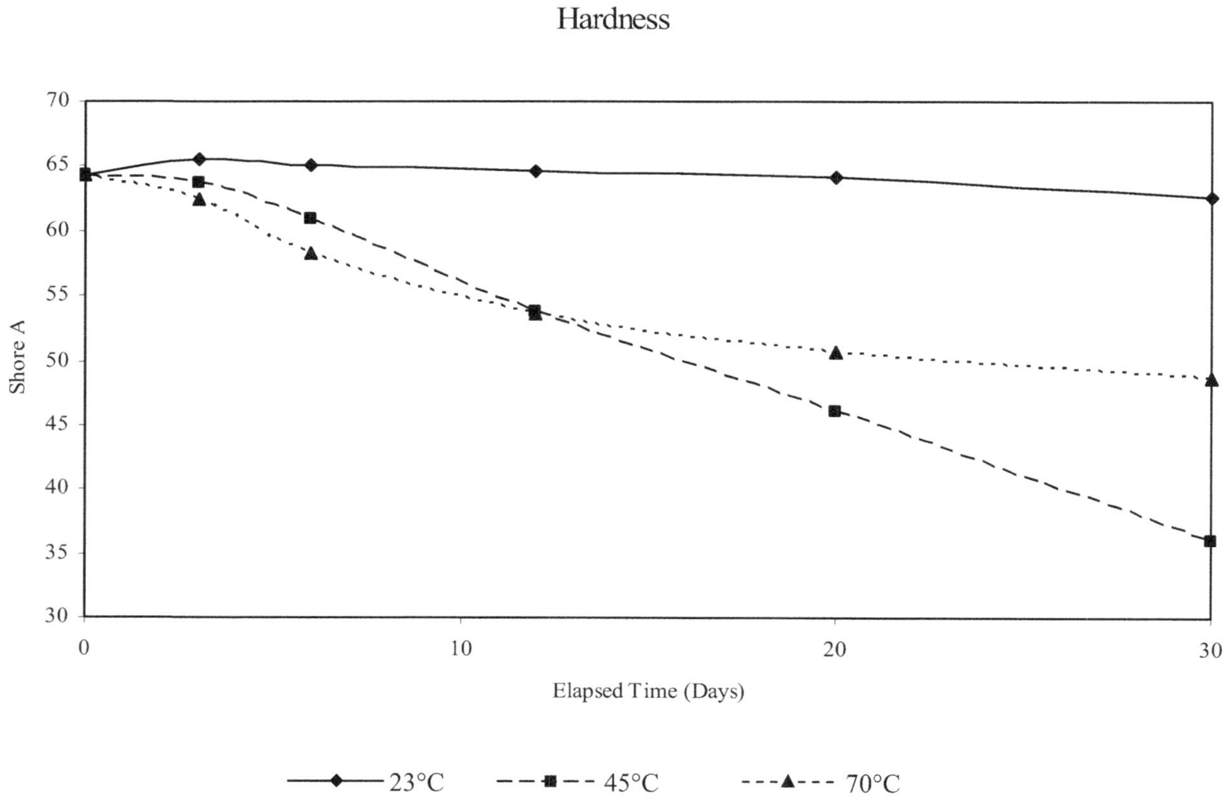

23°C at 30 days

45°C at 30 days

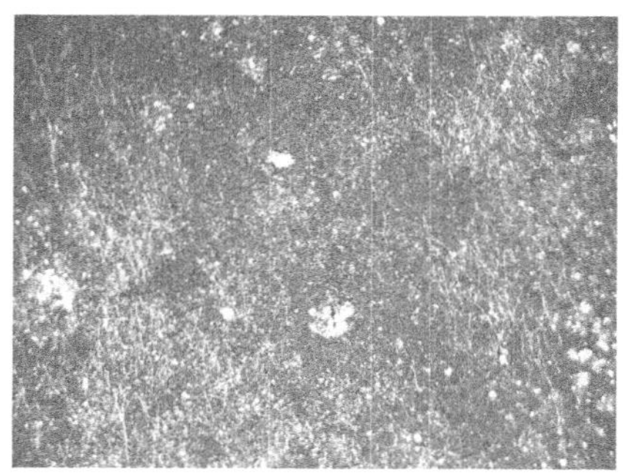

70°C at 30 days

Figure A.14 Temperature effect on hardness at 30 ppm chloramine and pH 8.3 of natural rubber with 40× magnified pictographs

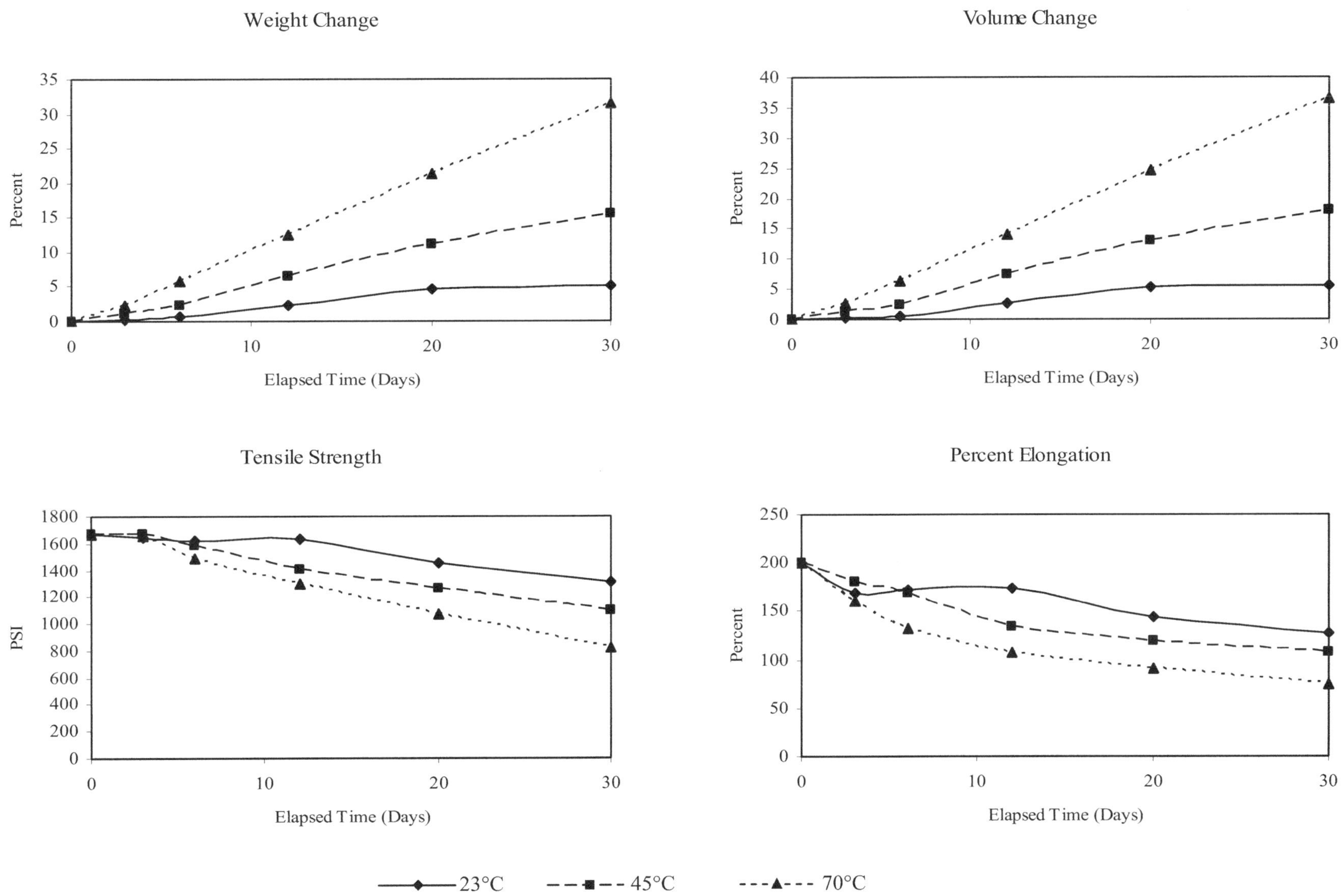

Figure A.15 Temperature effect on elastomeric properties at 30 ppm chloramine and pH 8.3 of EPDM-S

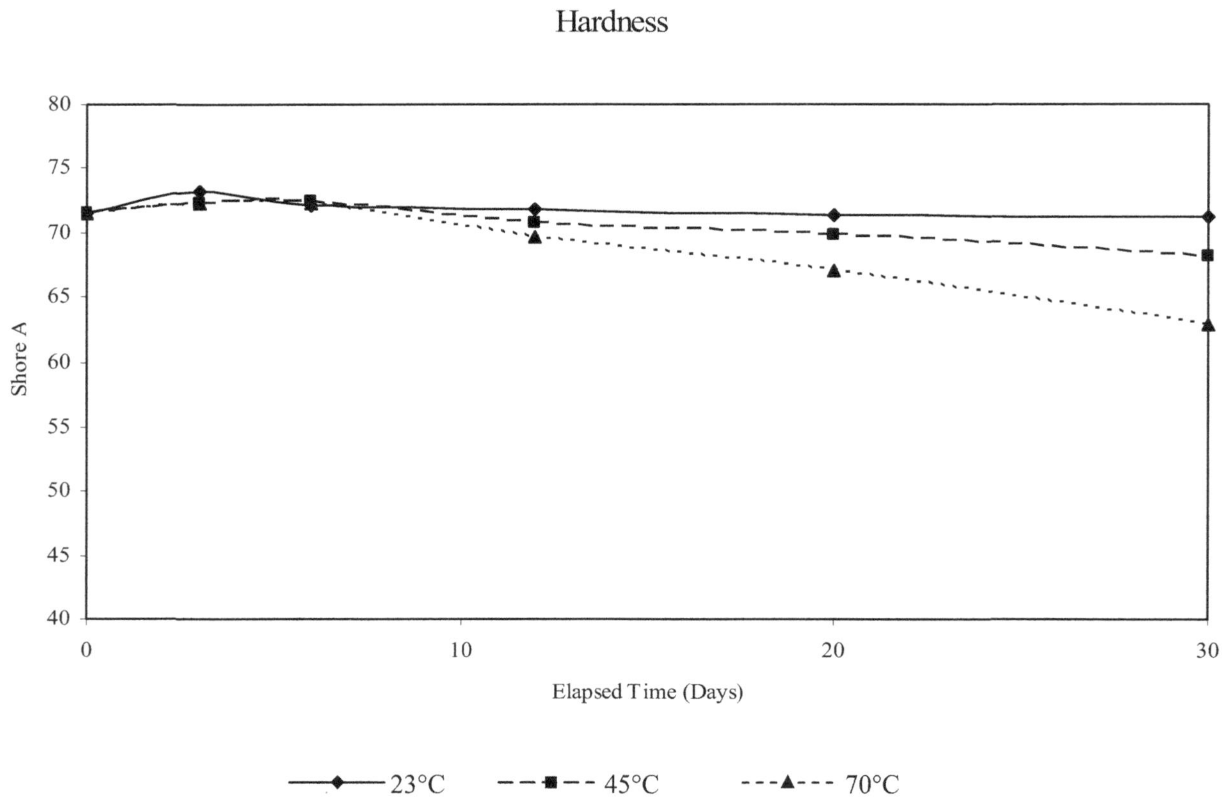

23°C at 30 days | 45°C at 30 days | 70°C at 30 days

Figure A.16 Temperature effect on hardness at 30 ppm chloramine and pH 8.3 of EPDM-S with 40× magnified pictographs

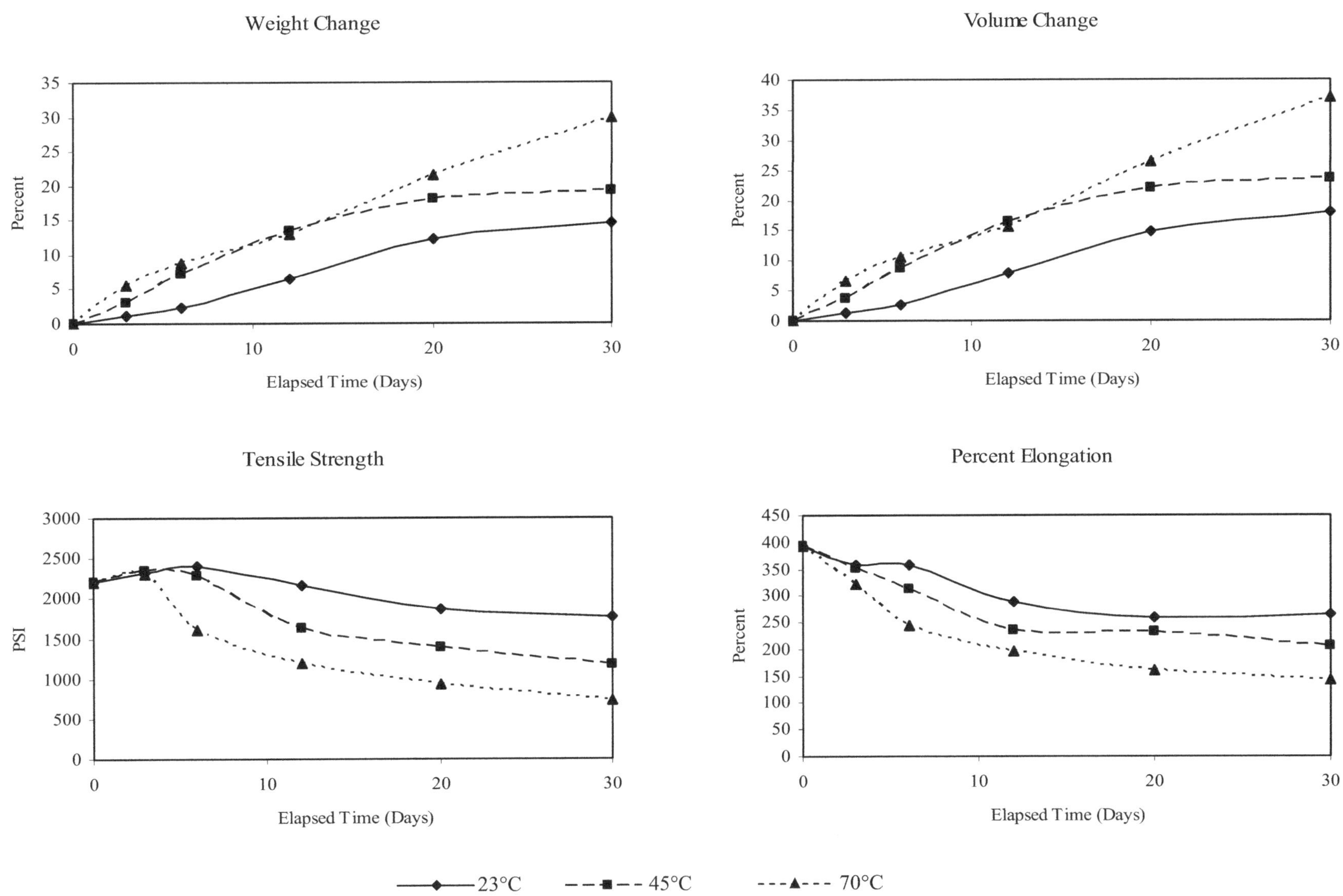

Figure A.17 Temperature effect on elastomeric properties at 30 ppm chloramine and pH 8.3 of nitrile (NBR)

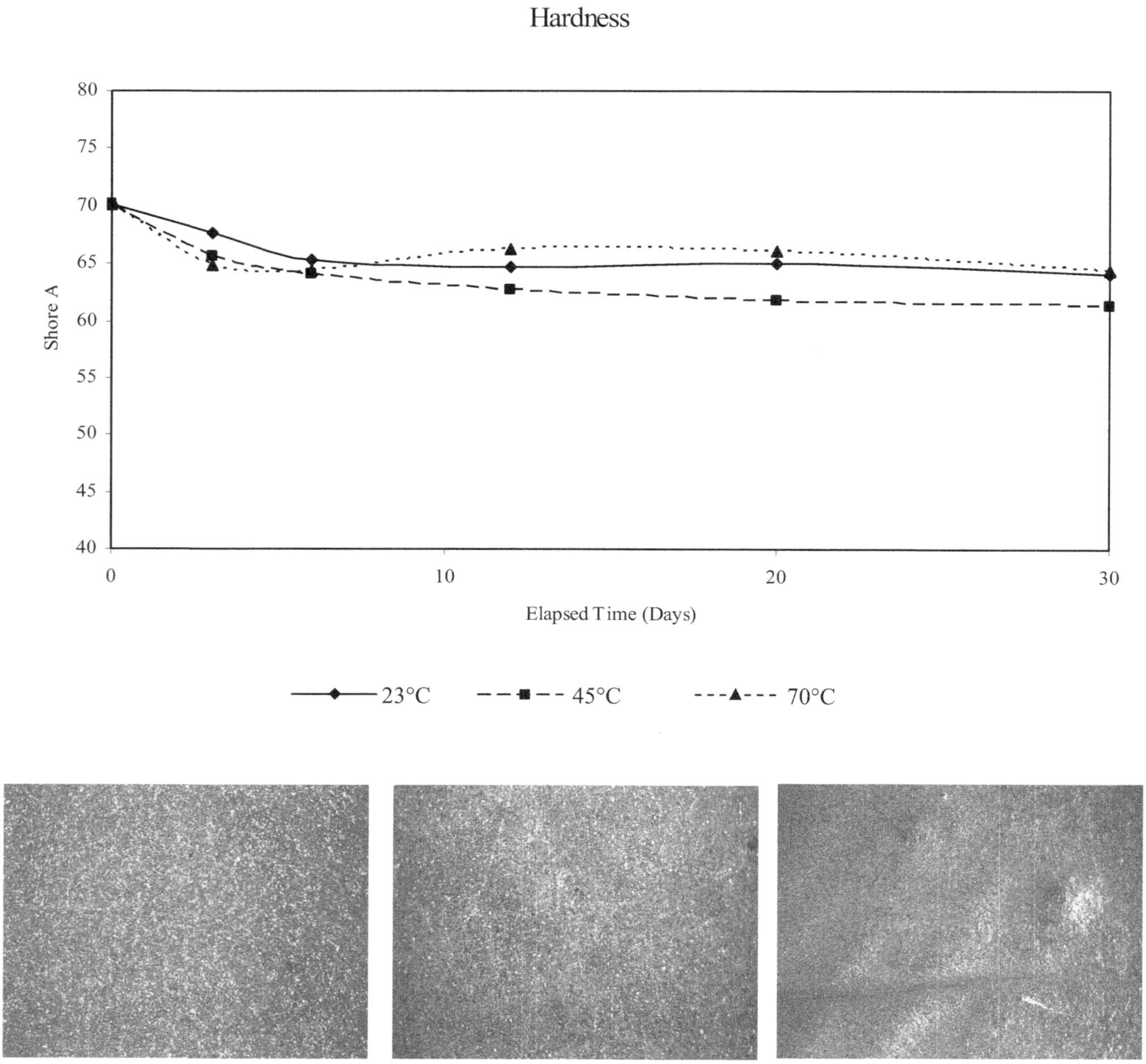

23°C at 30 days 45°C at 30 days 70°C at 30 days

Figure A.18 Temperature effect on hardness at 30 ppm chloramine and pH 8.3 of nitrile (NBR) with 40× magnified pictographs

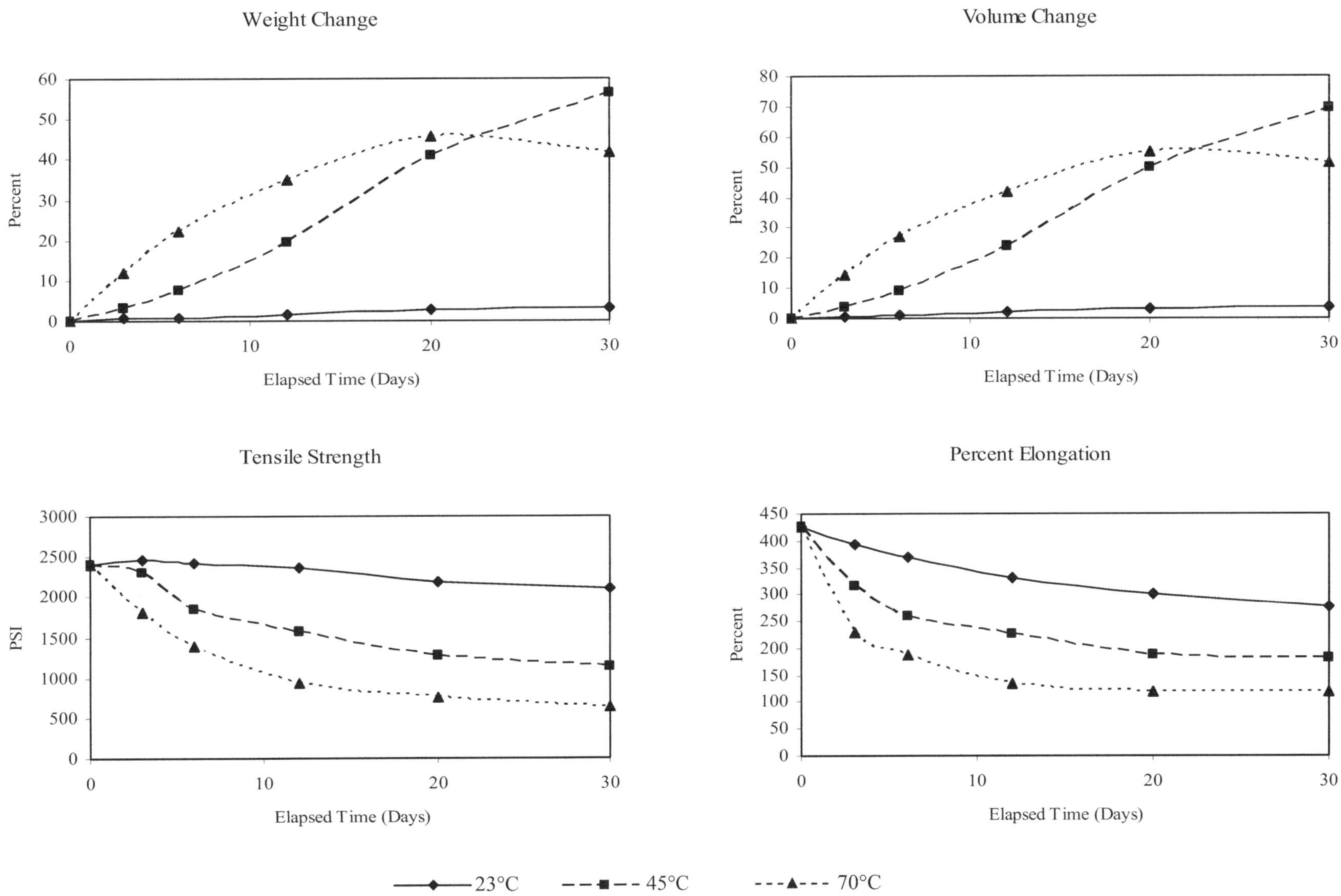

Figure A.19 Temperature effect on elastomeric properties at 30 ppm chloramine and pH 8.3 of SBR

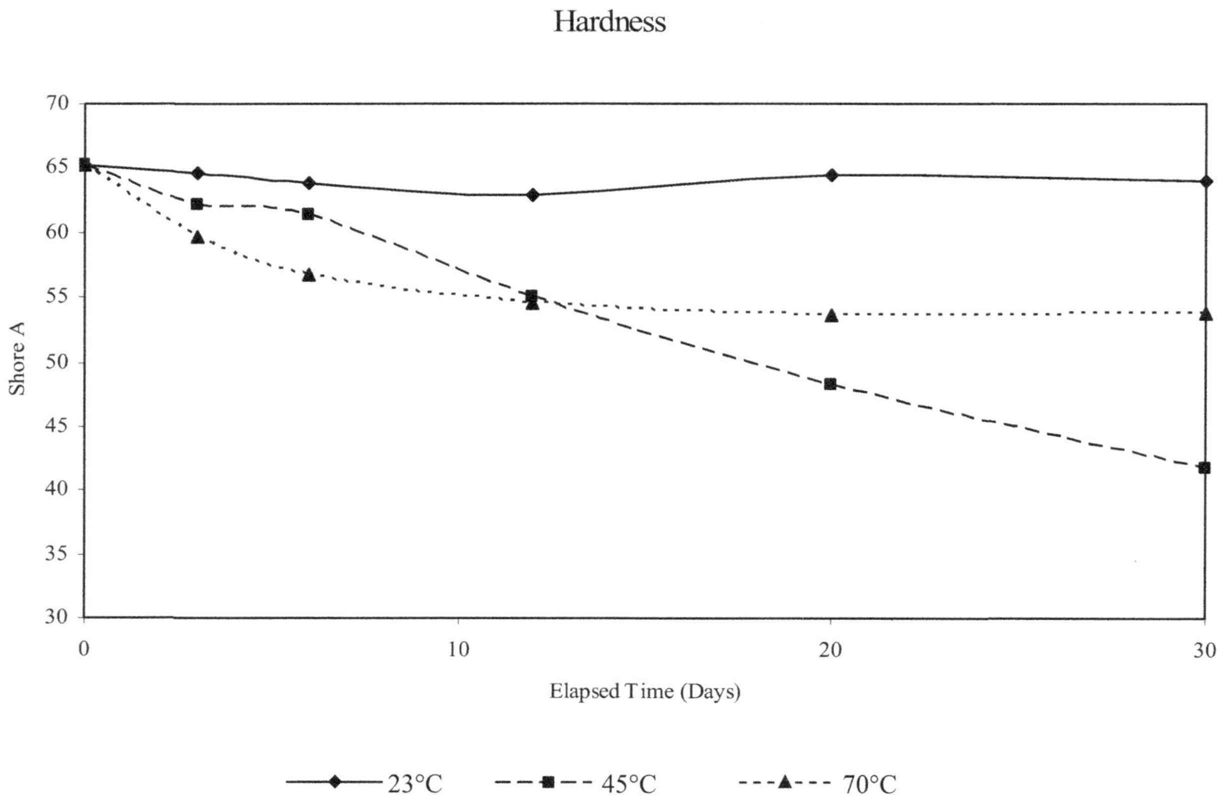

23°C at 30 days

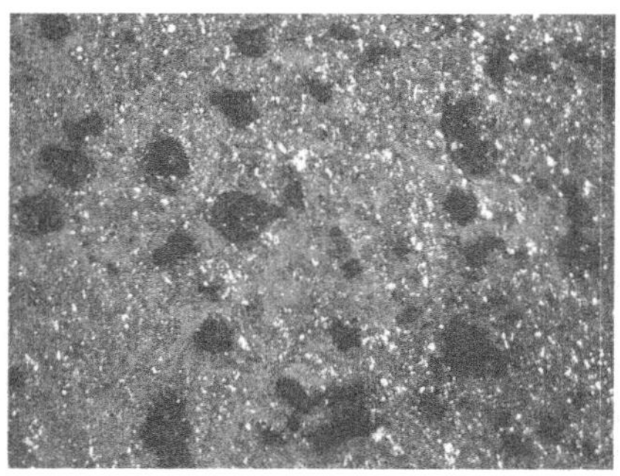

45°C at 30 days

70°C at 30 days

Figure A.20 Temperature effect on hardness at 30 ppm chloramine and pH 8.3 of SBR with 40× magnified pictographs

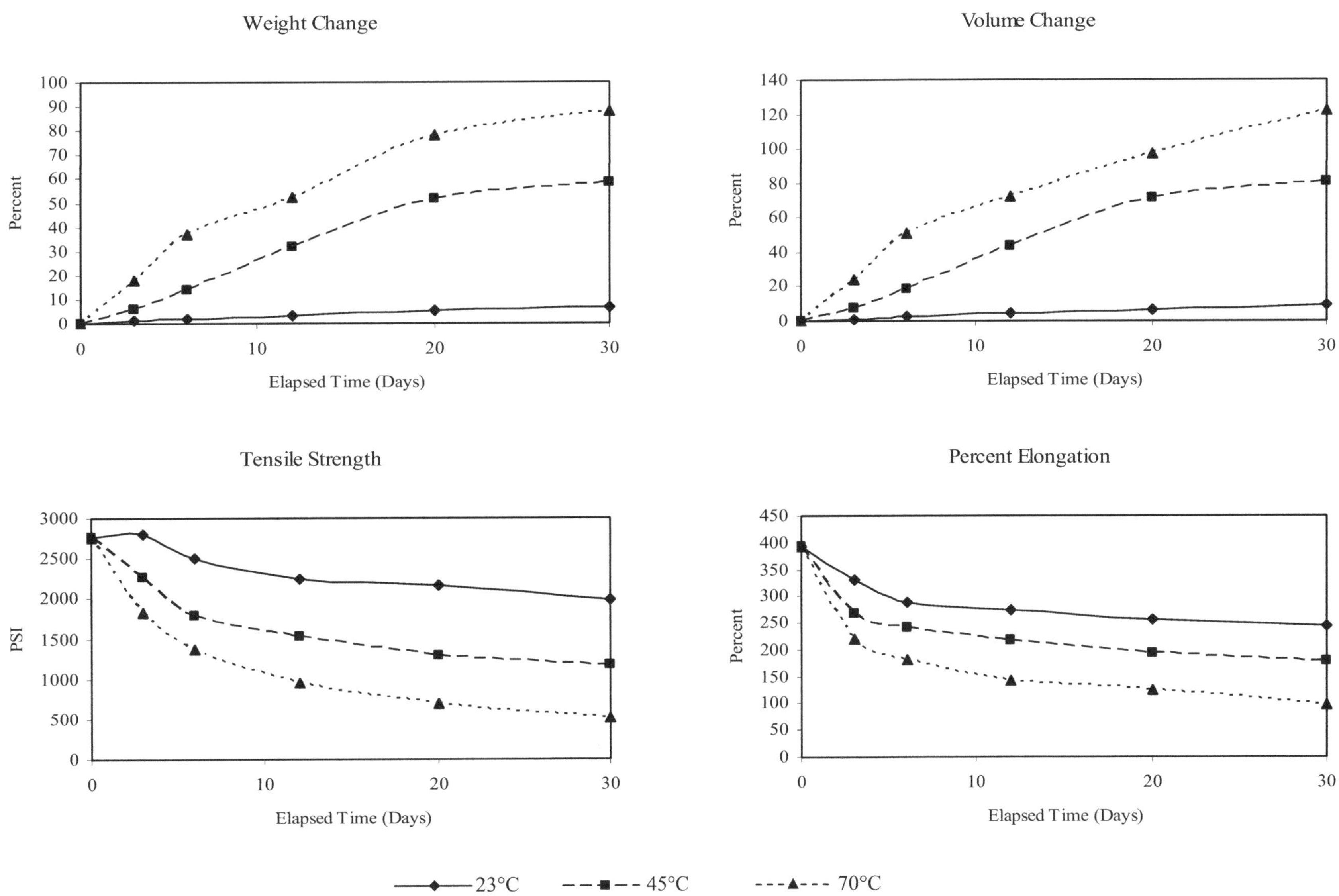

Figure A.21 Temperature effect on elastomeric properties at 30 ppm chloramine and pH 8.3 of neoprene

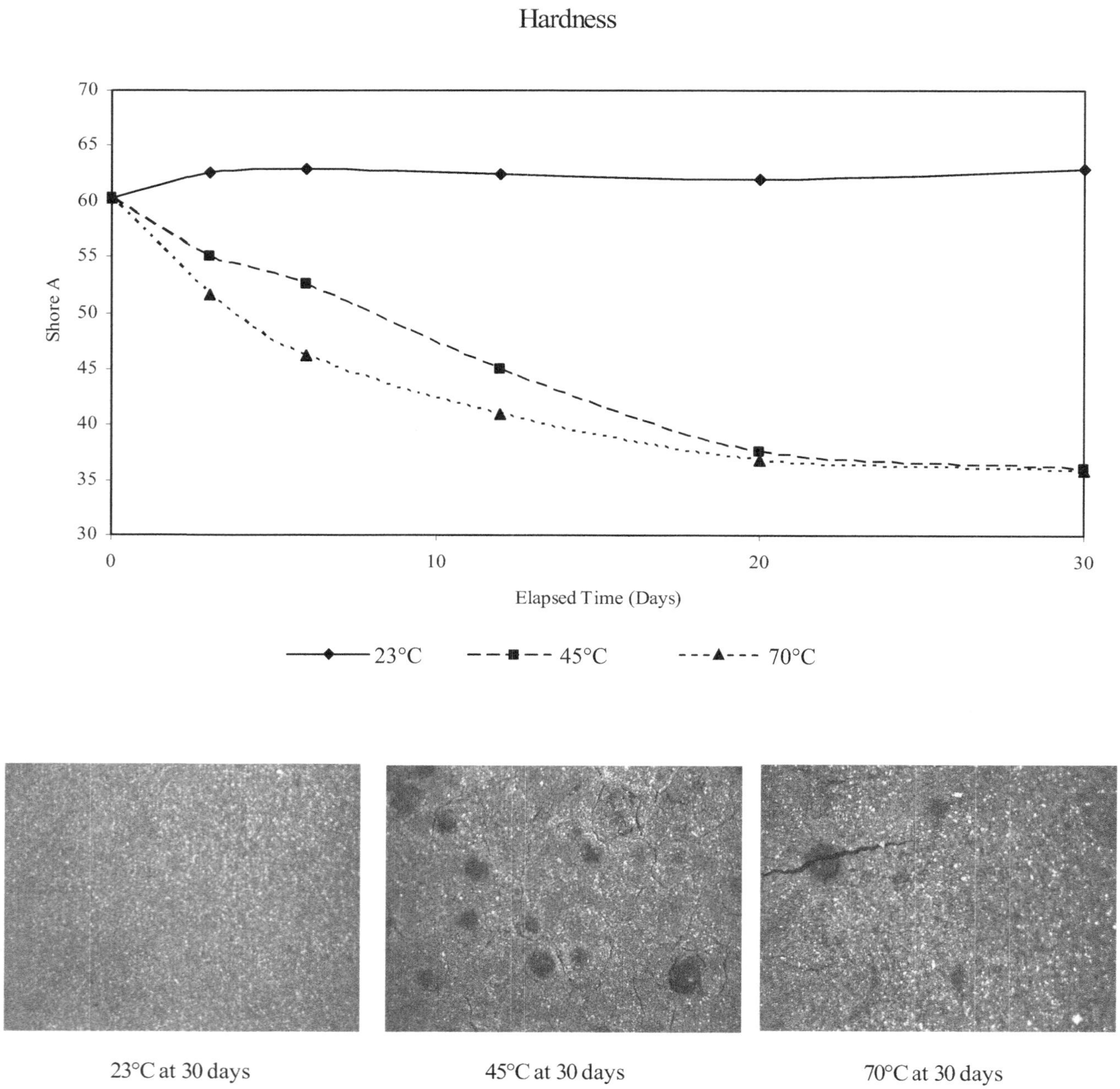

Figure A.22 Temperature effect on hardness at 30 ppm chloramine and pH 8.3 of neoprene with 40× magnified pictographs

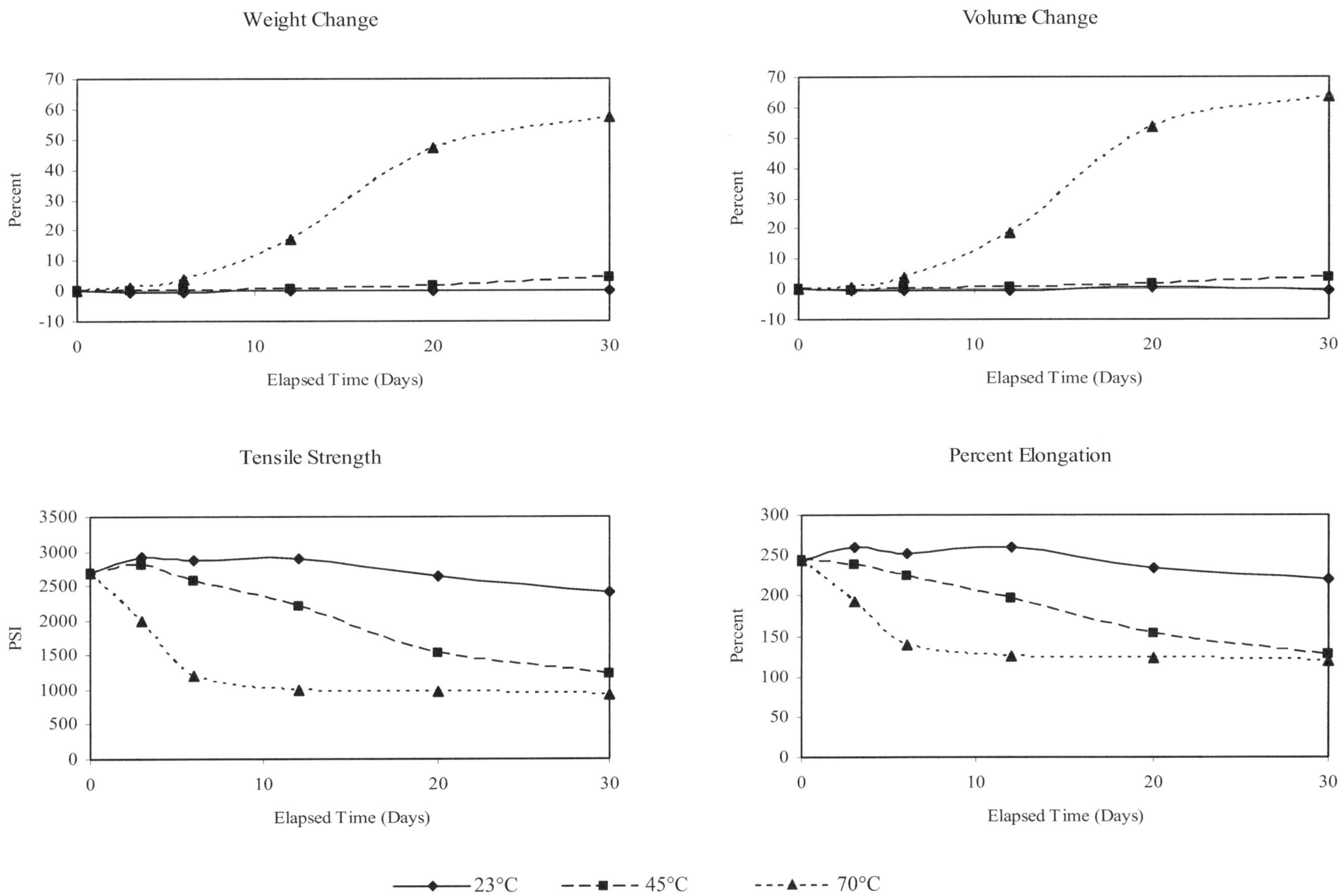

Figure A.23 Temperature effect on elastomeric properties at 30 ppm chloramine and pH 8.3 of EPDM-P

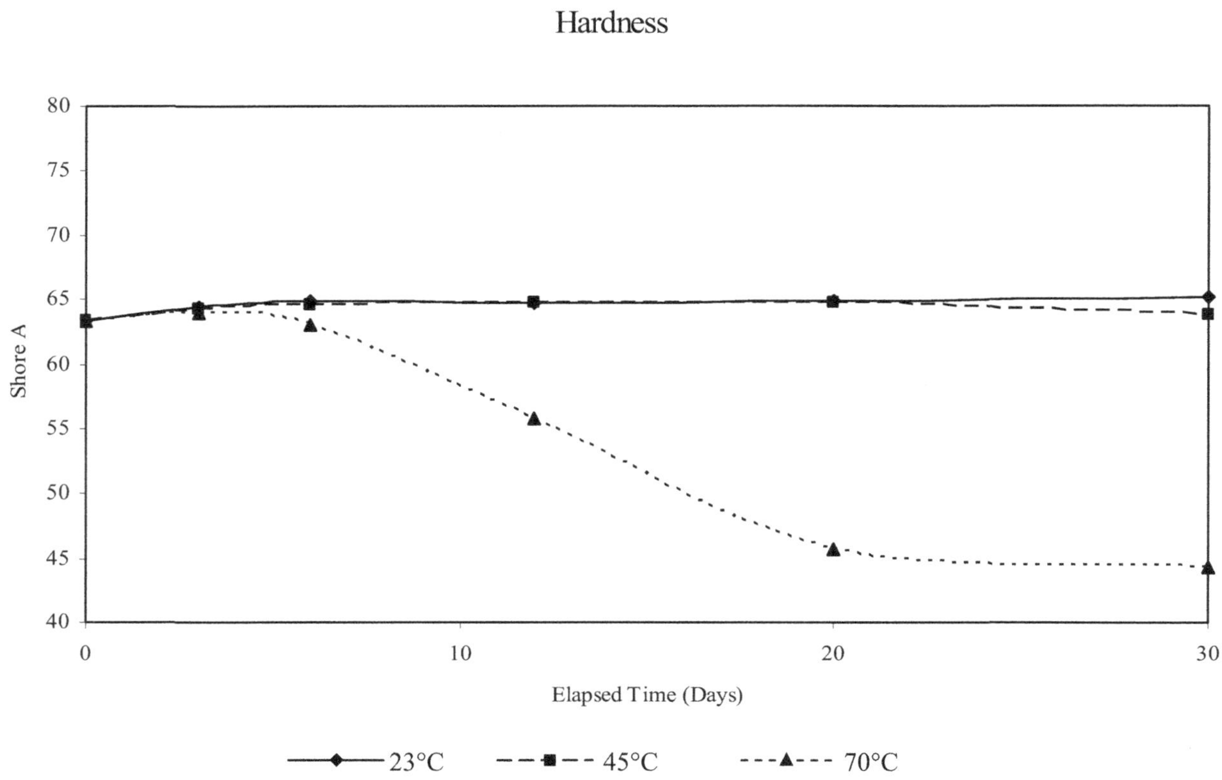

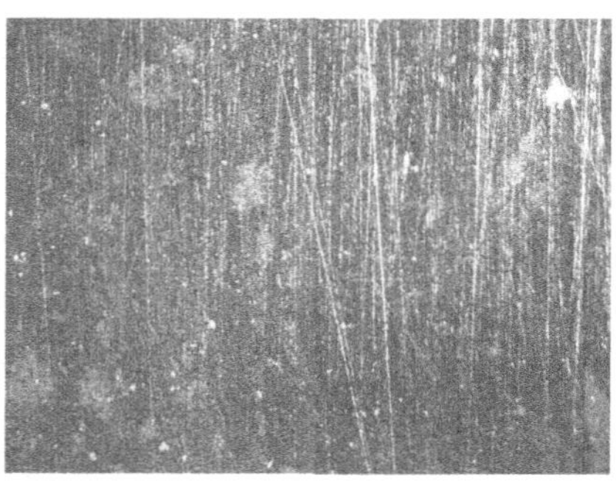

23°C at 30 days

45°C at 30 days

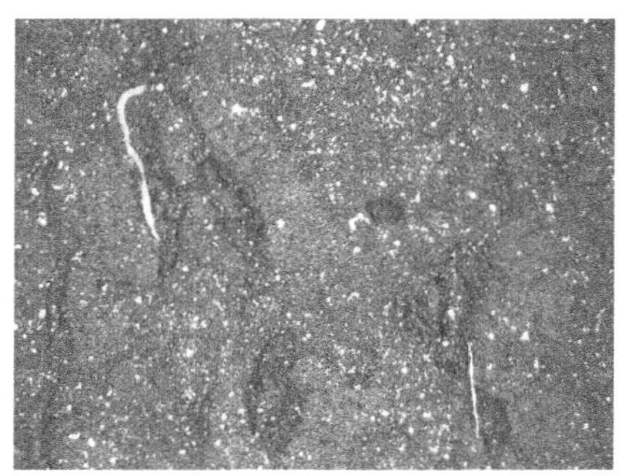

70°C at 30 days

Figure A.24 Temperature effect on hardness at 30 ppm chloramine and pH 8.3 of EPDM-P with 40× magnified pictographs

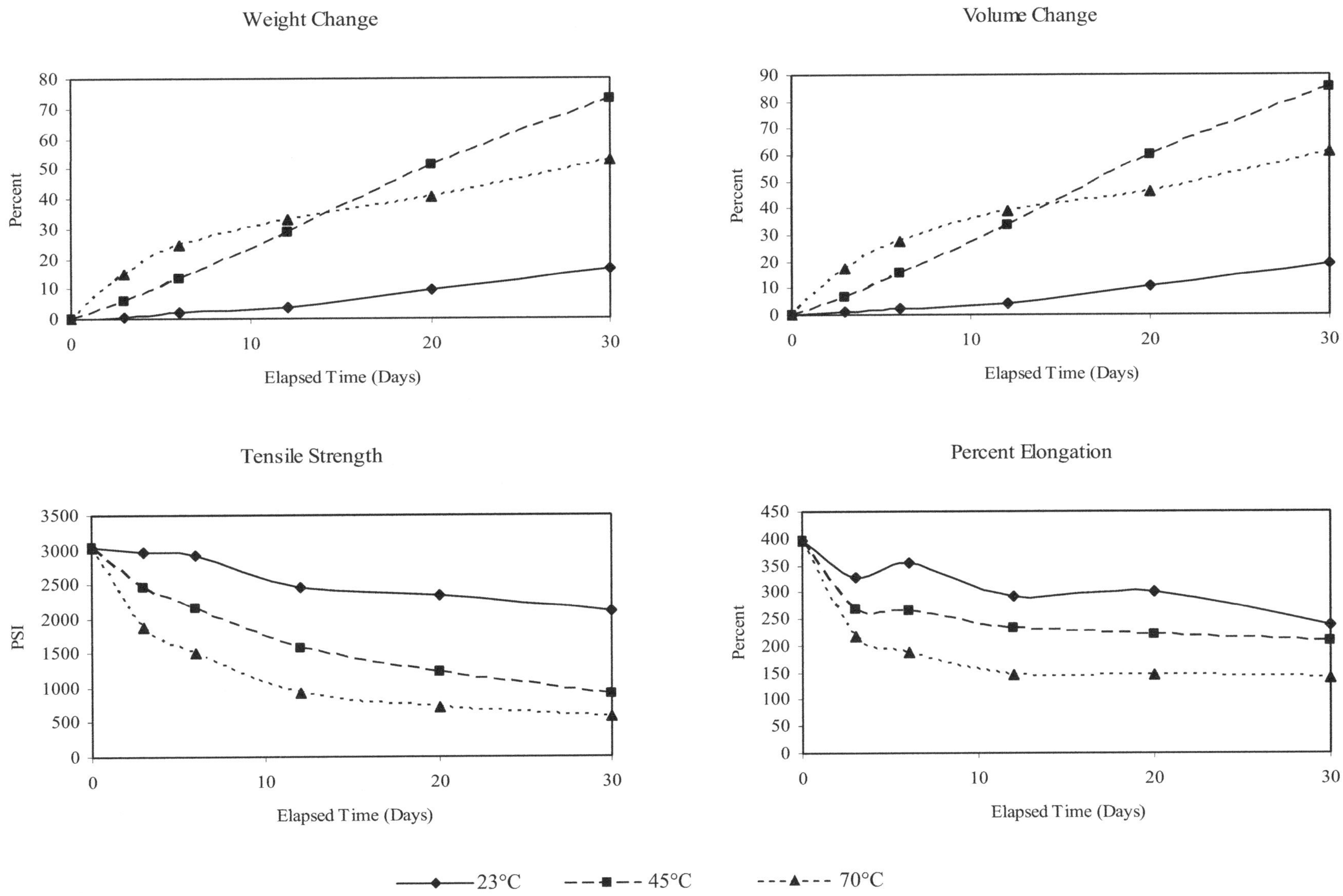

Figure A.25 Temperature effect on elastomeric properties at 60 ppm chloramine and pH 8.3 of natural rubber

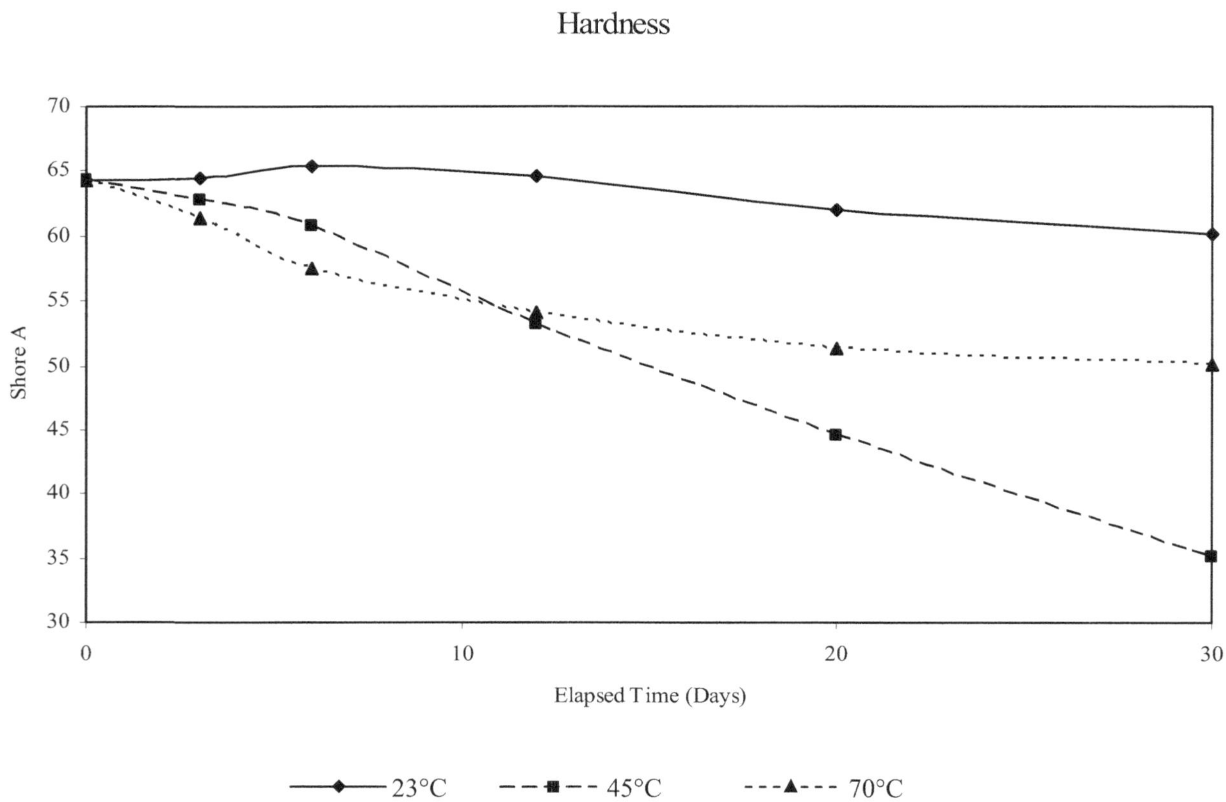

23°C at 30 days

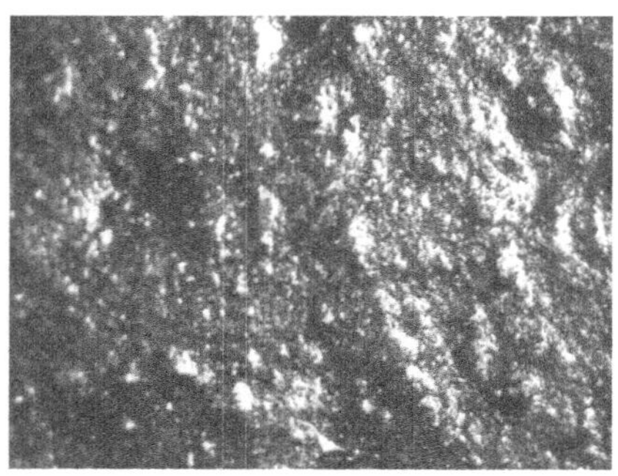

45°C at 30 days

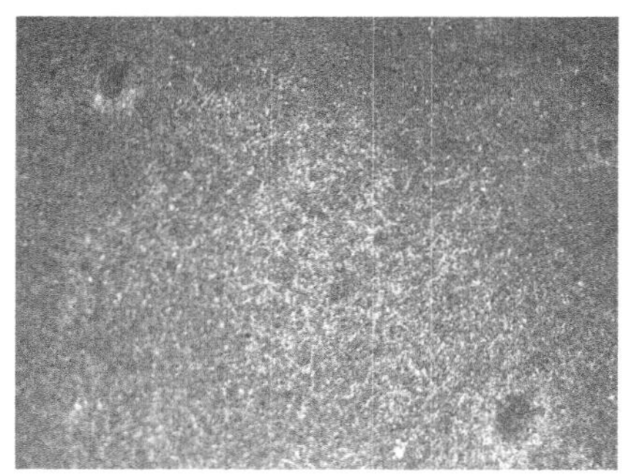

70°C at 30 days

Figure A.26 Temperature effect on hardness at 60 ppm chloramine and pH 8.3 of natural rubber with 40× magnified pictographs

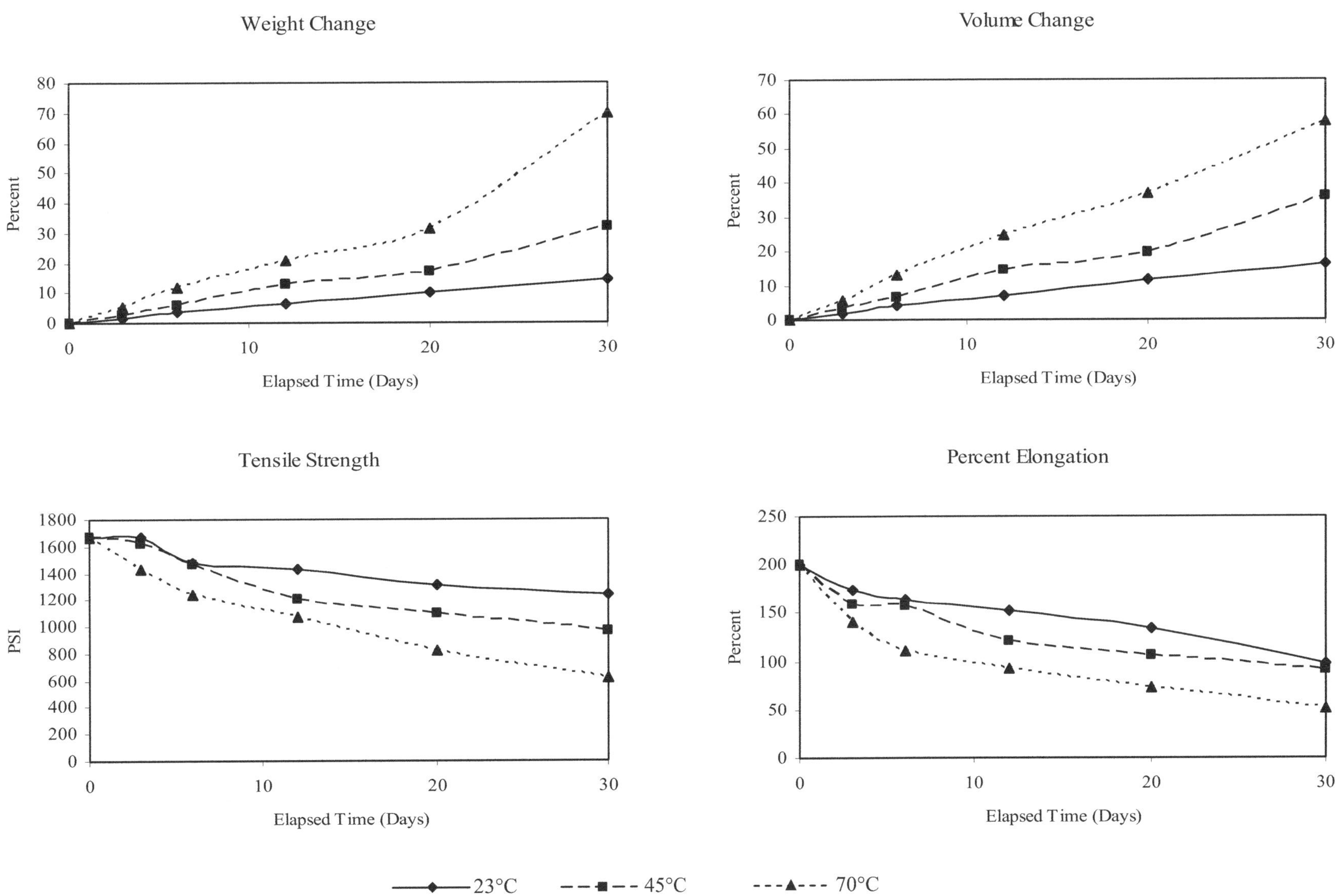

Figure A.27 Temperature effect on elastomeric properties at 60 ppm chloramine and pH 8.3 of EPDM-S

Hardness

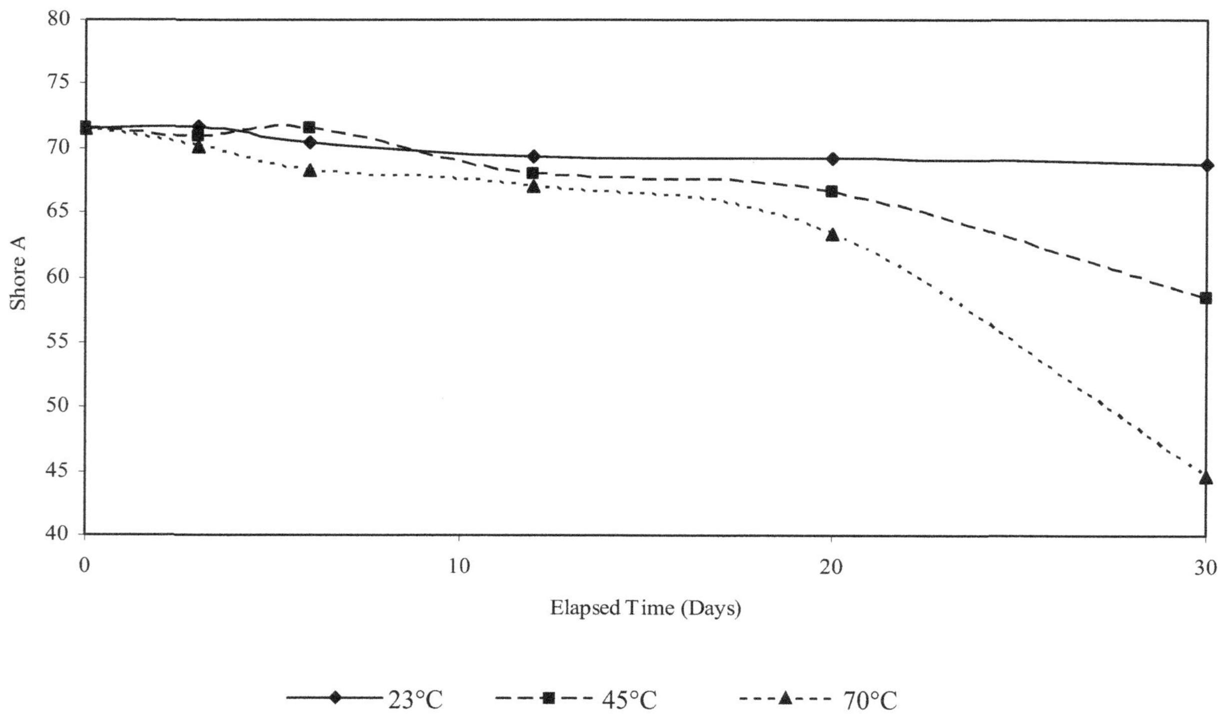

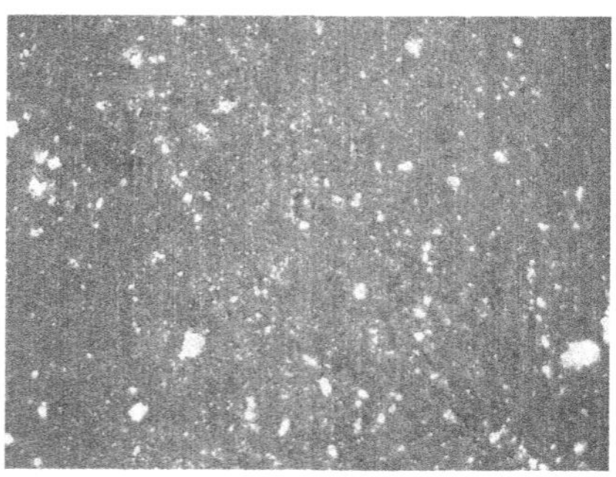

23°C at 30 days

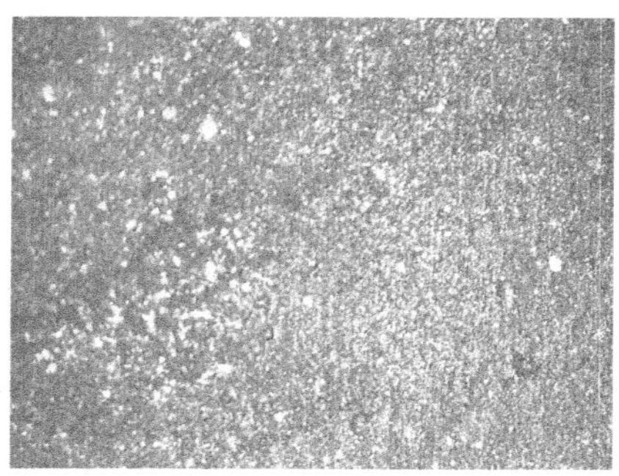

45°C at 30 days

70°C at 30 days

Figure A.28 Temperature effect on hardness at 60 ppm chloramine and pH 8.3 of EPDM-S with 40× magnified pictographs

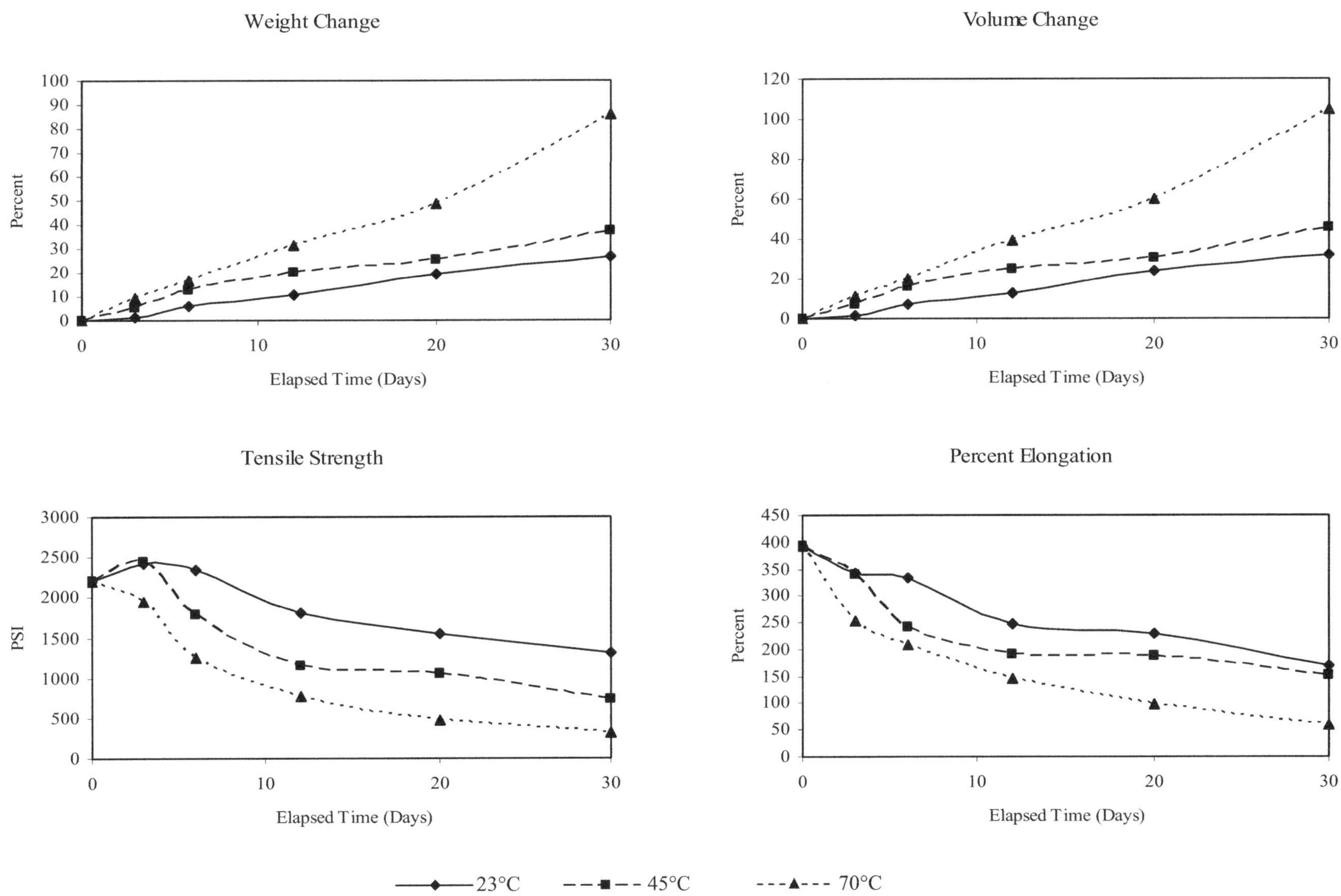

Figure A.29 Temperature effect on elastomeric properties at 60 ppm chloramine and pH 8.3 of nitrile (NBR)

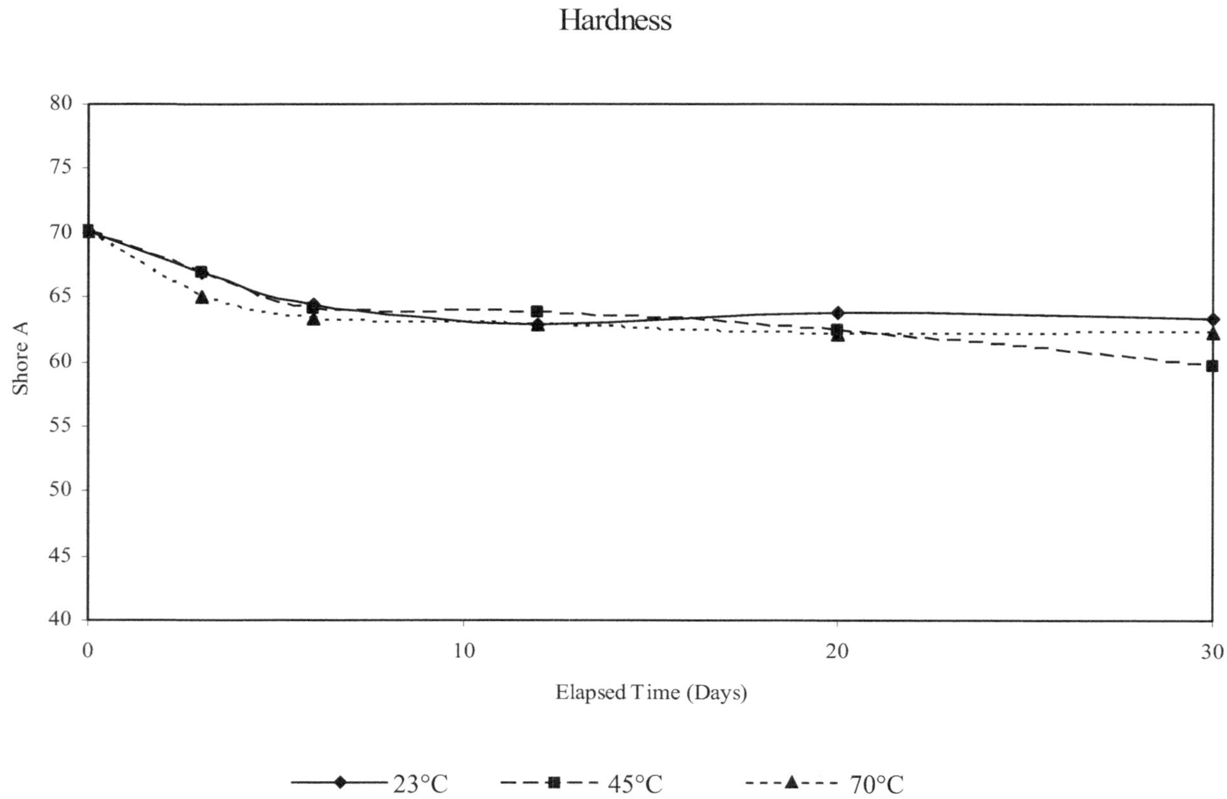

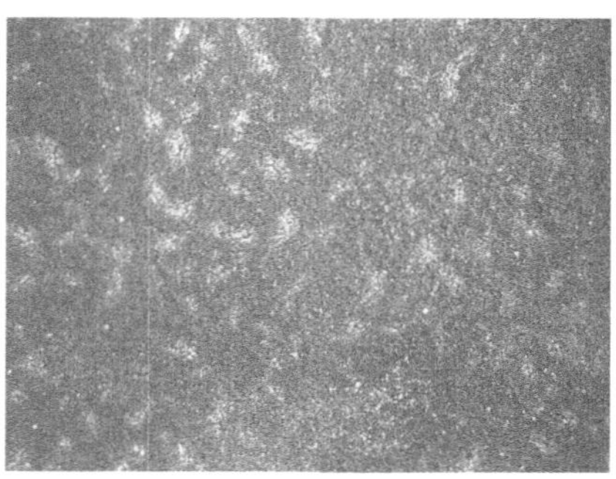

23°C at 30 days

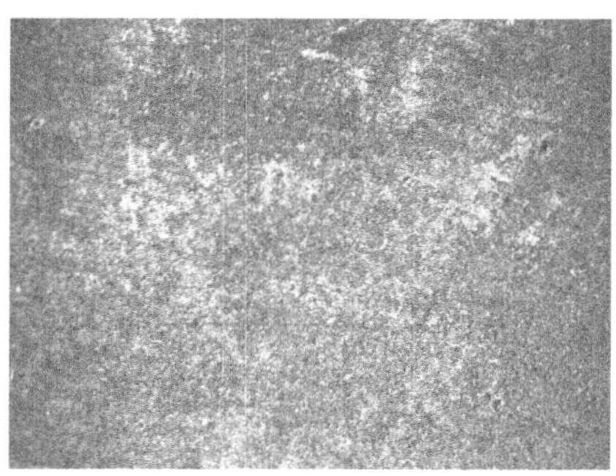

45°C at 30 days

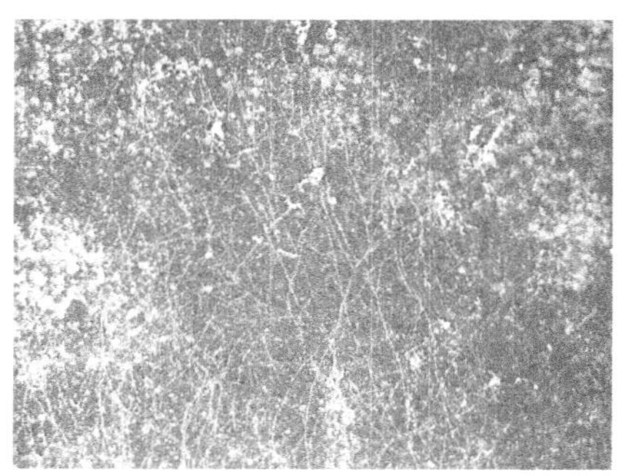

70°C at 30 days

Figure A.30 Temperature effect on hardness at 60 ppm chloramine and pH 8.3 of nitrile (NBR) with 40× magnified pictographs

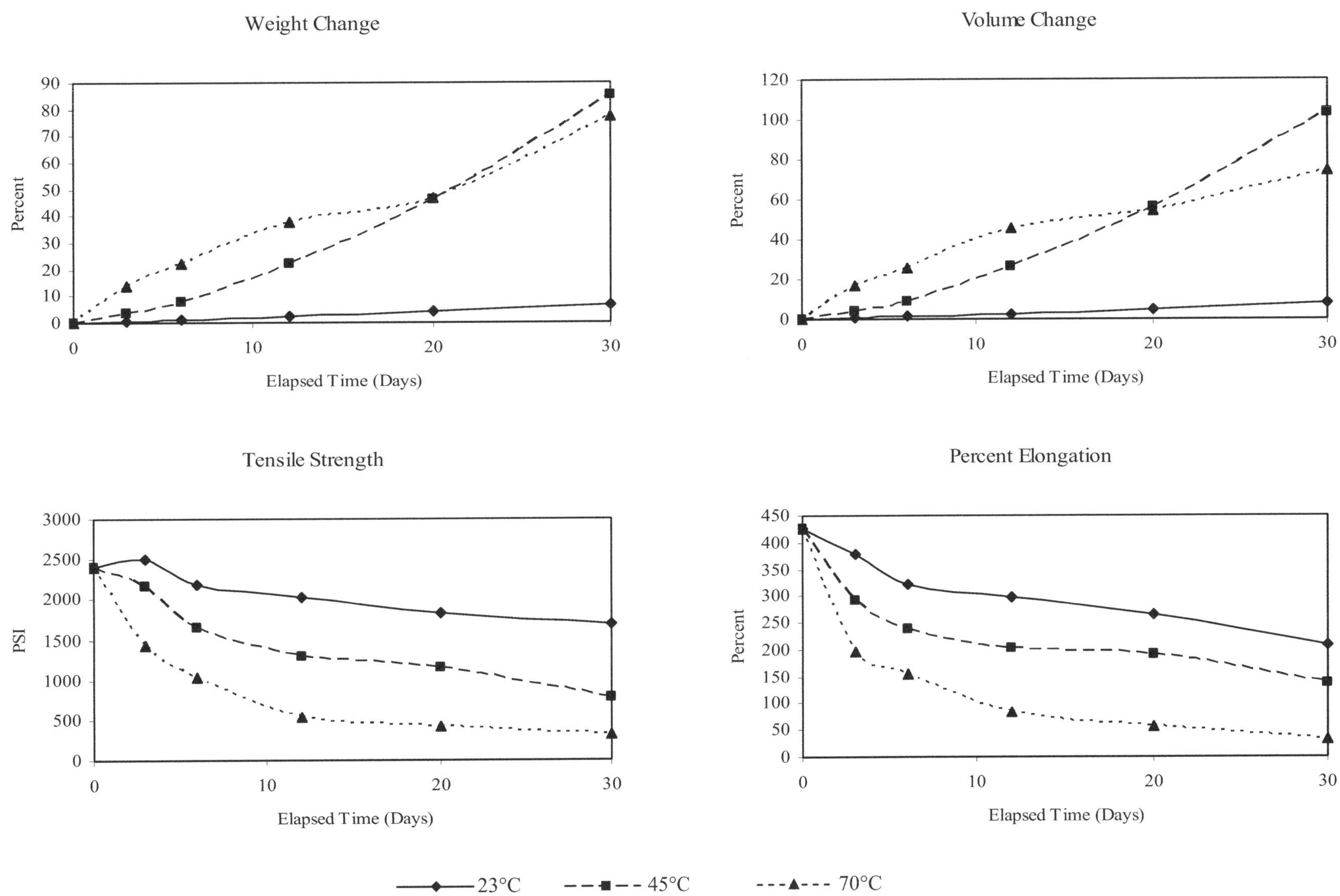

Figure A.31 Temperature effect on elastomeric properties at 60 ppm chloramine and pH 8.3 of SBR

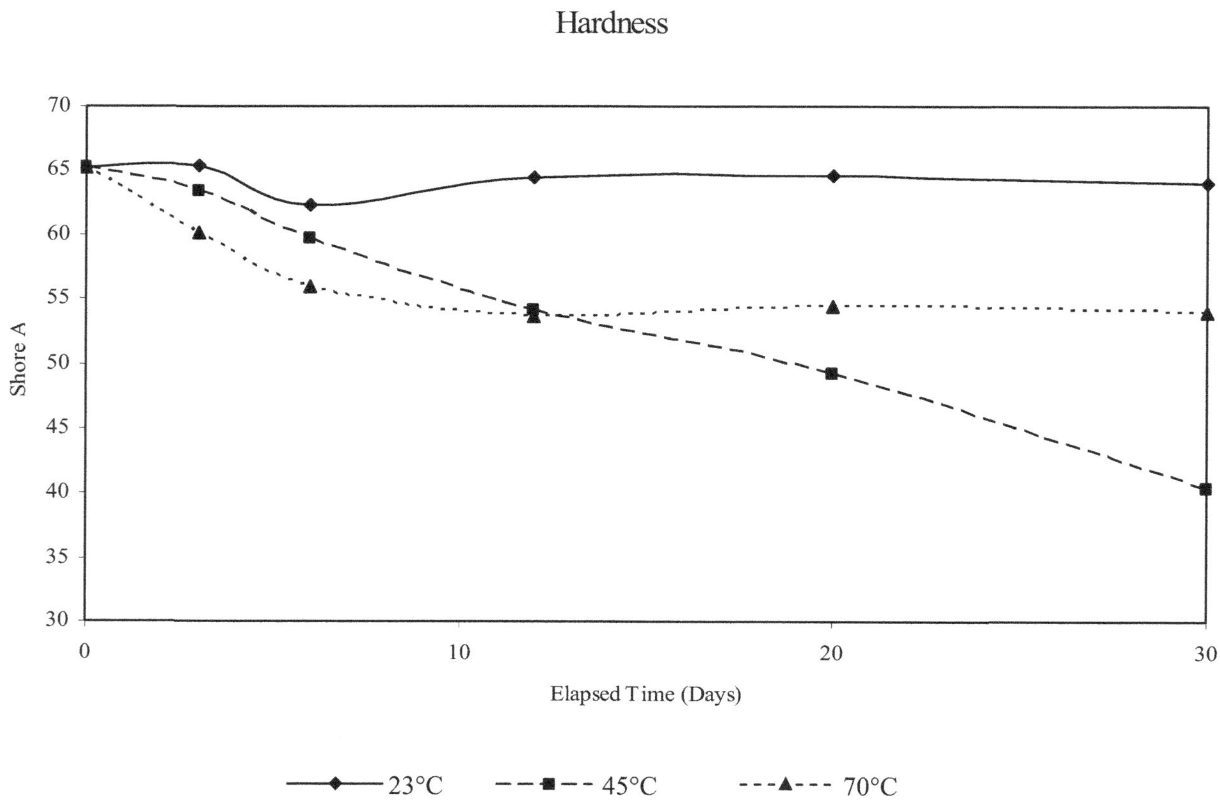

23°C at 30 days

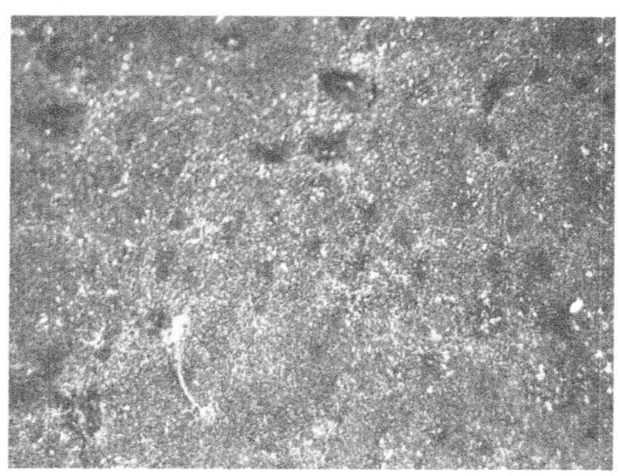

45°C at 30 days

70°C at 30 days

Figure A.32 Temperature effect on hardness at 60 ppm chloramine and pH 8.3 of SBR with 40× magnified pictographs

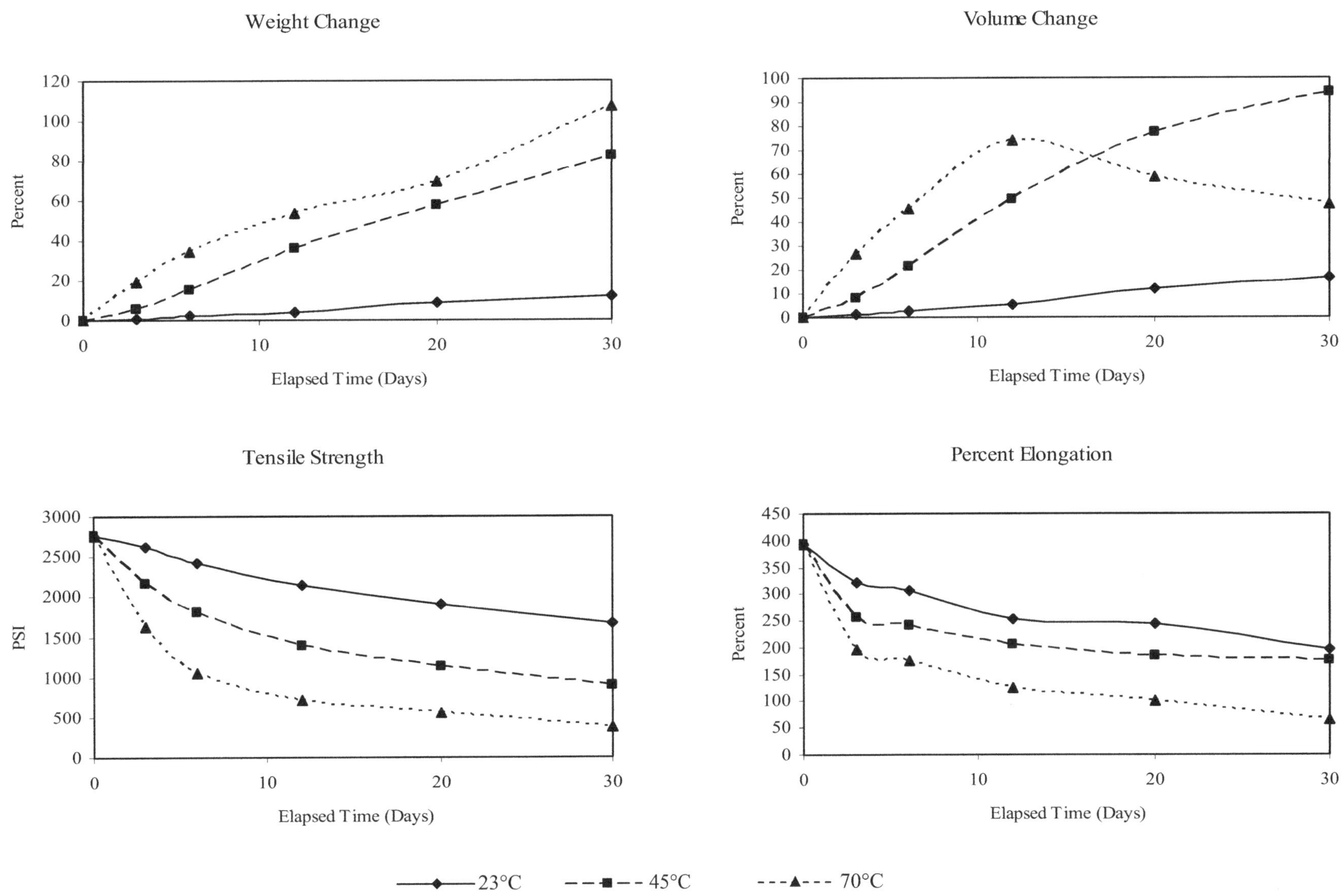

Figure A.33 Temperature effect on elastomeric properties at 60 ppm chloramine and pH 8.3 of neoprene

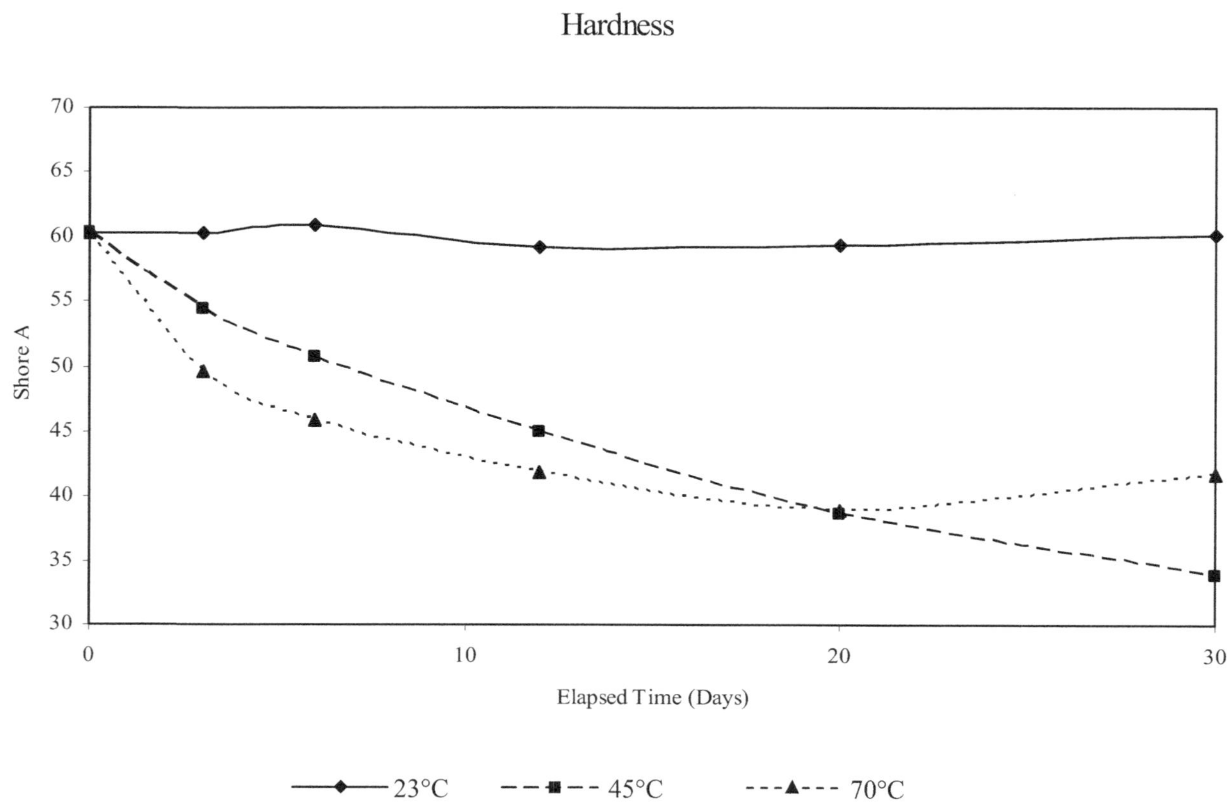

23°C at 30 days

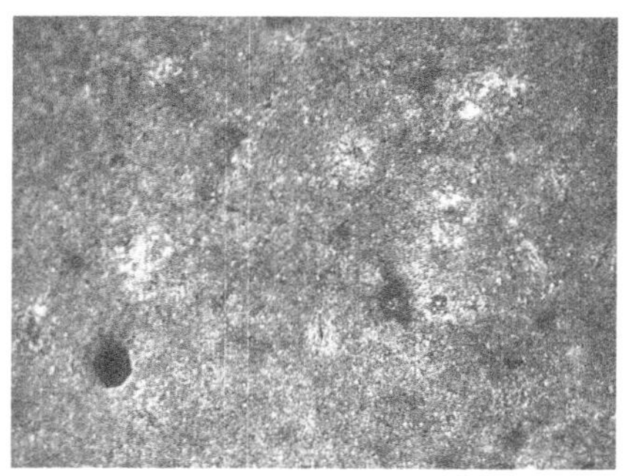

45°C at 30 days

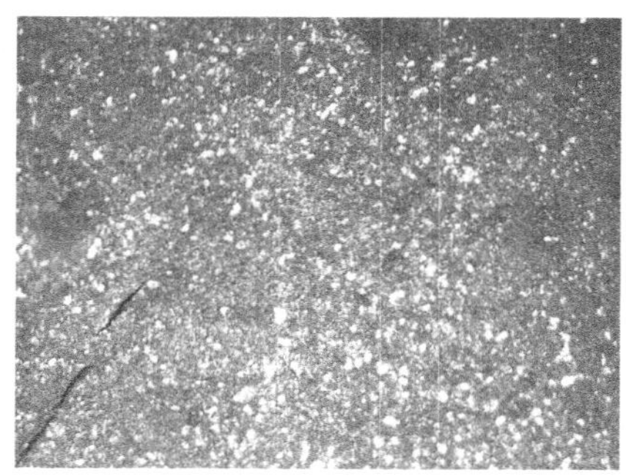

70°C at 30 days

Figure A.34 Temperature effect on hardness at 60 ppm chloramine and pH 8.3 of neoprene with 40× magnified pictographs

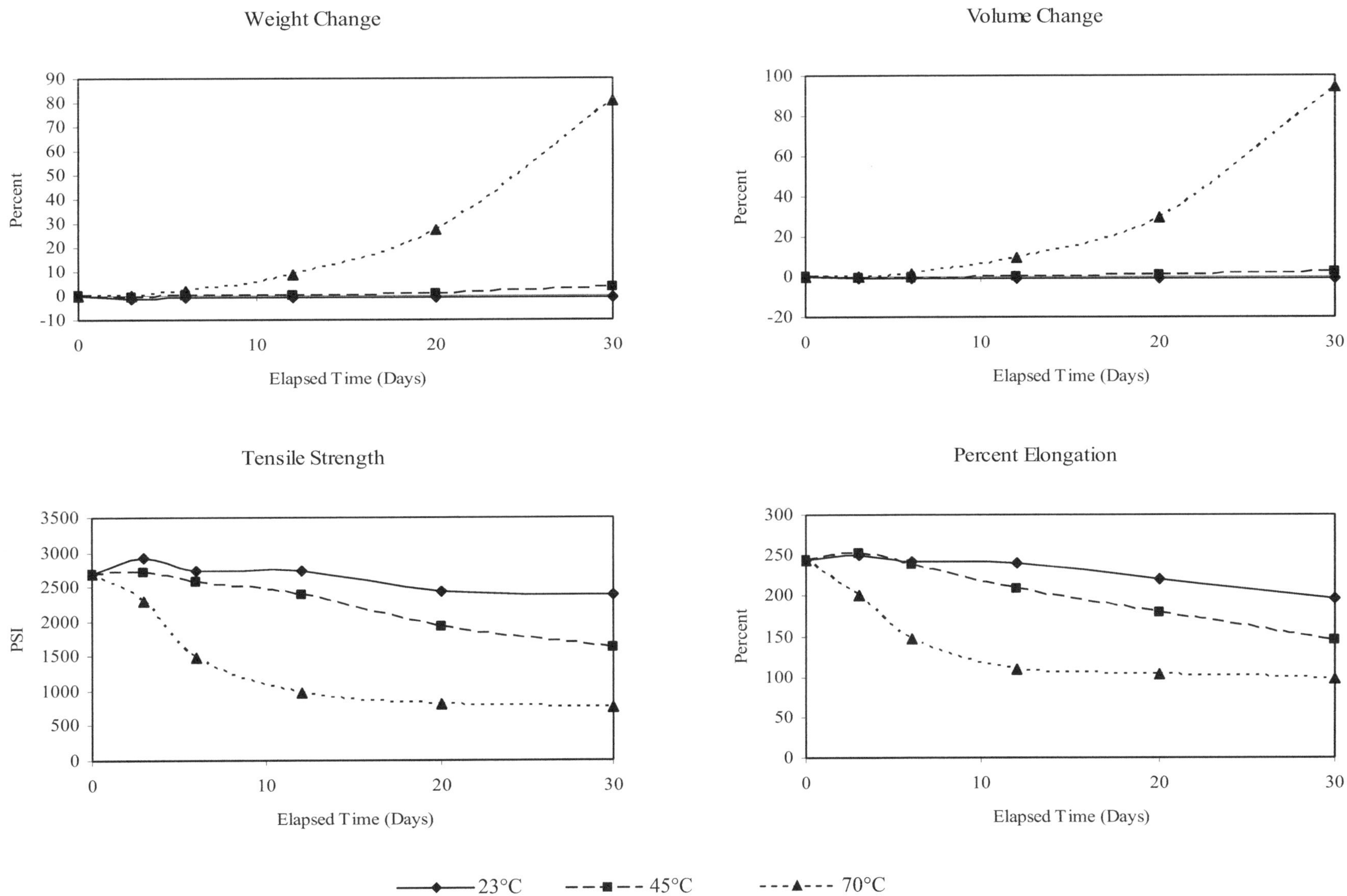

Figure A.35 Temperature effect on elastomeric properties at 60 ppm chloramine and pH 8.3 of EPDM-P

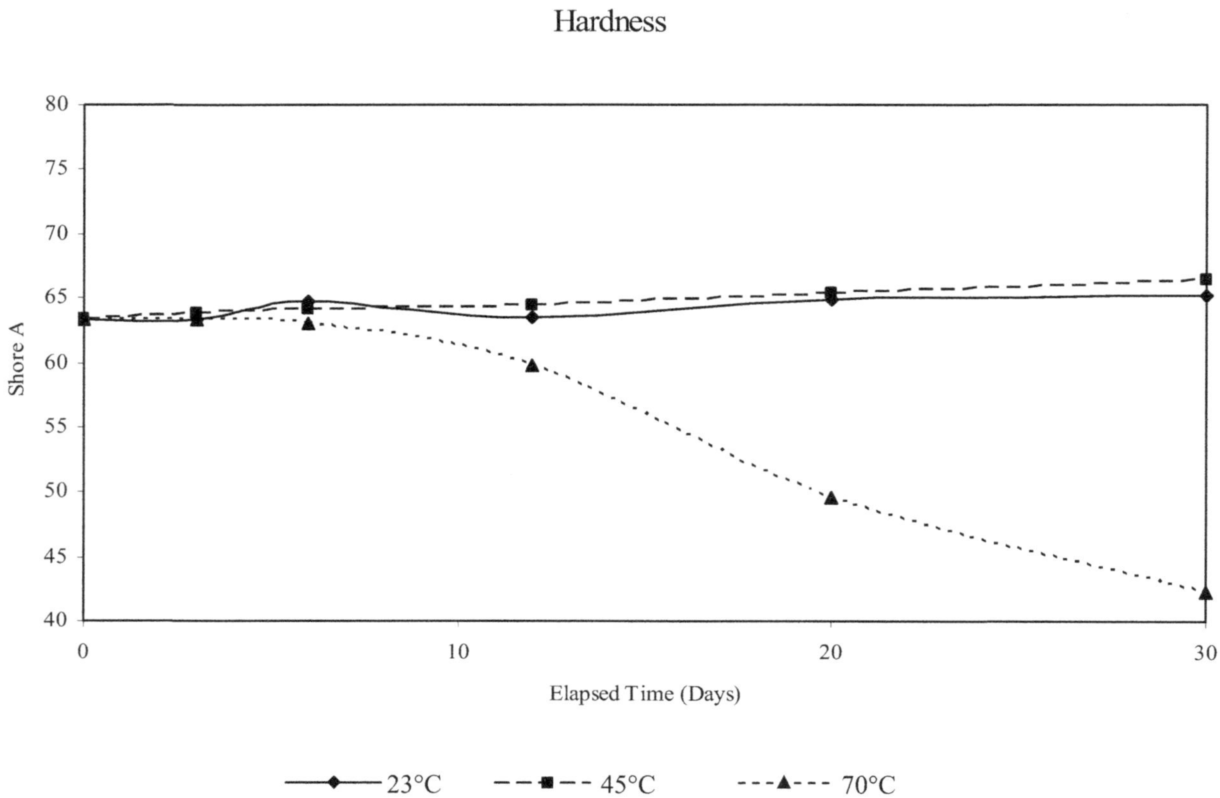

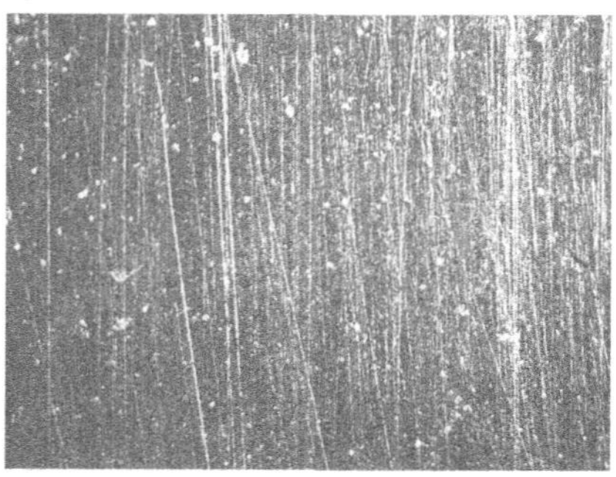

23°C at 30 days

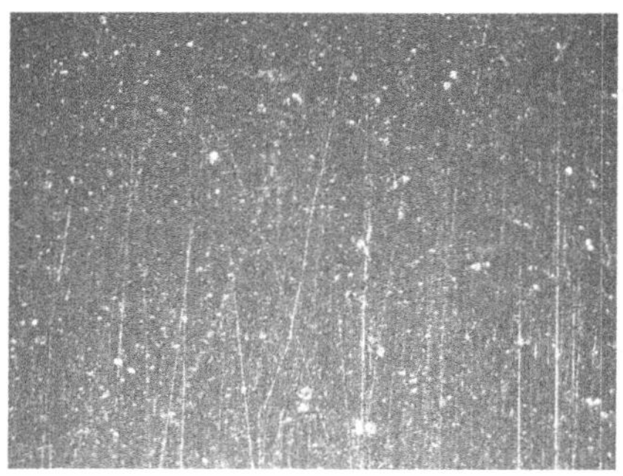

45°C at 30 days

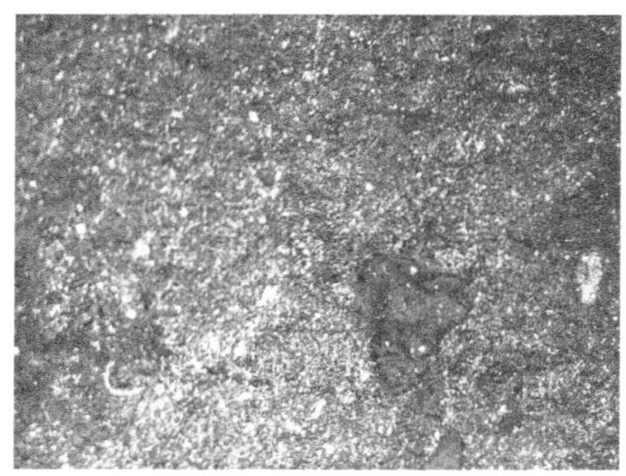

70°C at 30 days

Figure A.36 Temperature effect on hardness at 60 ppm chloramine and pH 8.3 of EPDM-P with 40× magnified pictographs

CONCENTRATION EFFECTS

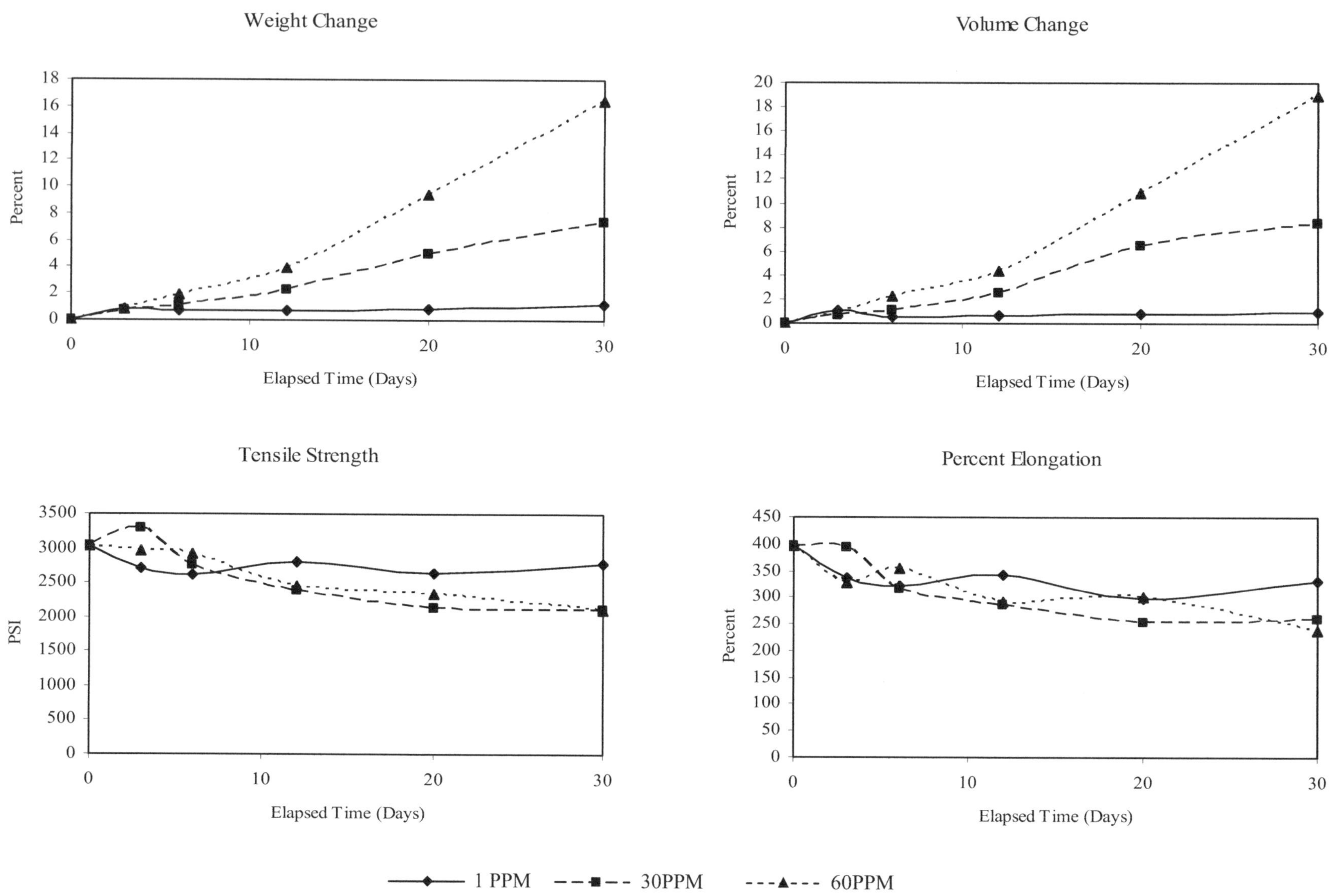

Figure A.37 Concentration of chloramine effect on elastomeric properties at 23°C and pH 8.3 of natural rubber

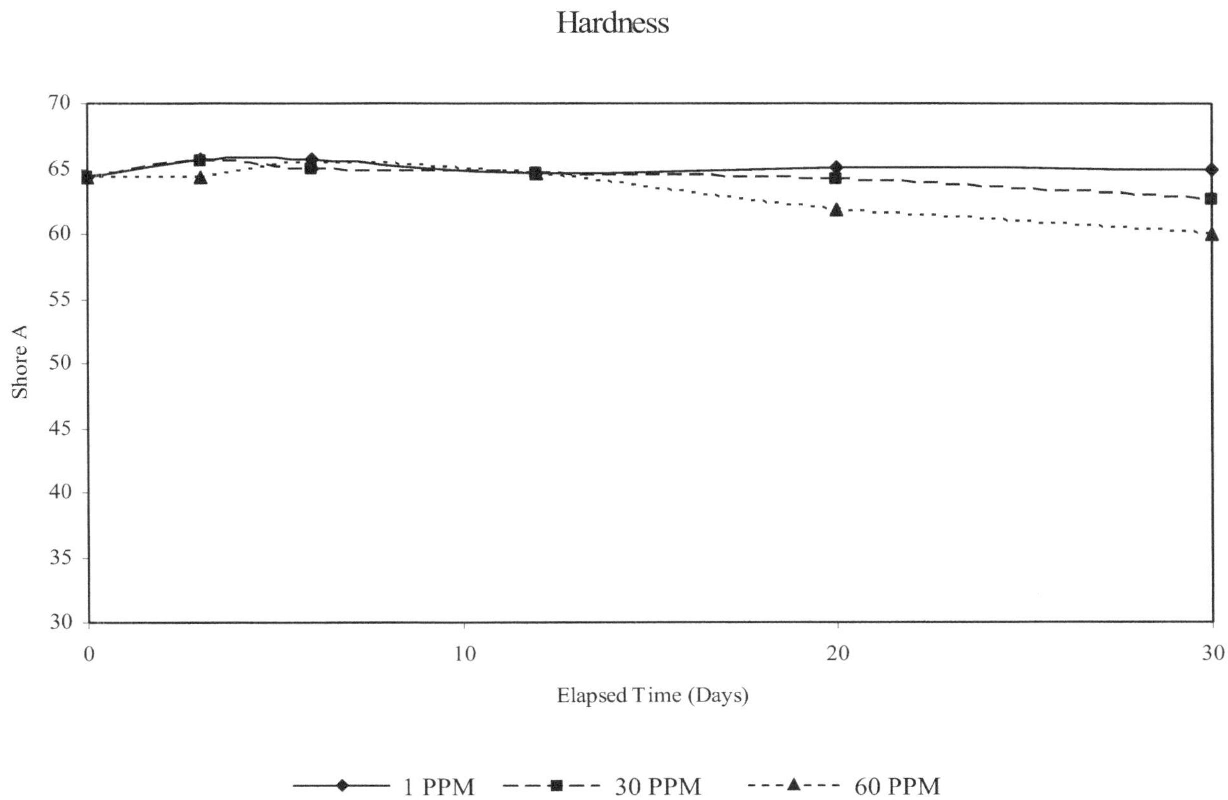

1 PPM at 30 days

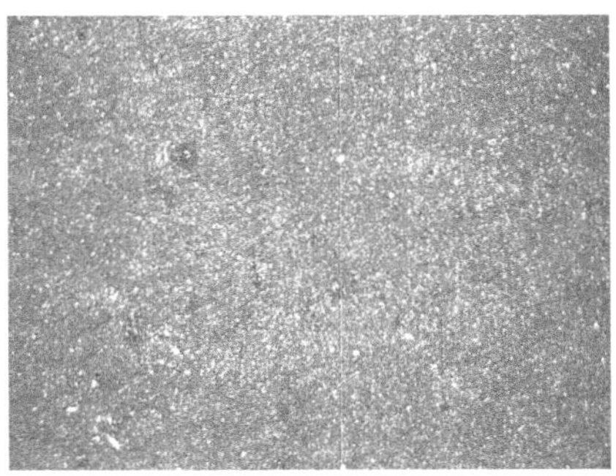

30 PPM at 30 days

60 PPM at 30 days

Figure A.38 Concentration of chloramine effect on hardness at 23°C and pH 8.3 of natural rubber with 40× magnified pictographs

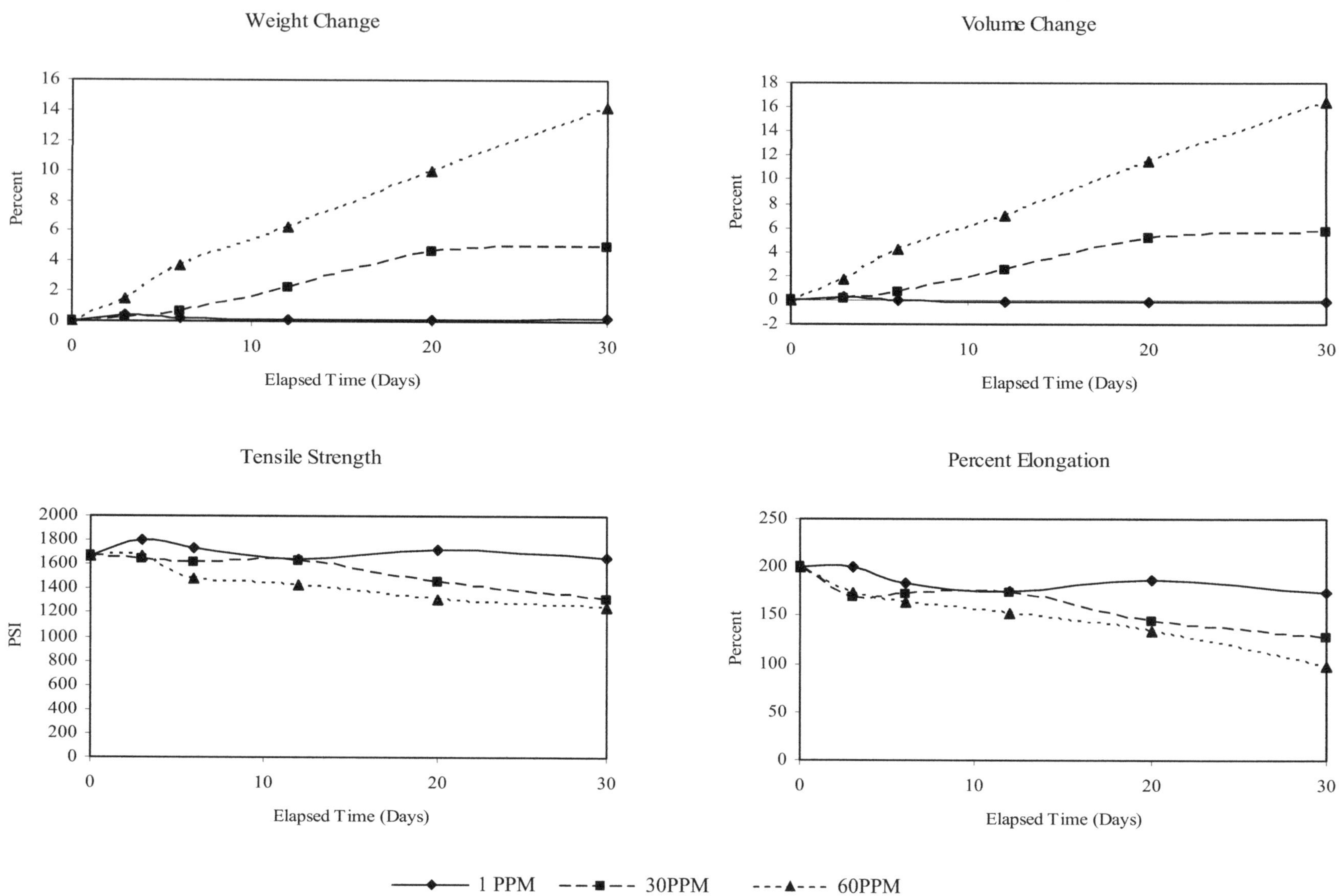

Figure A.39 Concentration of chloramine effect on elastomeric properties at 23°C and pH 8.3 of EPDM-S

Hardness

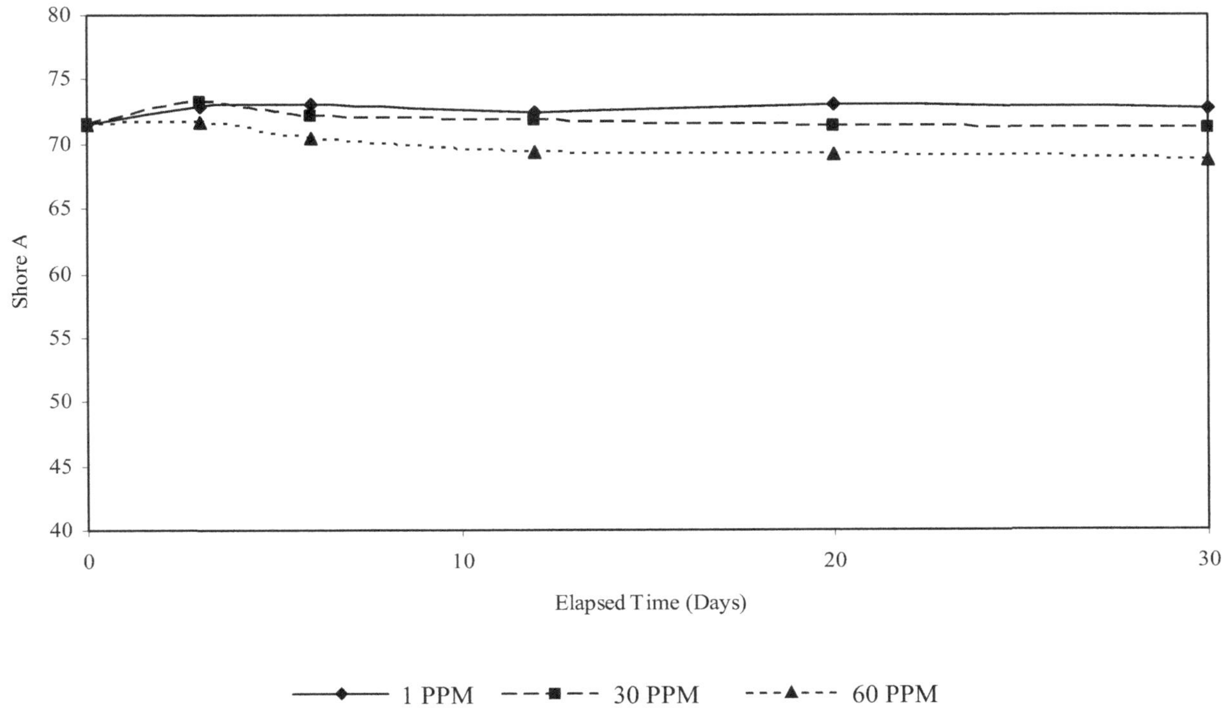

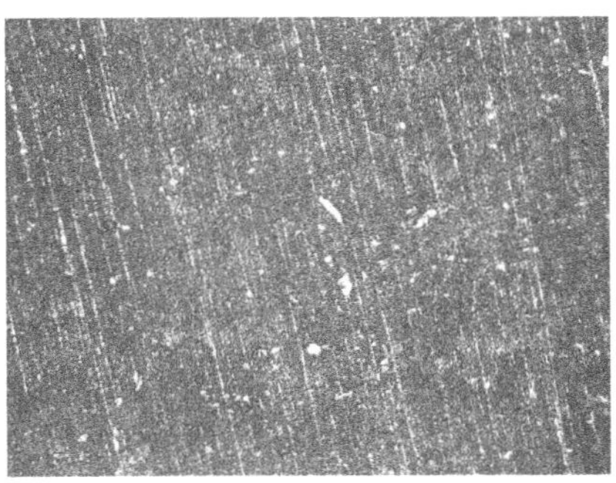

1 PPM at 30 days

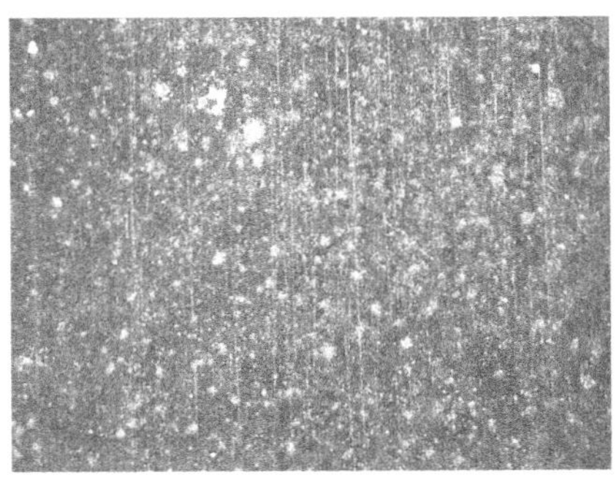

30 PPM at 30 days

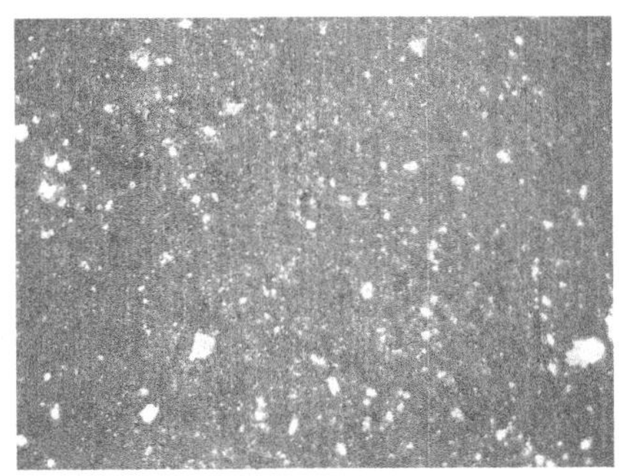

60 PPM at 30 days

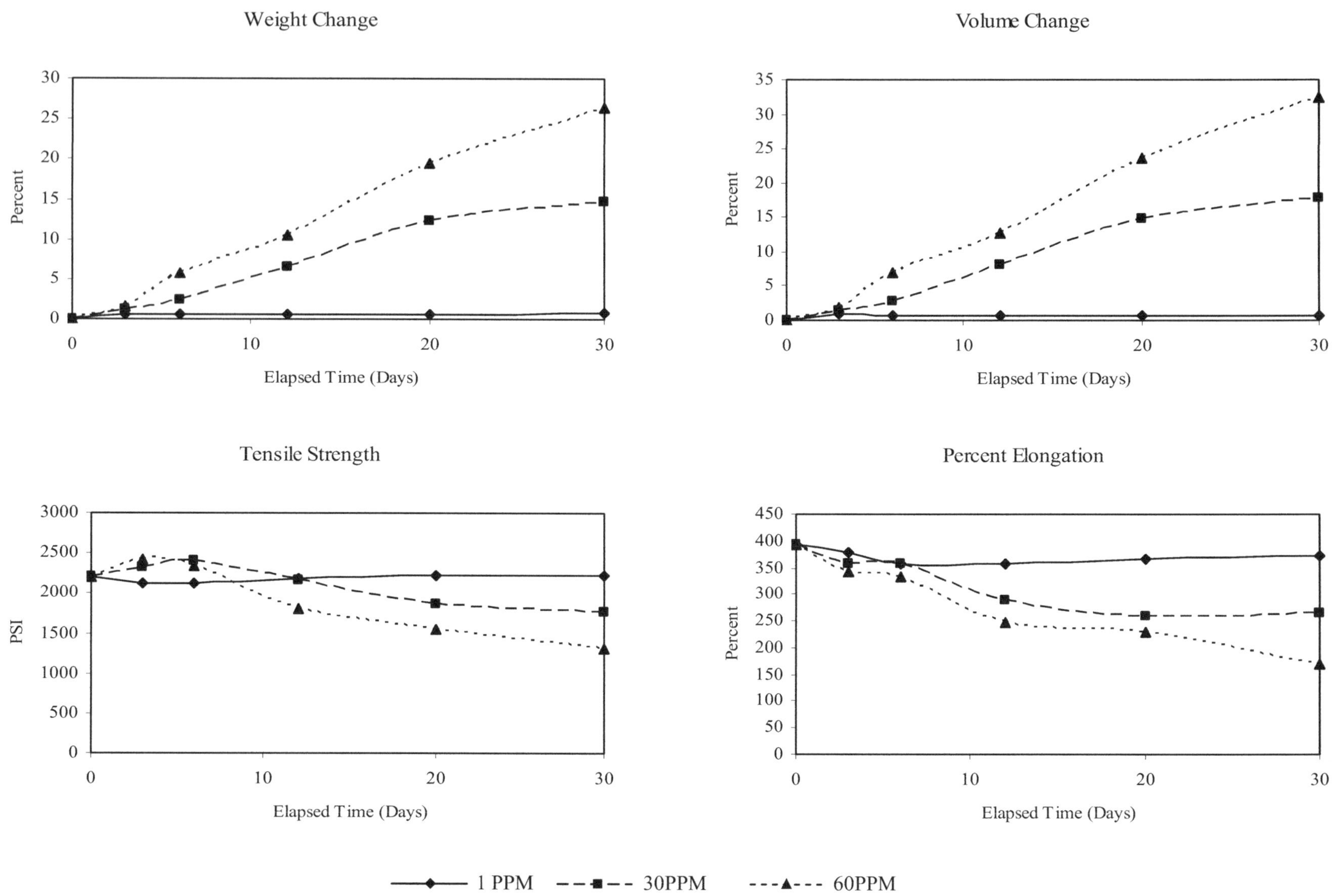

Figure A.41 Concentration of chloramine effect on elastomeric properties at 23°C and pH 8.3 of nitrile (NBR)

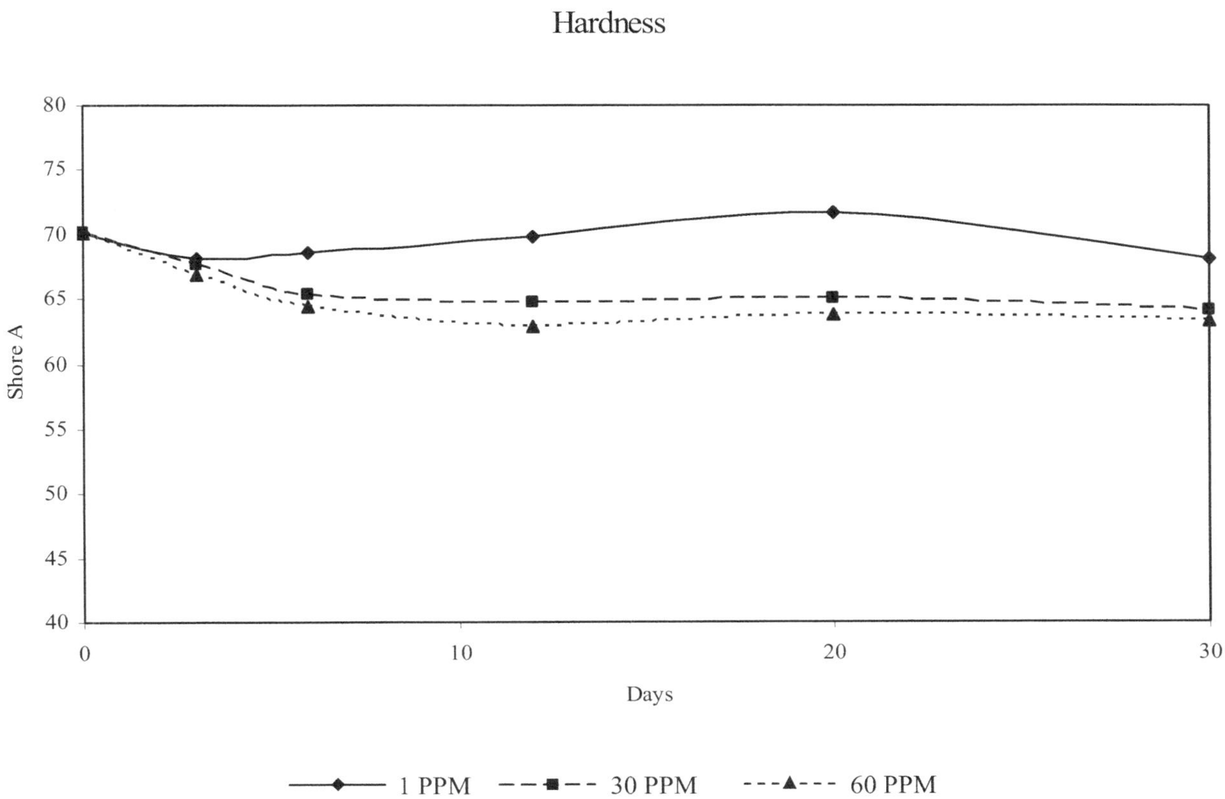

1 PPM at 30 days

30 PPM at 30 days

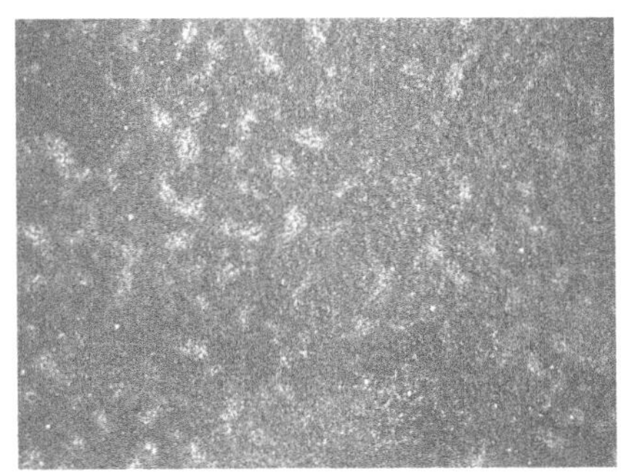

60 PPM at 30 days

Figure A.42 Concentration of chloramine effect on hardness at 23°C and pH 8.3 of nitrile (NBR) with 40× magnified pictographs

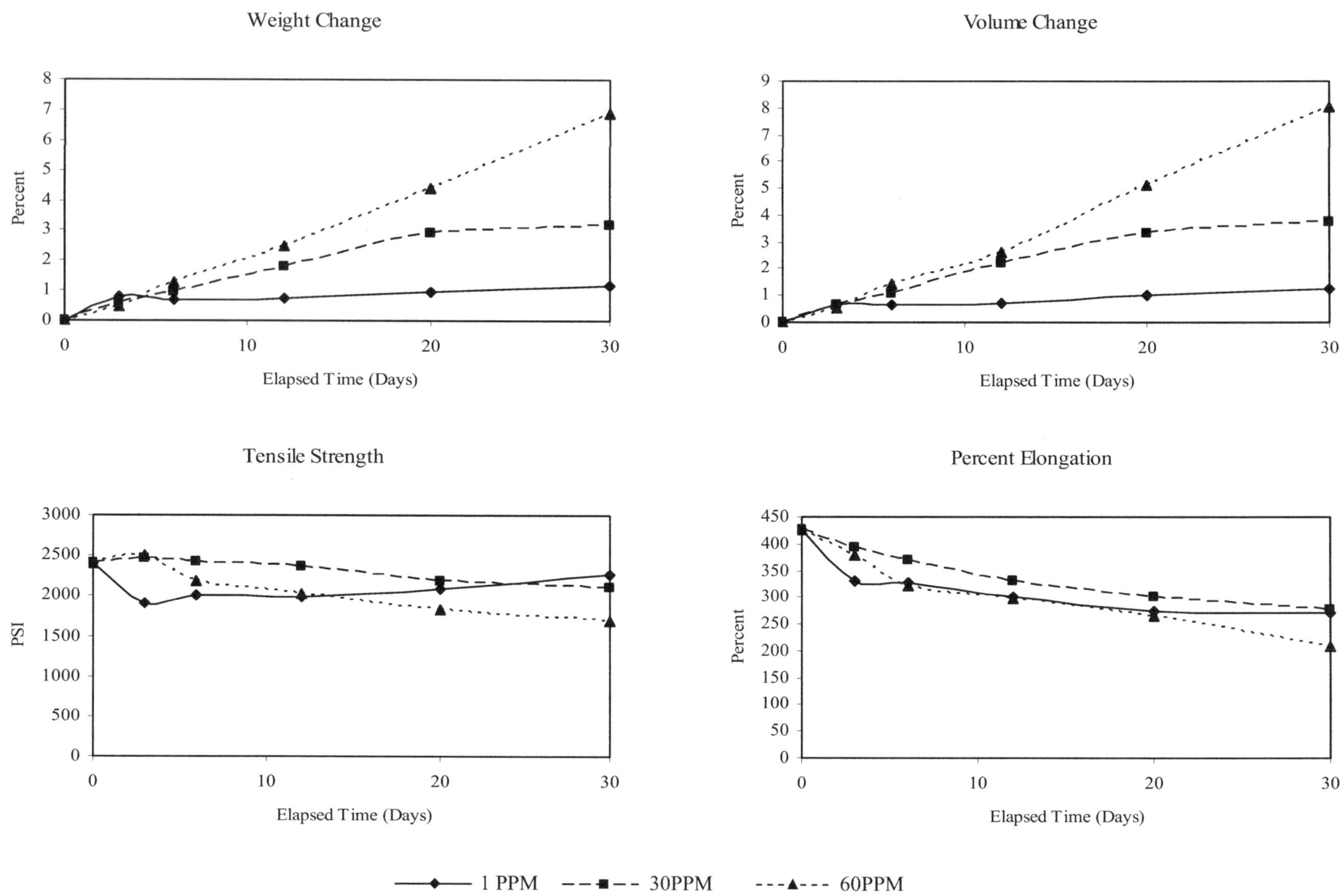

Figure A.43 Concentration of chloramine effect on elastomeric properties at 23°C and pH 8.3 of SBR

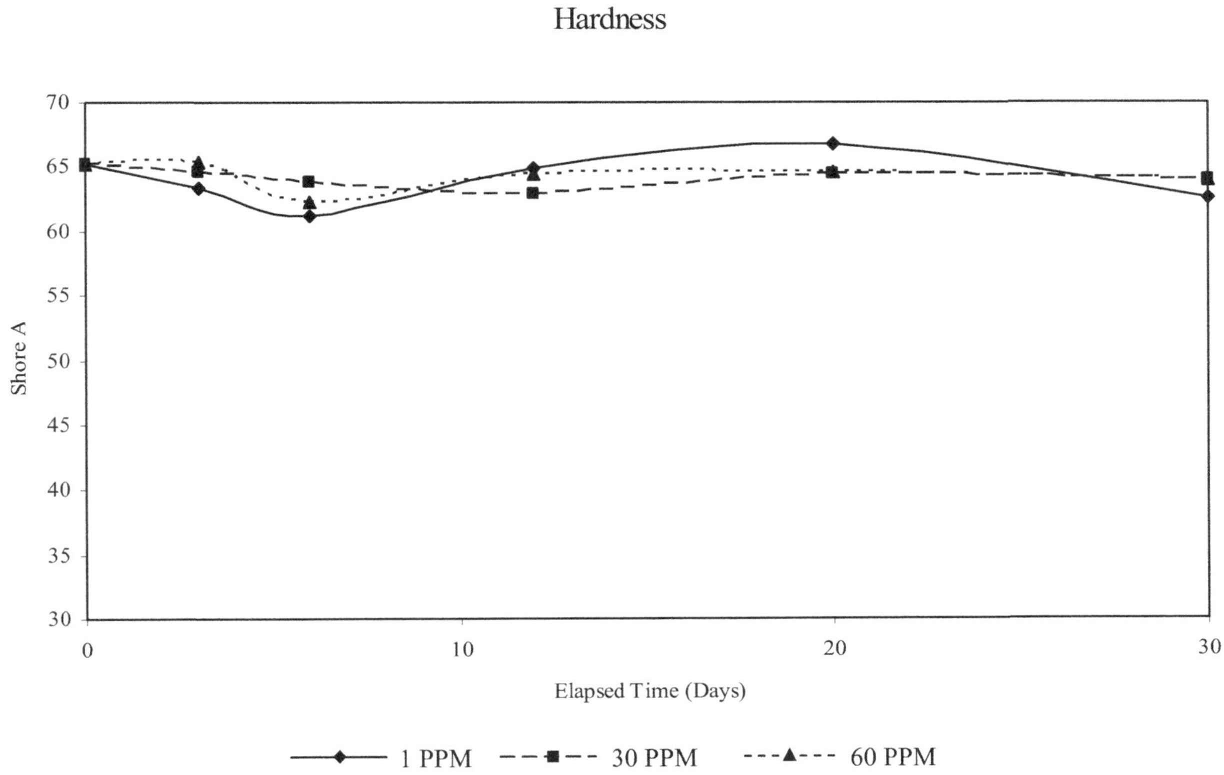

1 PPM at 30 days

30 PPM at 30 days

60 PPM at 30 days

Figure A.44 Concentration of chloramine effect on hardness at 23°C and pH 8.3 of SBR with 40× magnified pictographs

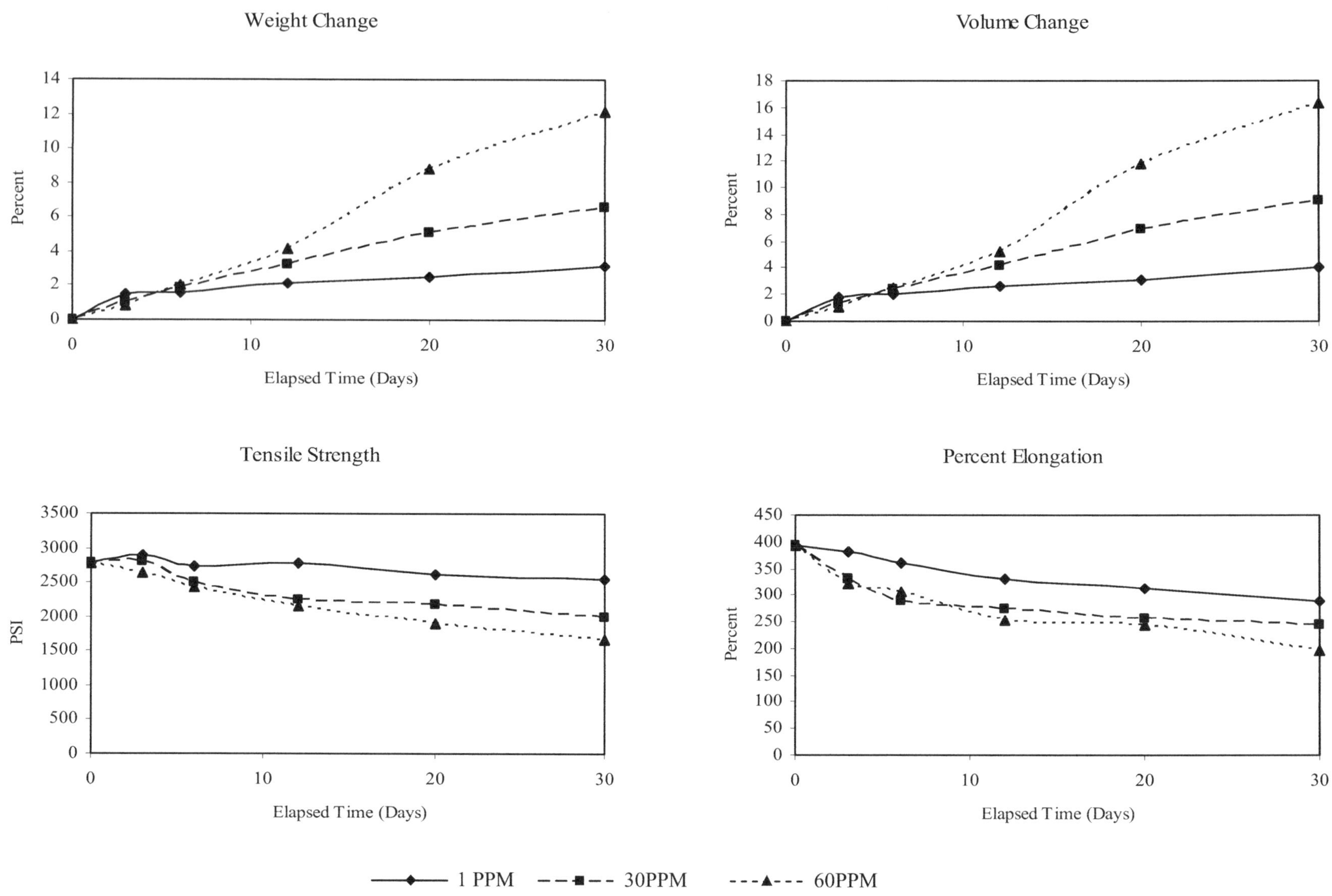

Figure A.45 Concentration of chloramine effect on elastomeric properties at 23°C and pH 8.3 of neoprene

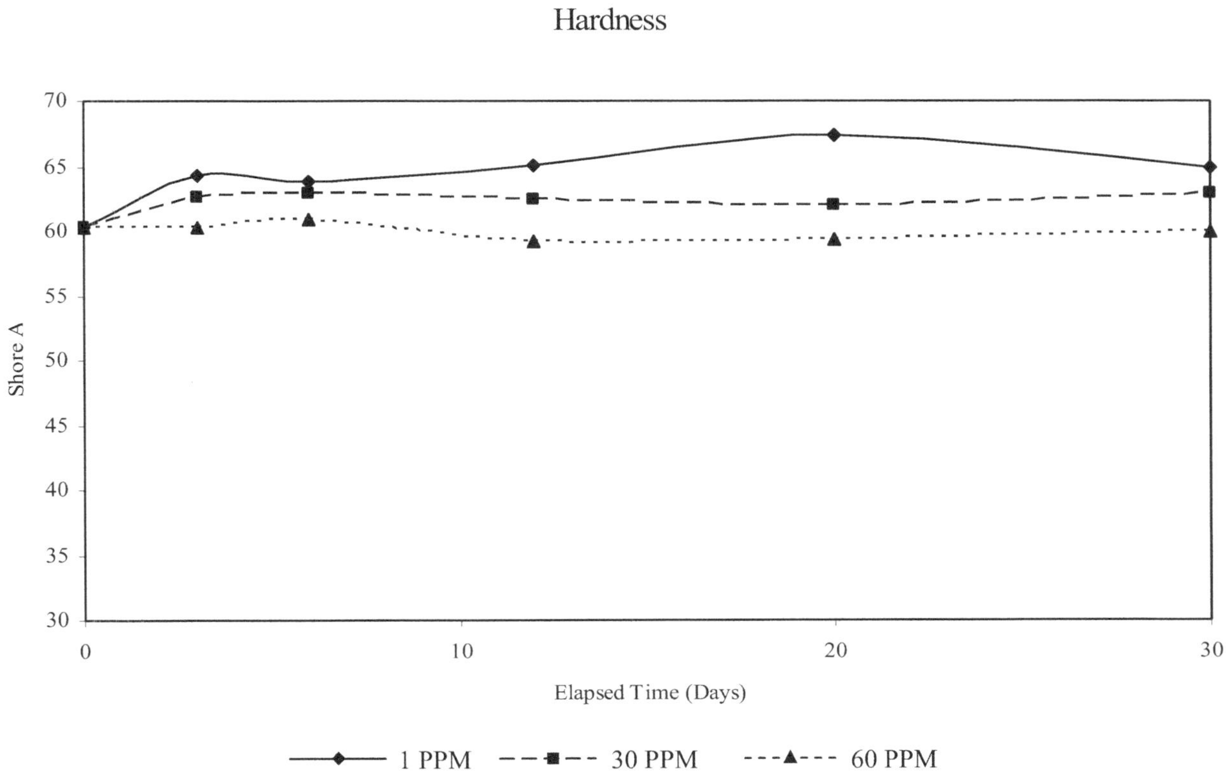

1 PPM at 30 days

30 PPM at 30 days

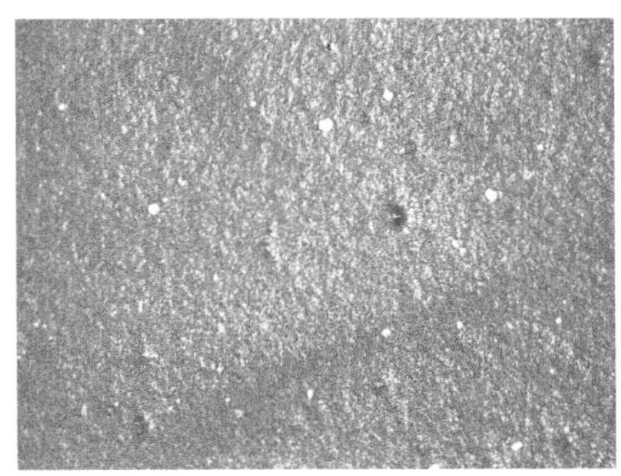

60 PPM at 30 days

Figure A.46 Concentration of chloramine effect on hardness at 23°C and pH 8.3 of neoprene with 40× magnified pictographs

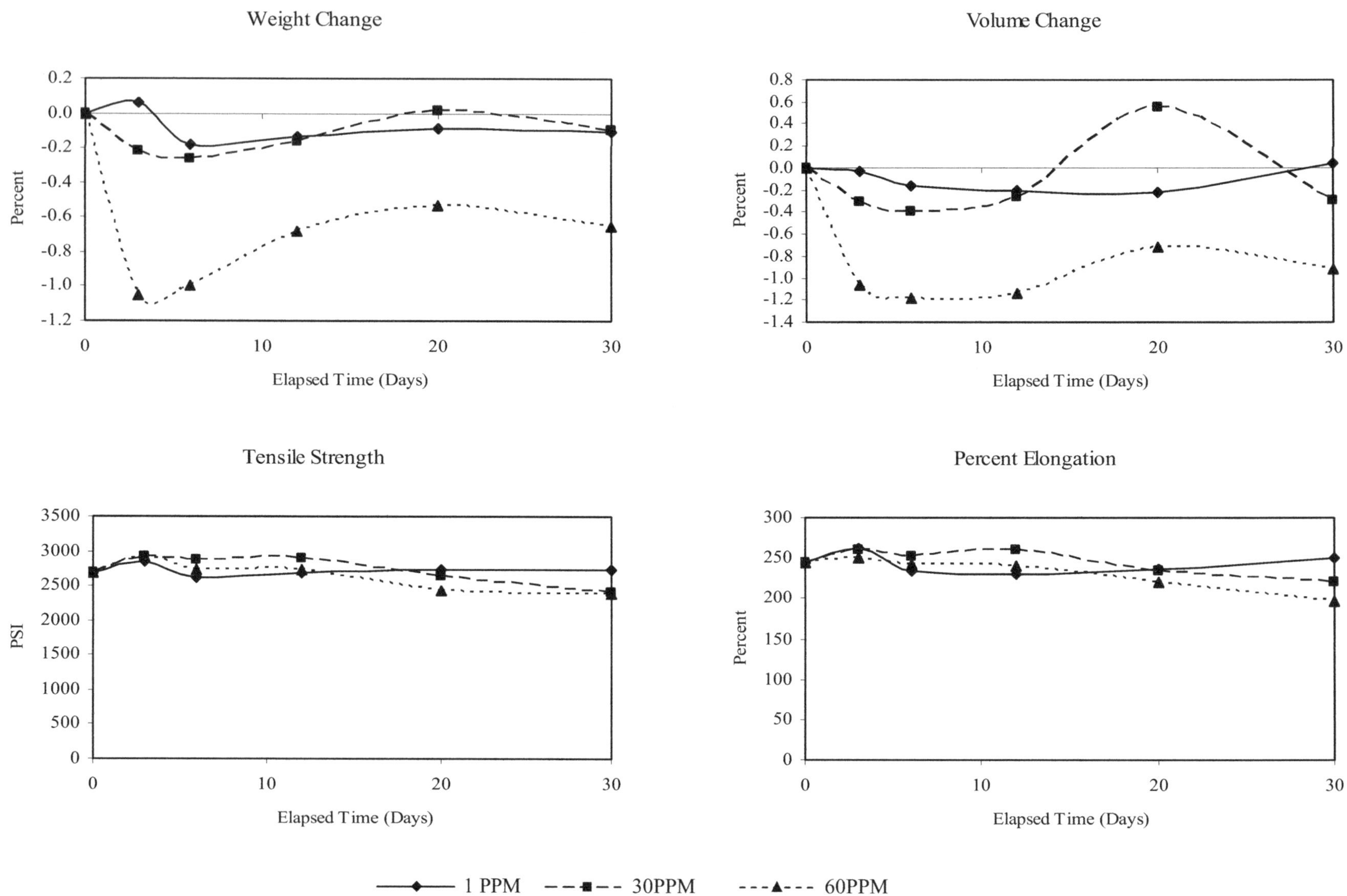

Figure A.47 Concentration of chloramine effect on elastomeric properties at 23°C and pH 8.3 of EPDM-P

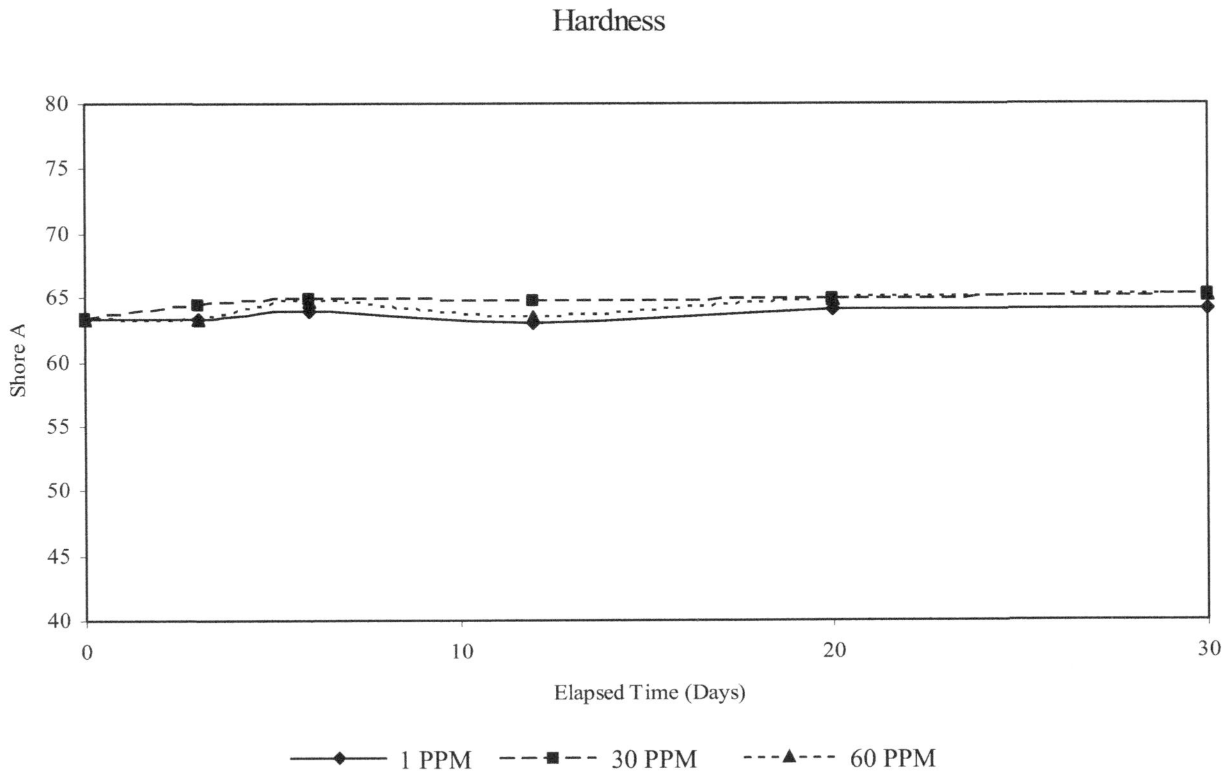

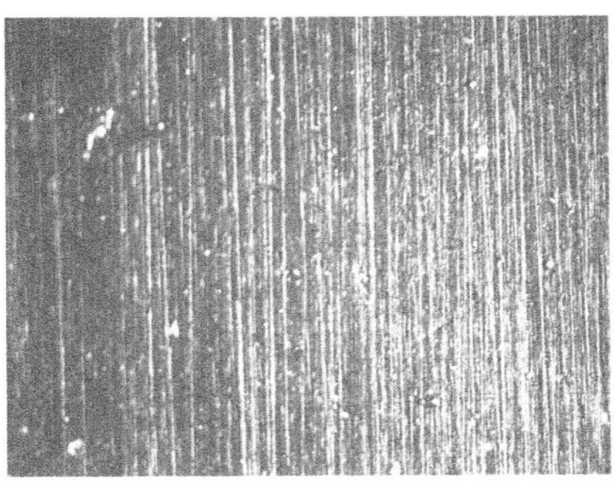

1 PPM at 30 days

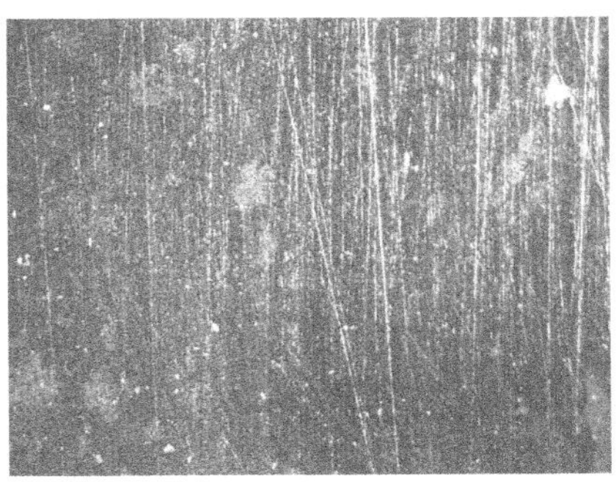

30 PPM at 30 days

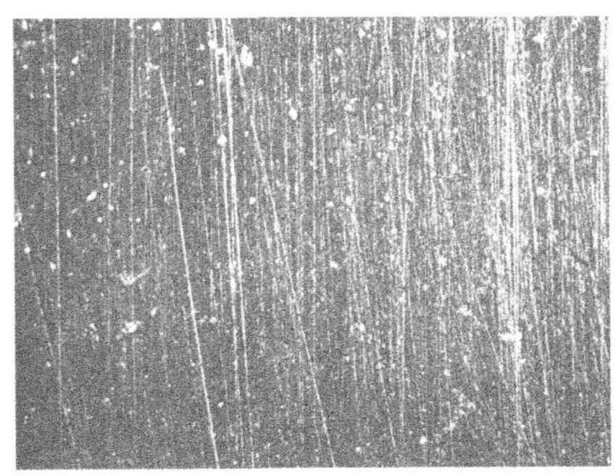

60 PPM at 30 days

Figure A.48 Concentration of chloramine effect on hardness at 23°C and pH 8.3 of EPDM-P with 40× magnified pictographs

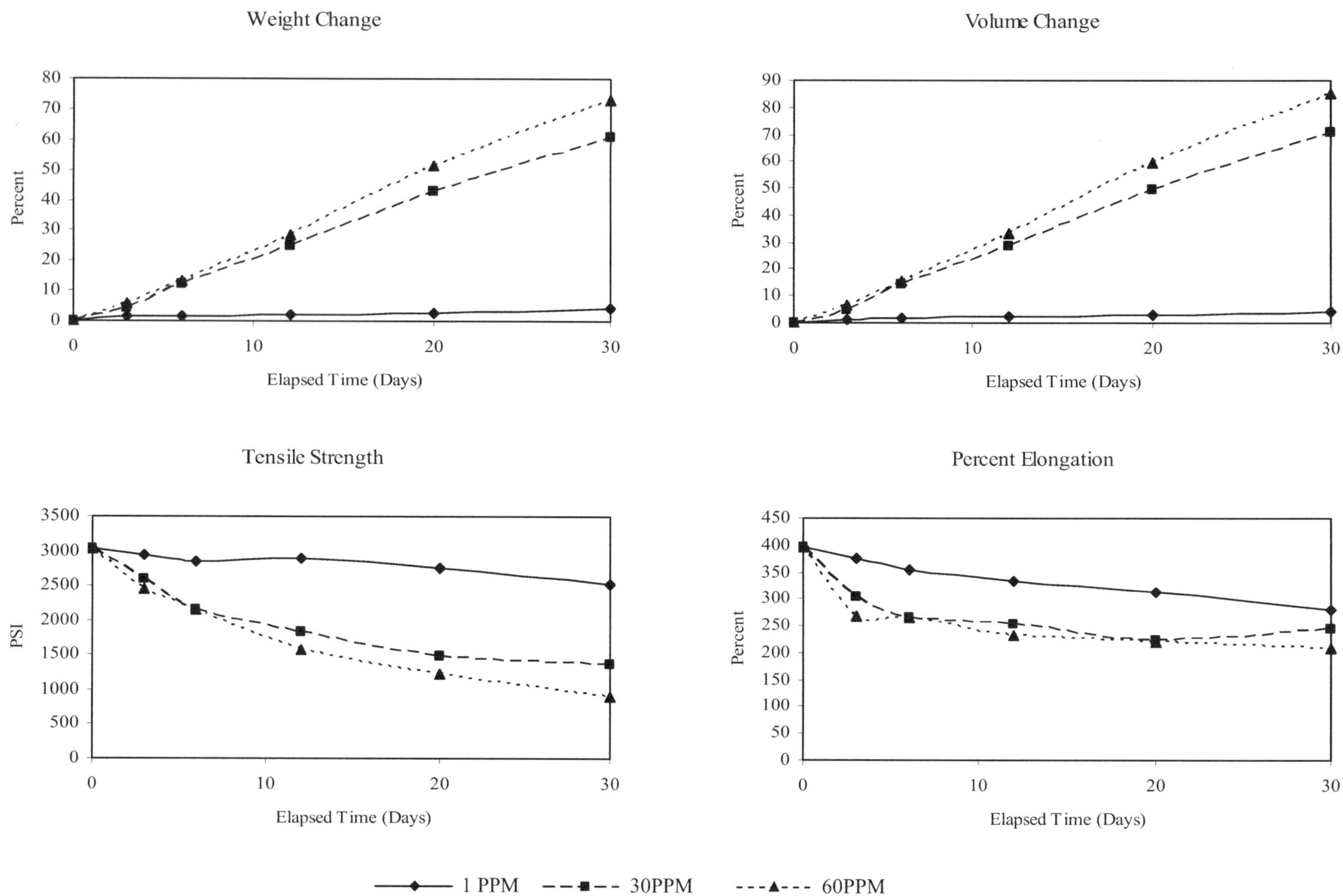

Figure A.49 Concentration of chloramine effect on elastomeric properties at 45°C and pH 8.3 of natural rubber

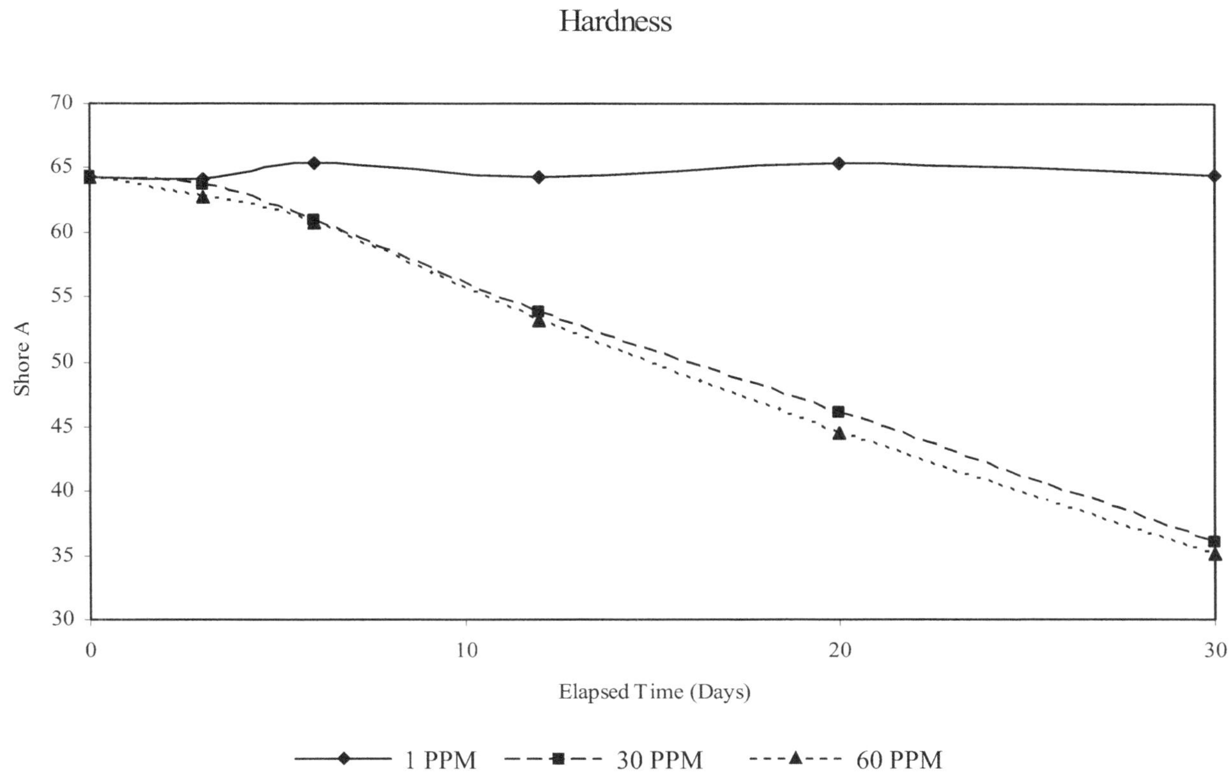

1 PPM at 30 days

30 PPM at 30 days

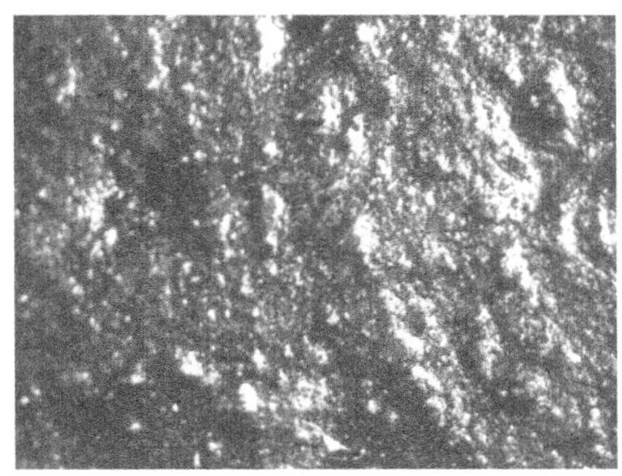

60 PPM at 30 days

Figure A.50 Concentration of chloramine effect on hardness at 45°C and pH 8.3 of natural rubber with 40× magnified pictographs

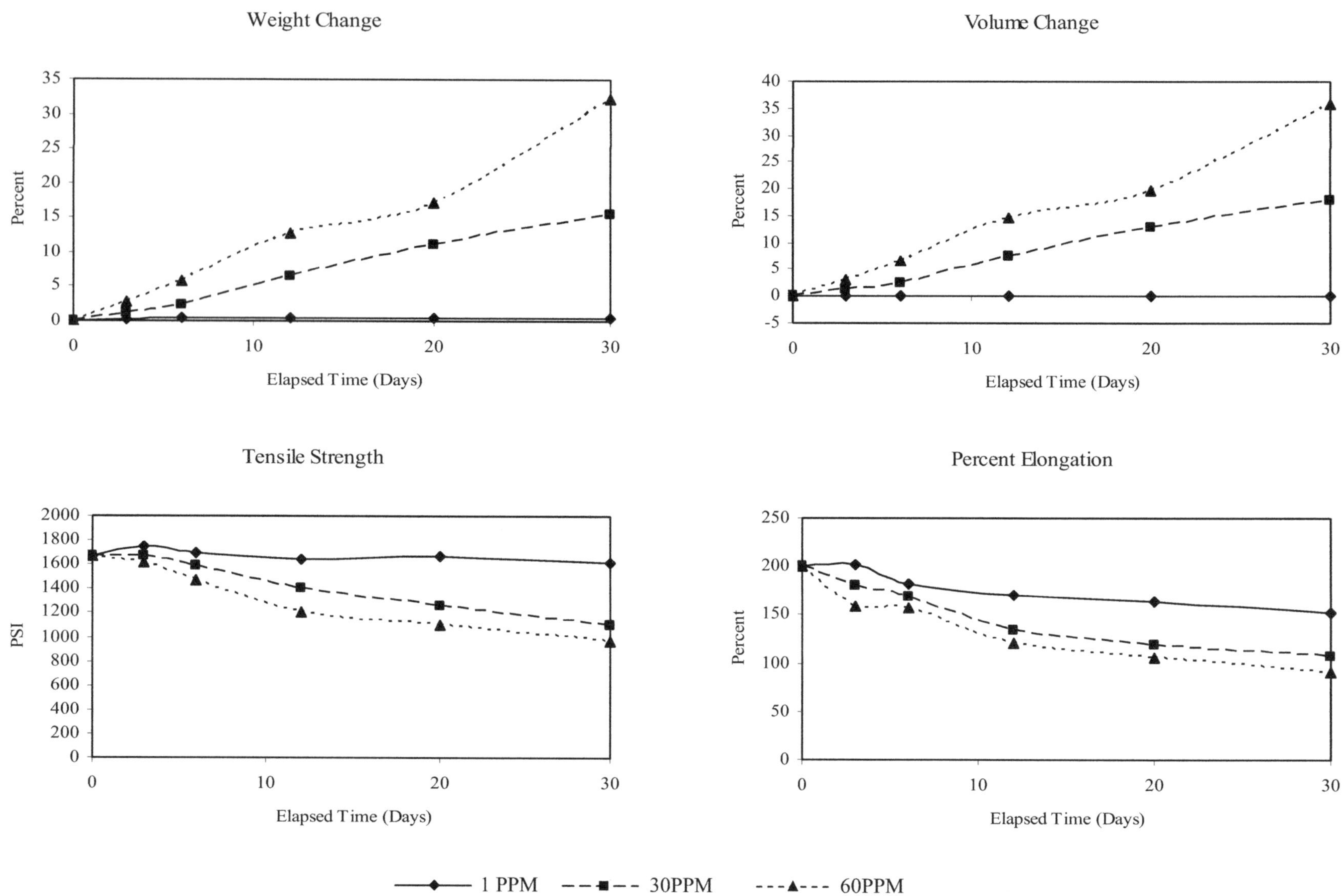

Figure A.51 Concentration of chloramine effect on elastomeric properties at 45°C and pH 8.3 of EPDM-S

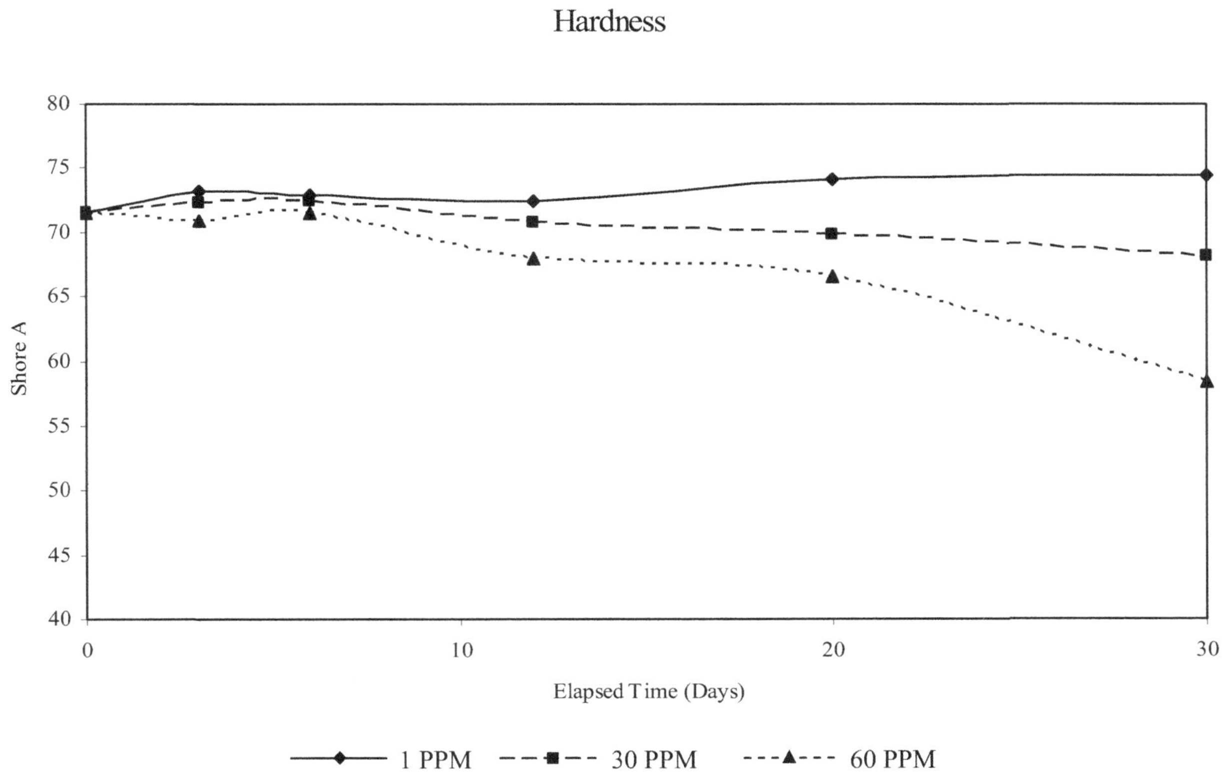

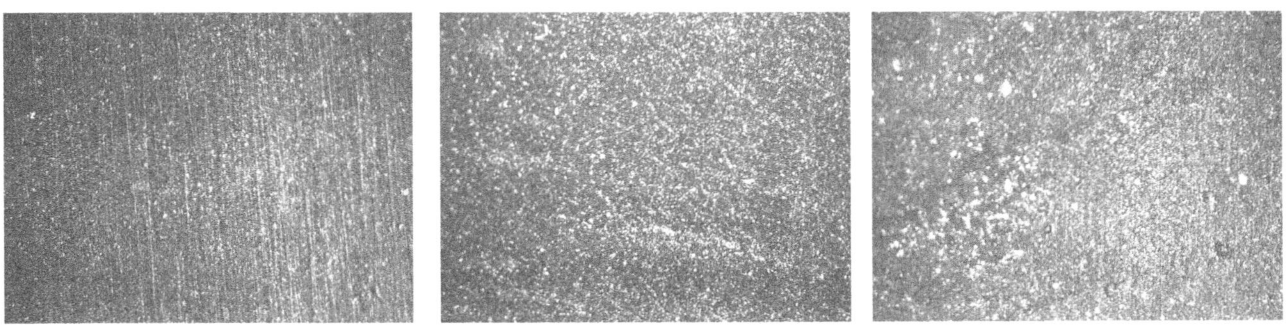

1 PPM at 30 days | 30 PPM at 30 days | 60 PPM at 30 days

Figure A.52 Concentration of chloramine effect on hardness at 45°C and pH 8.3 of EPDM-S with 40× magnified pictographs

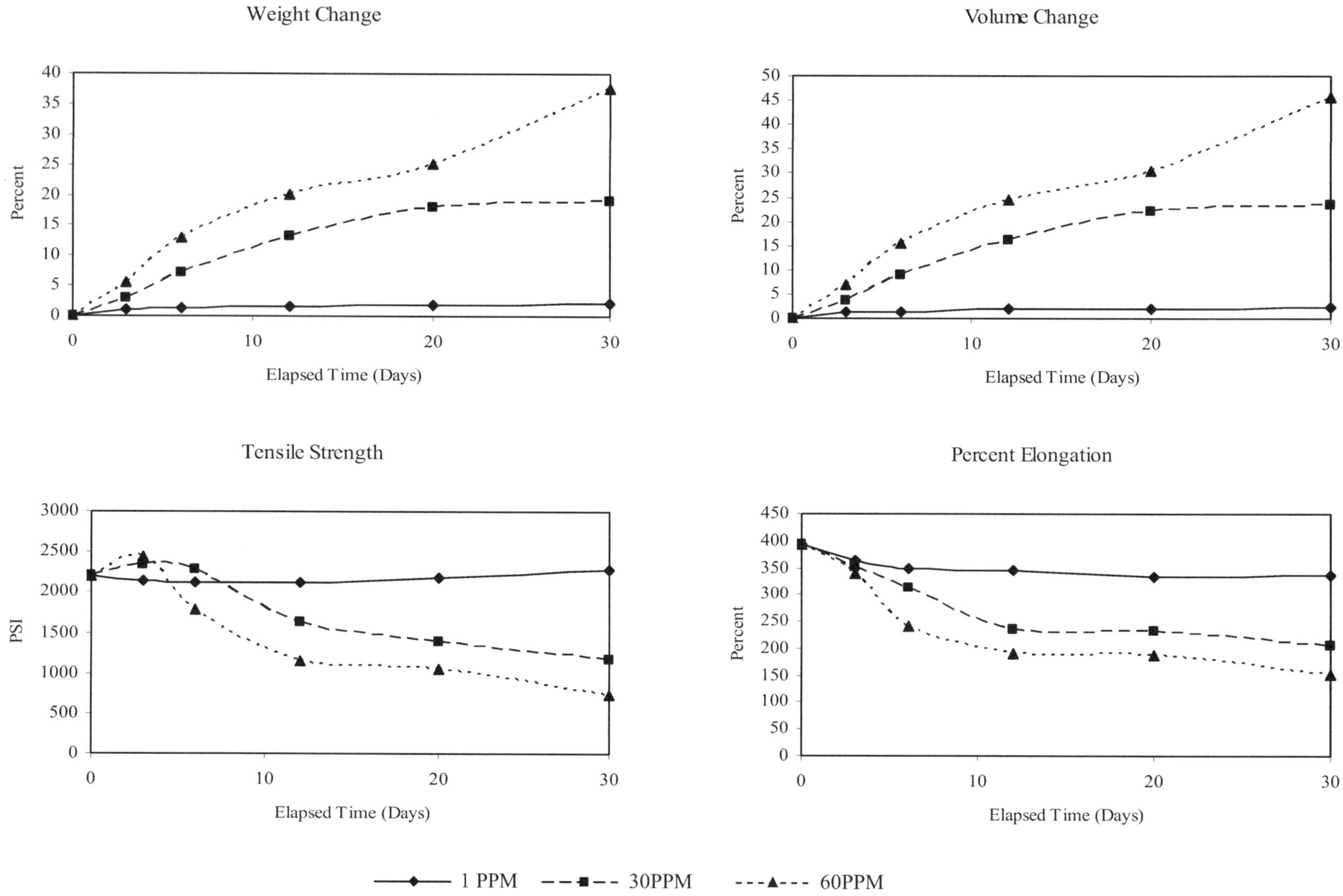

Figure A.53 Concentration of chloramine effect on elastomeric properties at 45°C and pH 8.3 of nitrile (NBR)

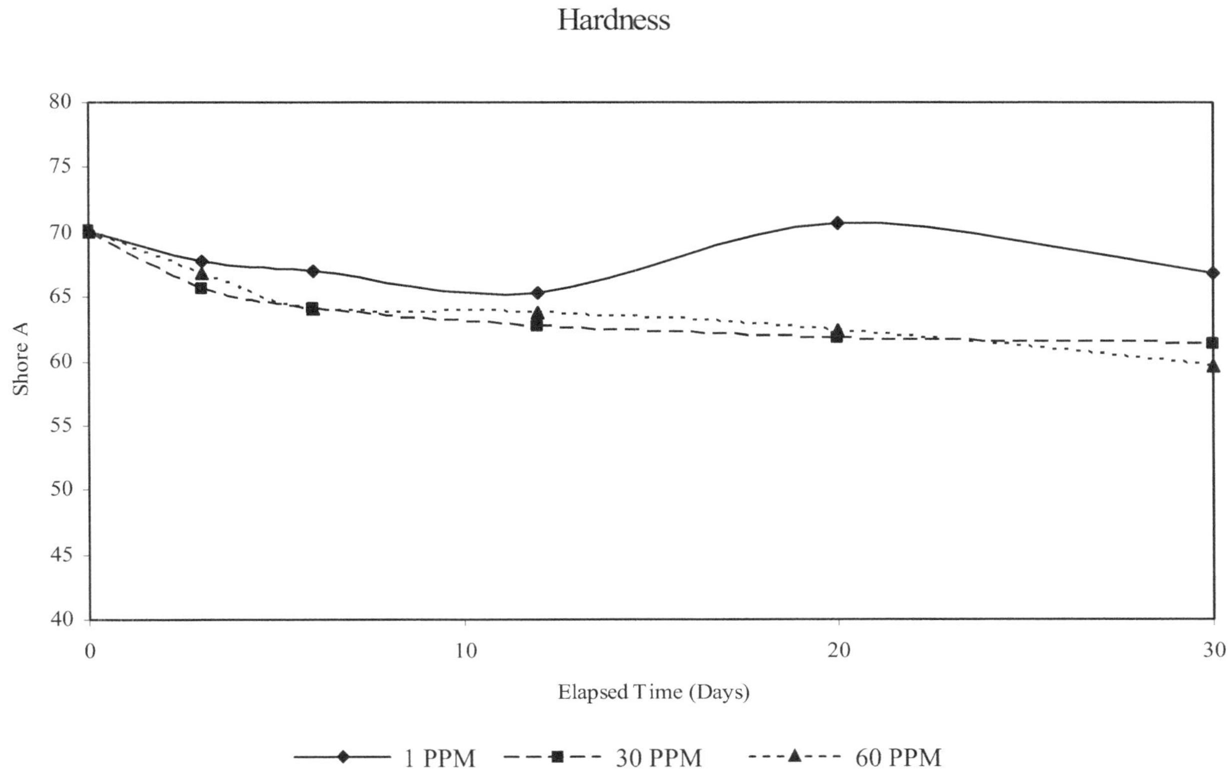

1 PPM at 30 days

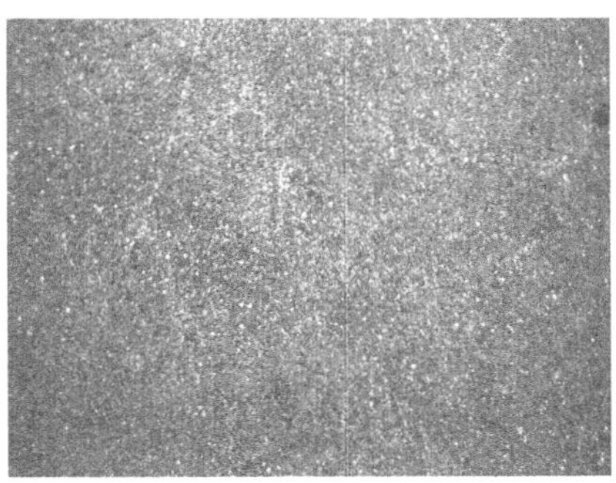

30 PPM at 30 days

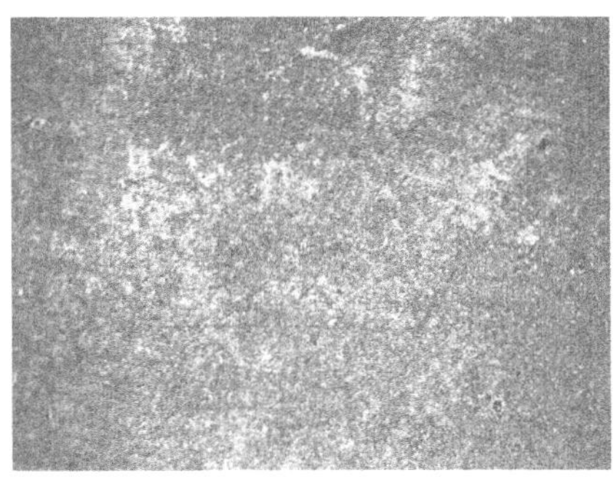

60 PPM at 30 days

Figure A.54 Concentration of chloramine effect on hardness at 45°C and pH 8.3 of nitrile (NBR) with 40× magnified pictographs

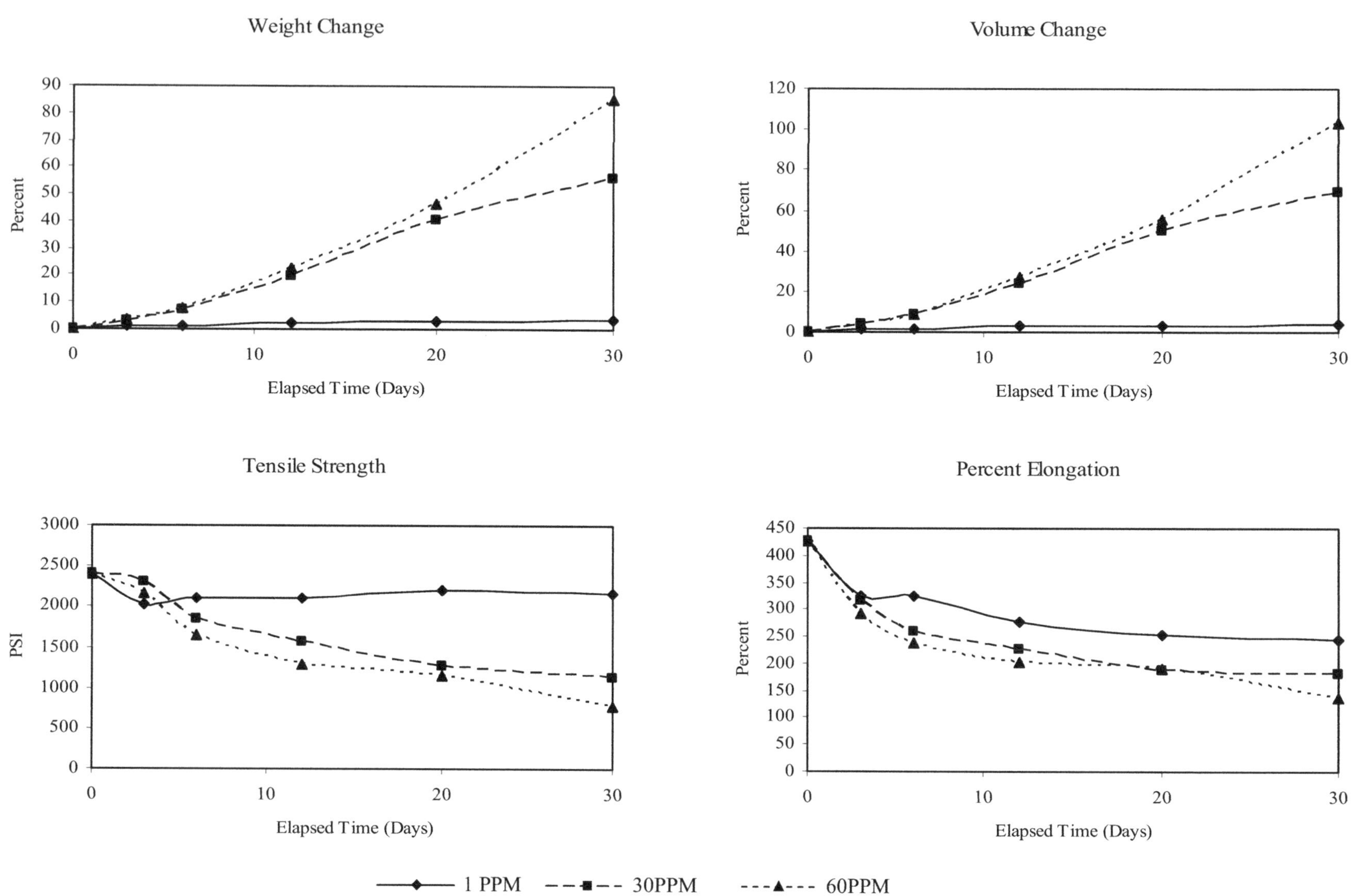

Figure A.55 Concentration of chloramine effect on elastomeric properties at 45°C and pH 8.3 of SBR

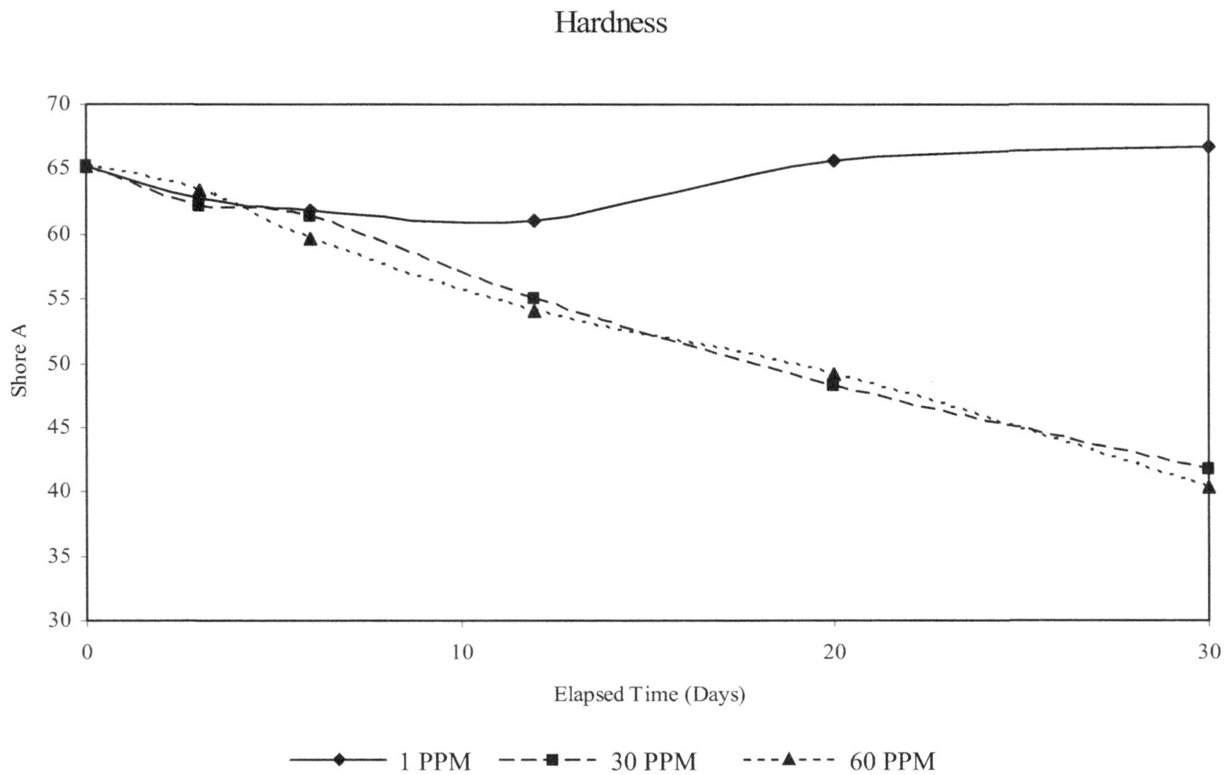

1 PPM at 30 days

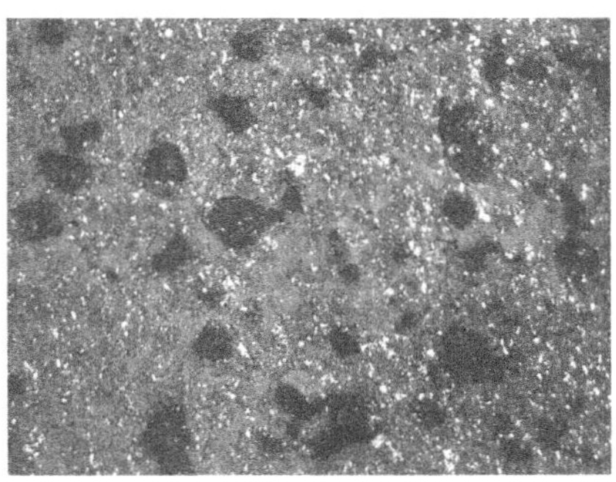

30 PPM at 30 days

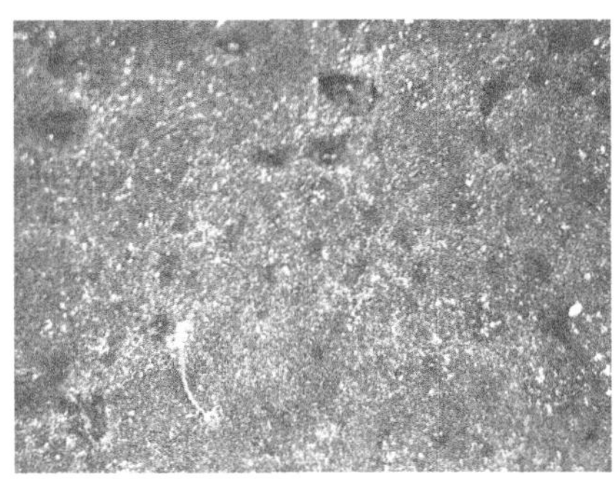

60 PPM at 30 days

Figure A.56 Concentration of chloramine effect on hardness at 45°C and pH 8.3 of SBR with 40× magnified pictographs

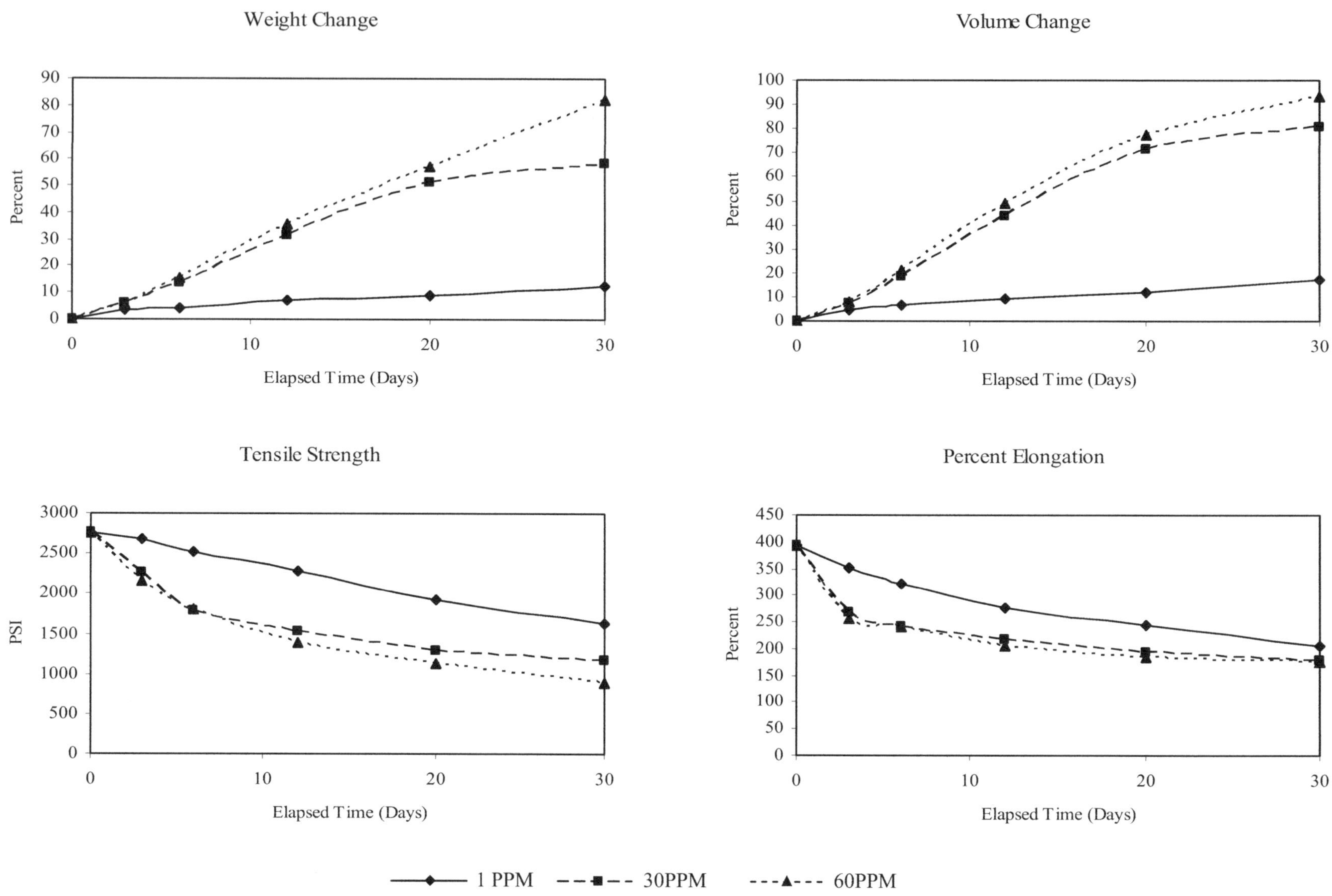

Figure A.57 Concentration of chloramine effect on elastomeric properties at 45°C and pH 8.3 of neoprene

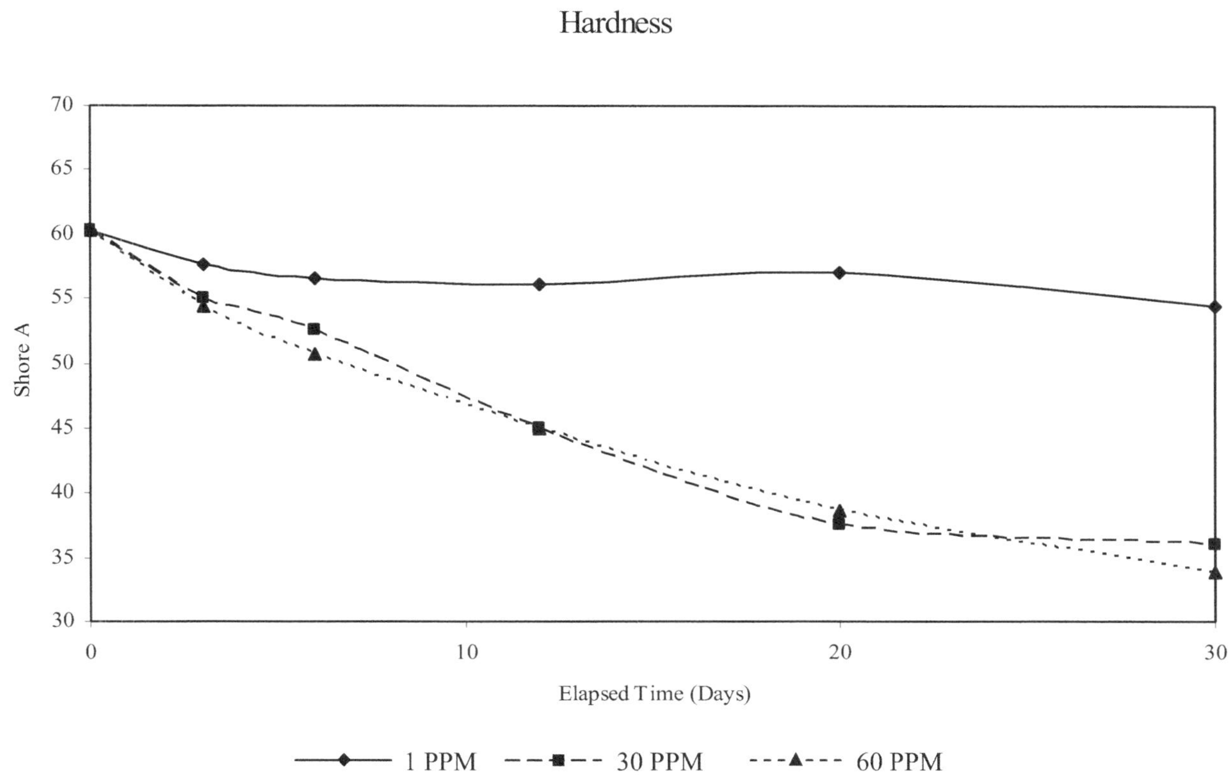

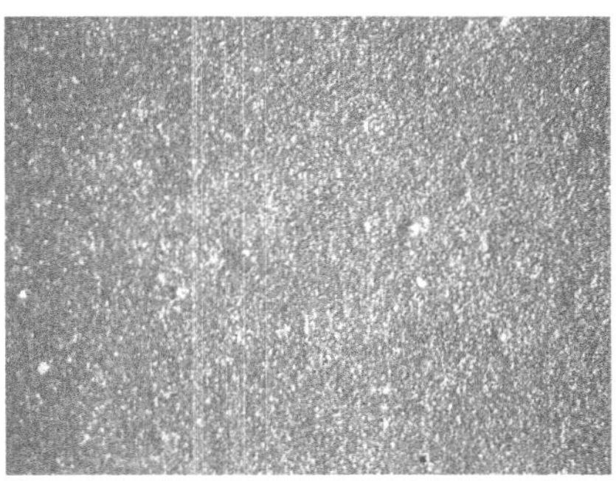

1 PPM at 30 days

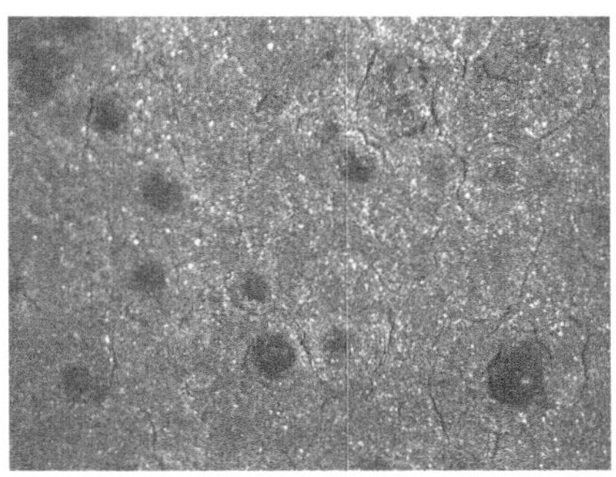

30 PPM at 30 days

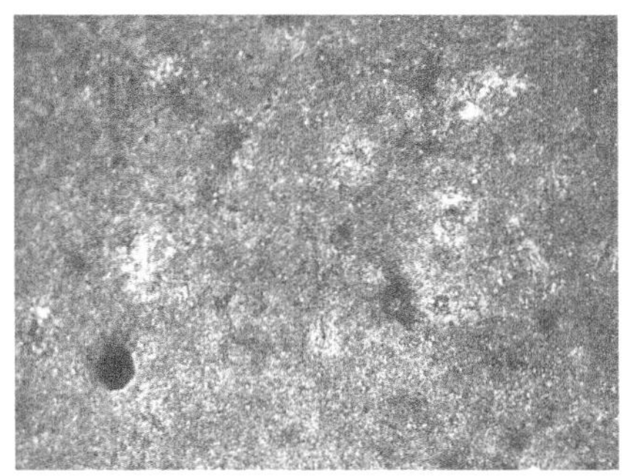

60 PPM at 30 days

Figure A.58 Concentration of chloramine effect on hardness at 45°C and pH 8.3 of neoprene with 40× magnified pictographs

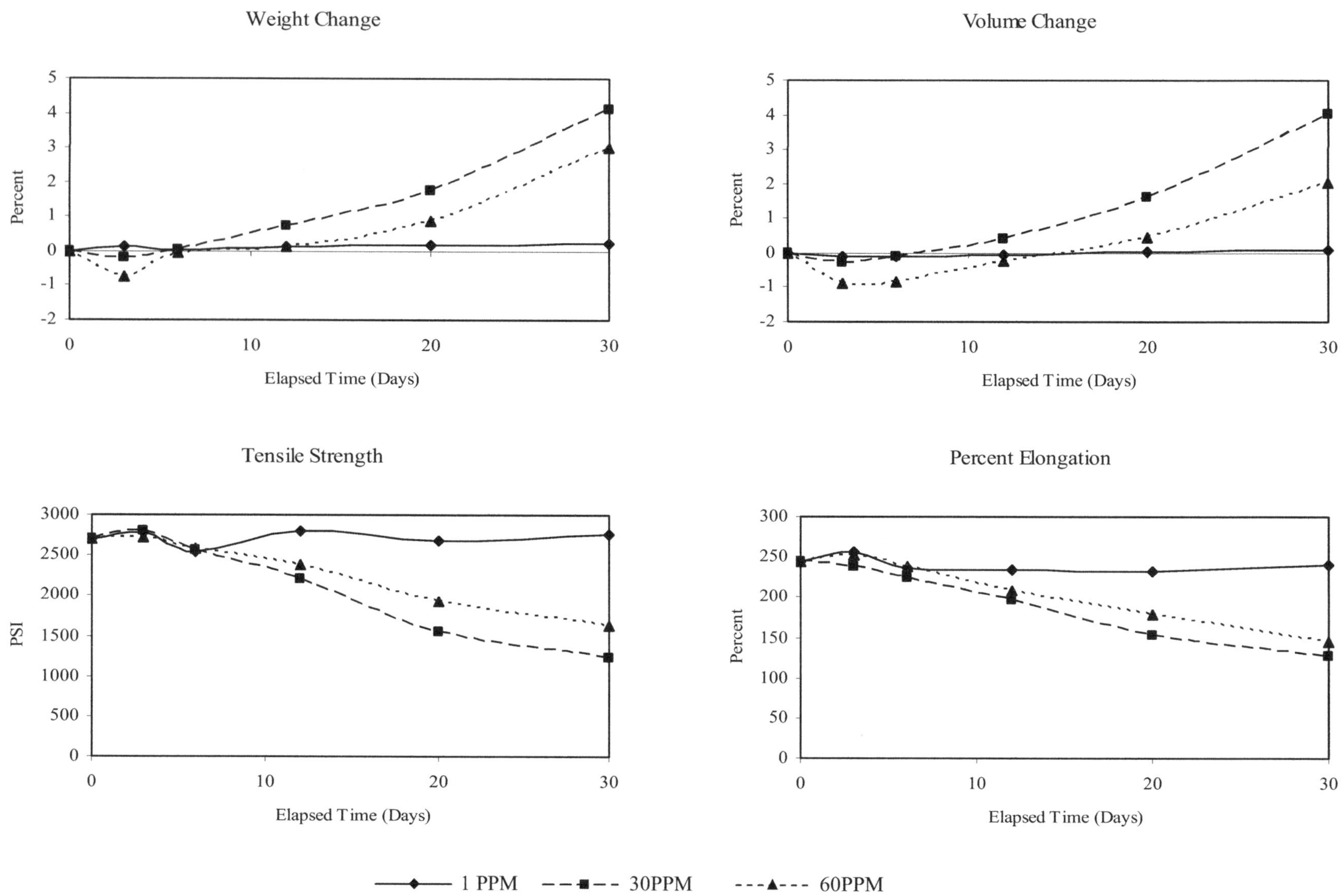

Figure A.59 Concentration of chloramine effect on elastomeric properties at 45°C and pH 8.3 of EPDM-P

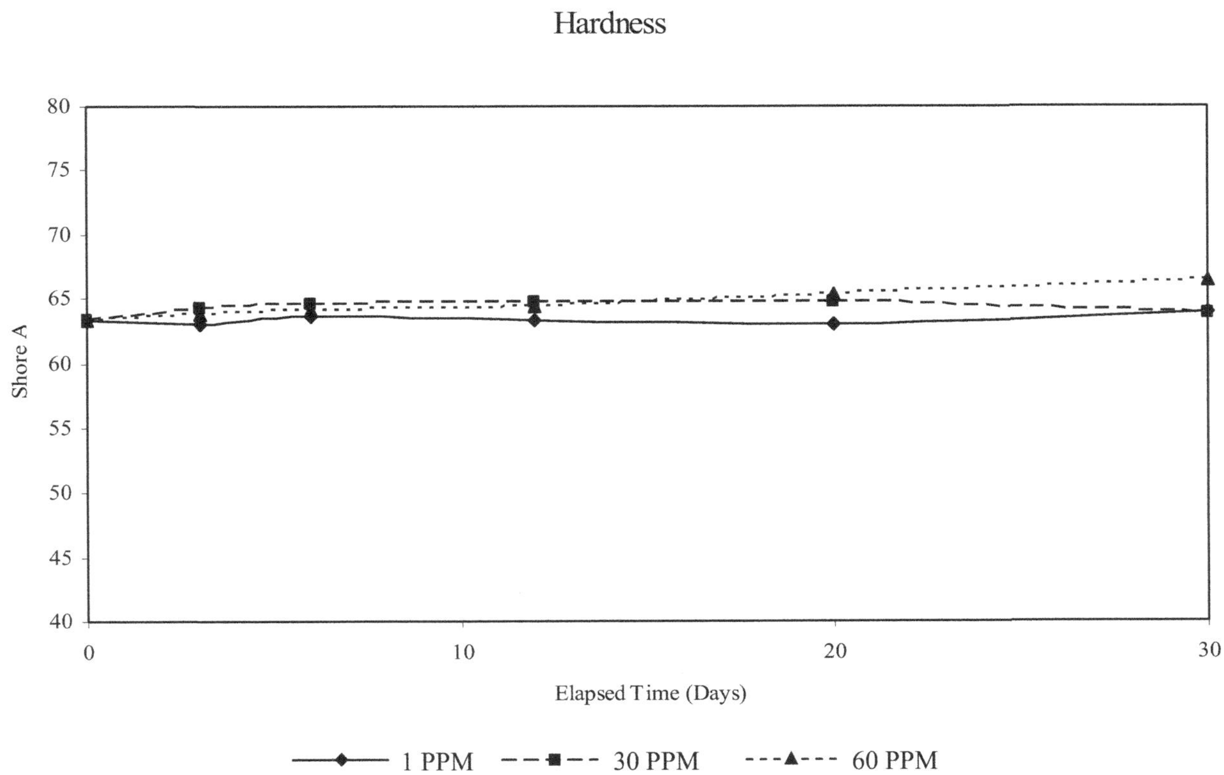

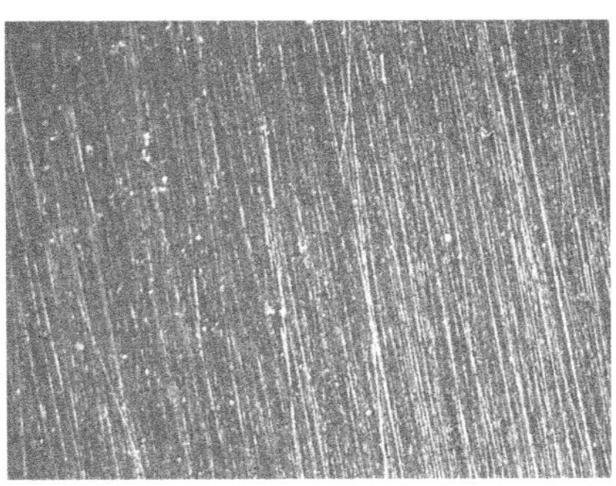

1 PPM at 30 days

30 PPM at 30 days

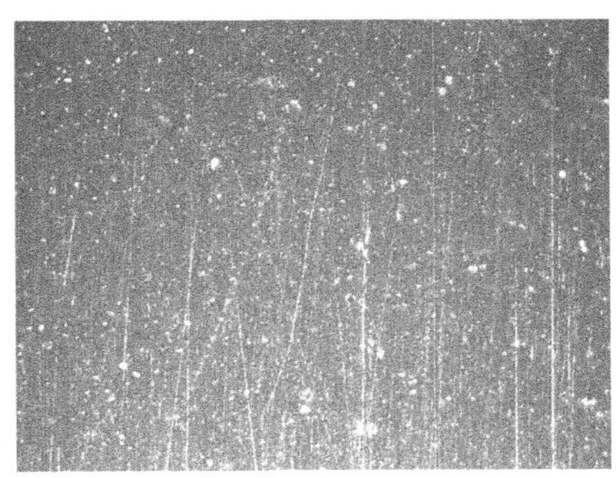

60 PPM at 30 days

Figure A.60 Concentration of chloramine effect on hardness at 45°C and pH 8.3 of EPDM-P with 40× magnified pictographs

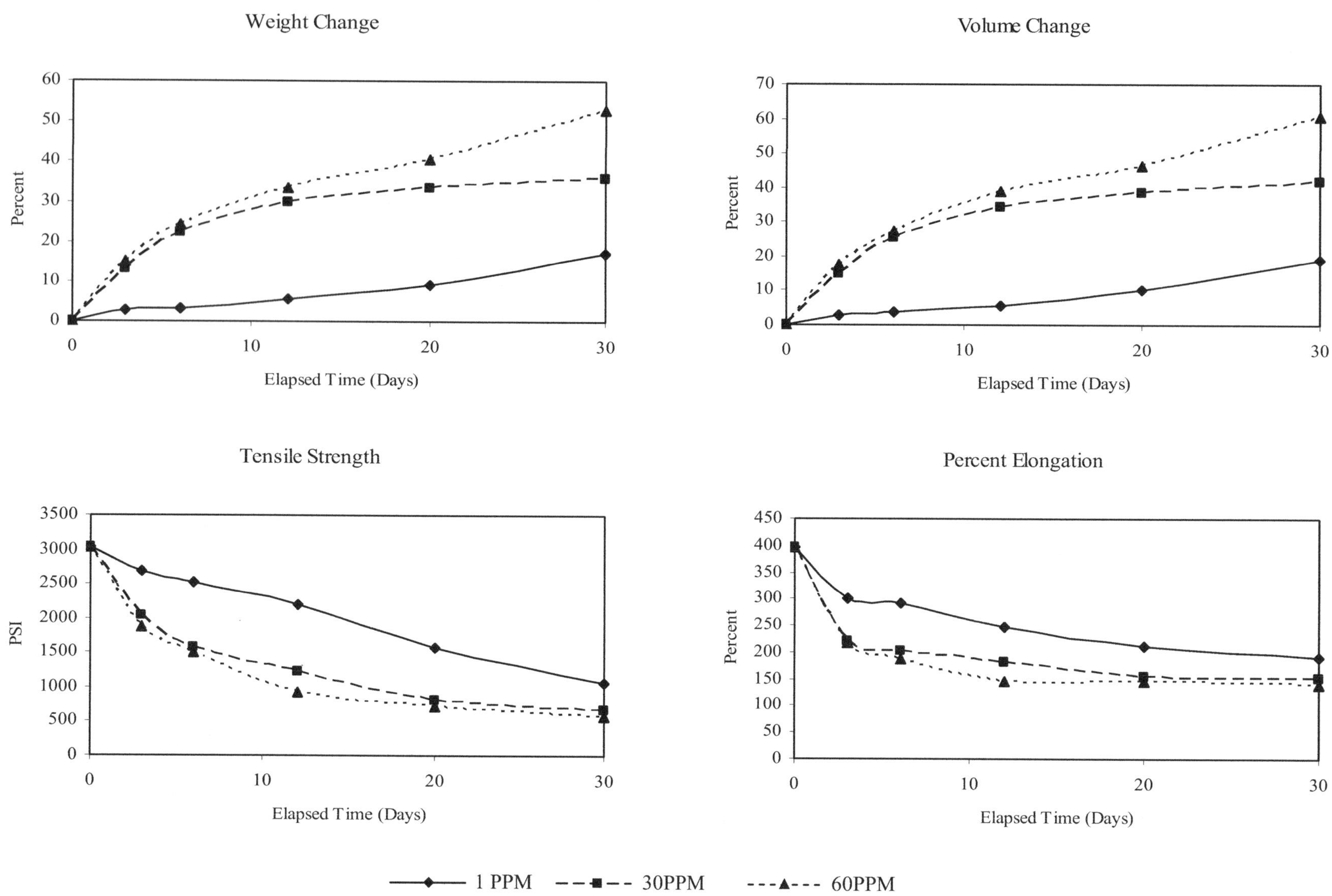

Figure A.61 Concentration of chloramine effect on elastomeric properties at 70°C and pH 8.3 of natural rubber

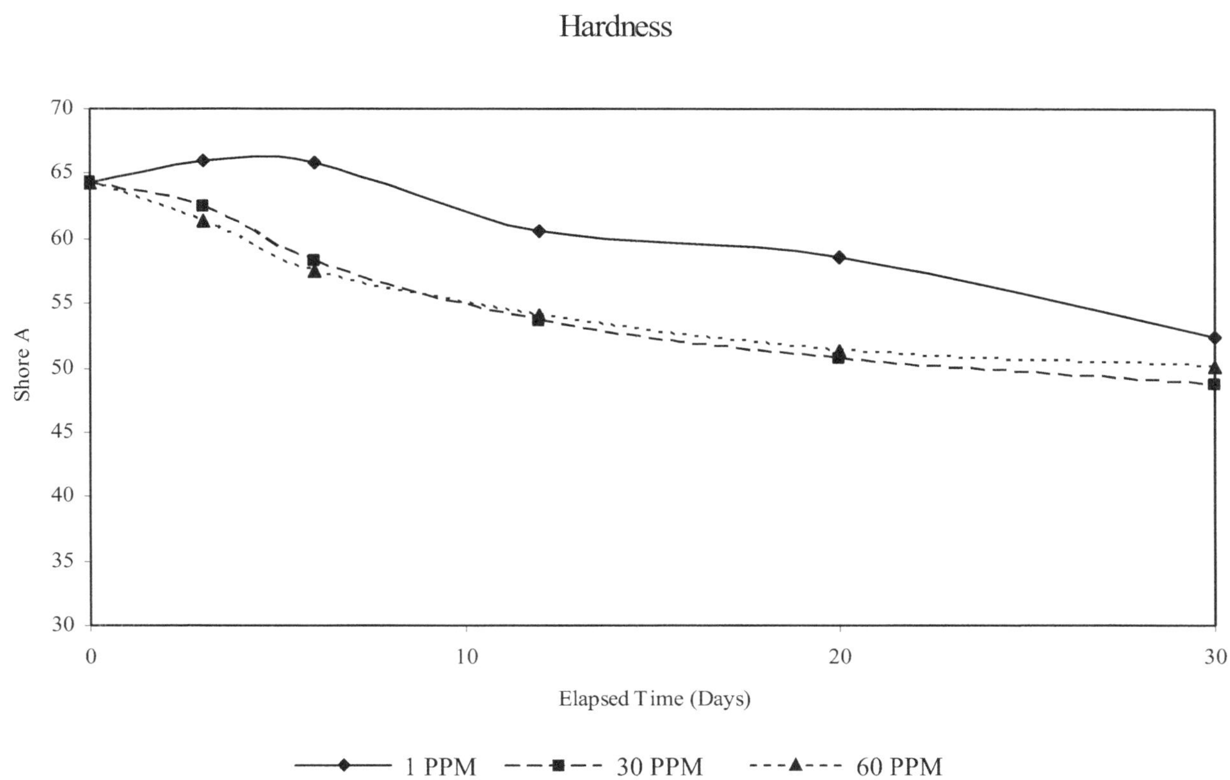

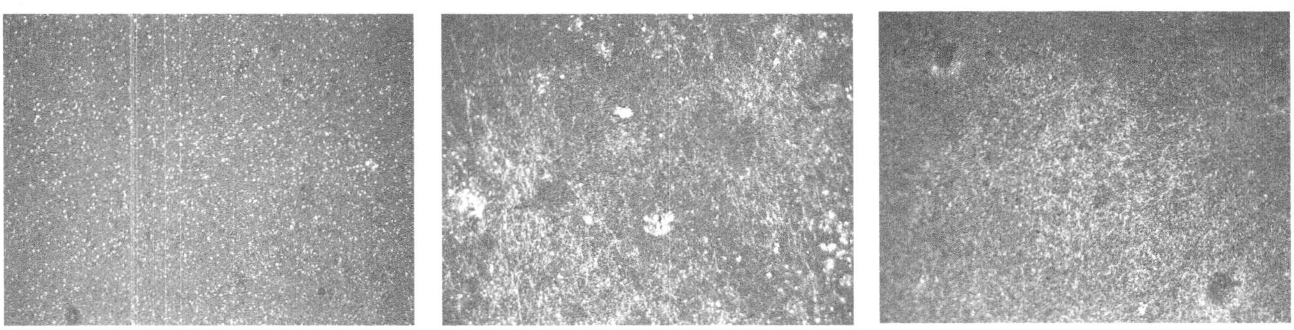

1 PPM at 30 days

30 PPM at 30 days

60 PPM at 30 days

Figure A.62 Concentration of chloramine effect on hardness at 70°C and pH 8.3 of natural rubber with 40× magnified pictographs

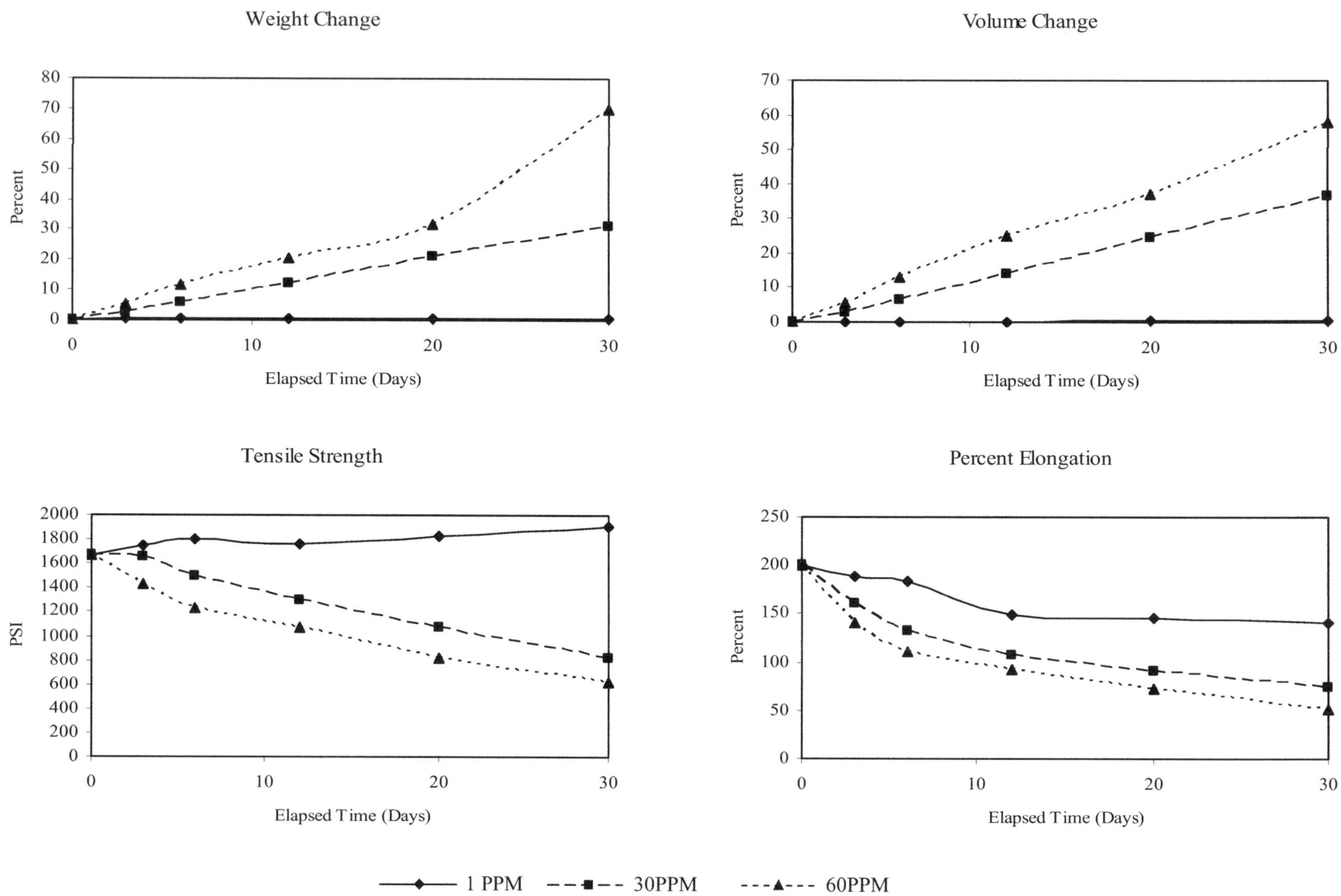

Figure A.63 Concentration of chloramine effect on elastomeric properties at 70°C and pH 8.3 of EPDM-S

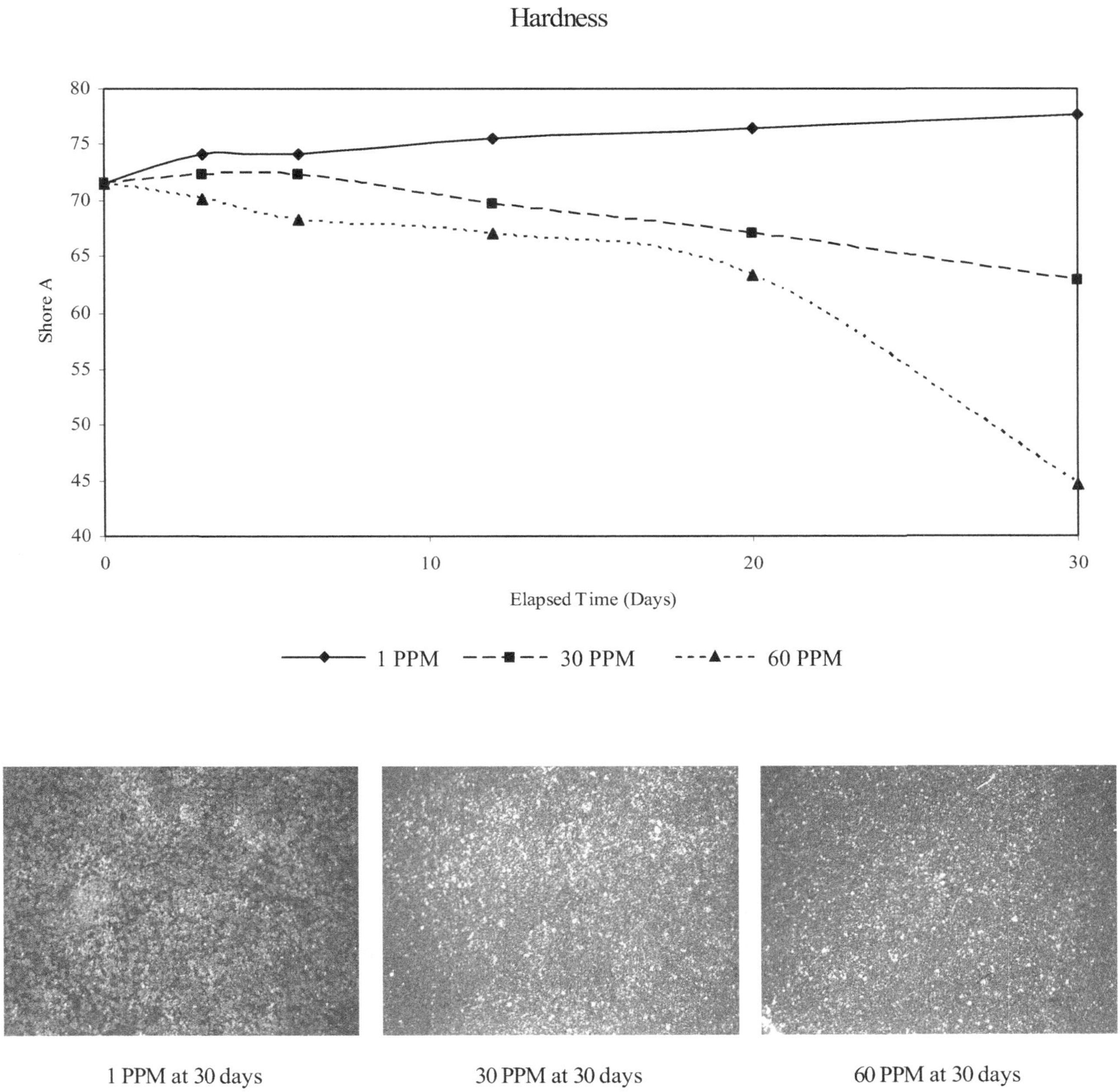

Figure A.64 Concentration of chloramine effect on hardness at 70°C and pH 8.3 of EPDM-S with 40× magnified pictographs

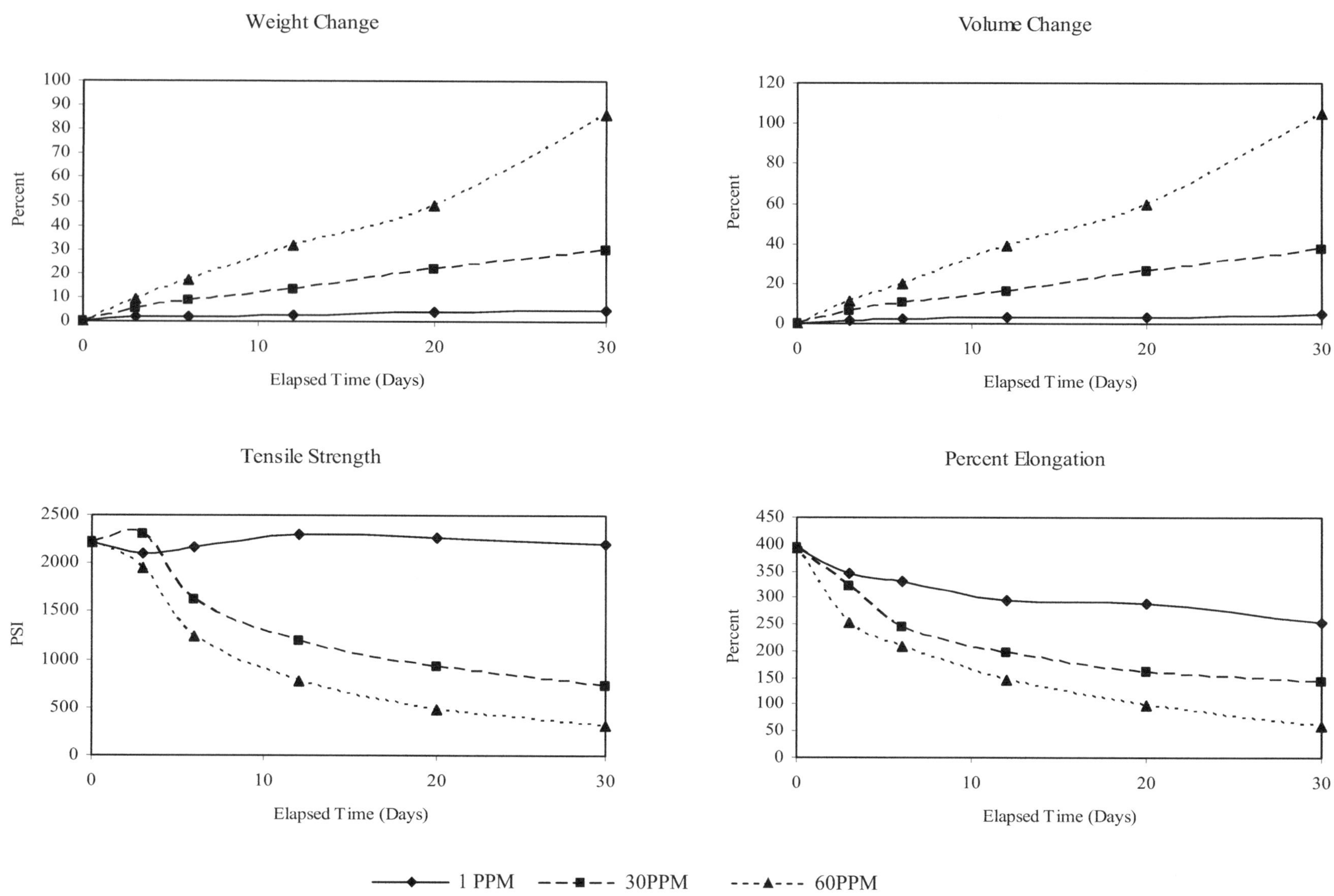

Figure A.65 Concentration of chloramine effect on elastomeric properties at 70°C and pH 8.3 of nitrile (NBR)

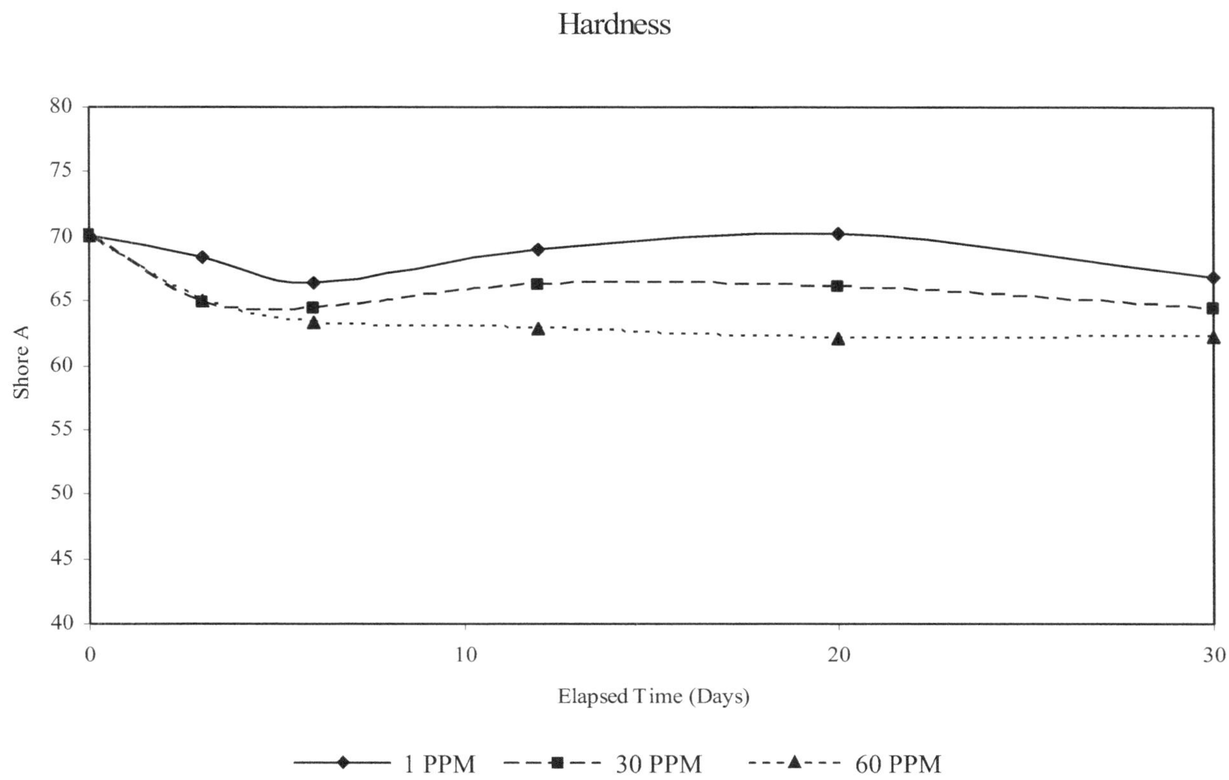

1 PPM at 30 days

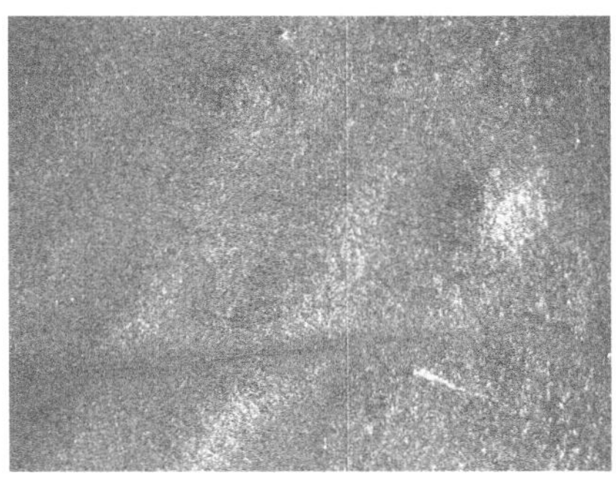

30 PPM at 30 days

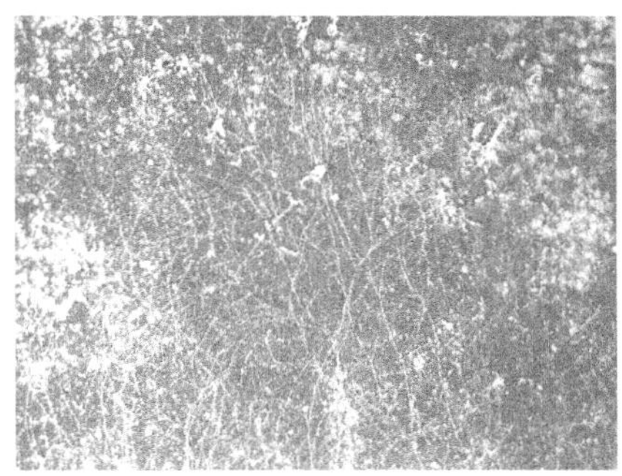

60 PPM at 30 days

Figure A.66 Concentration of chloramine effect on hardness at 70°C and pH 8.3 of nitrile (NBR) with 40× magnified pictographs

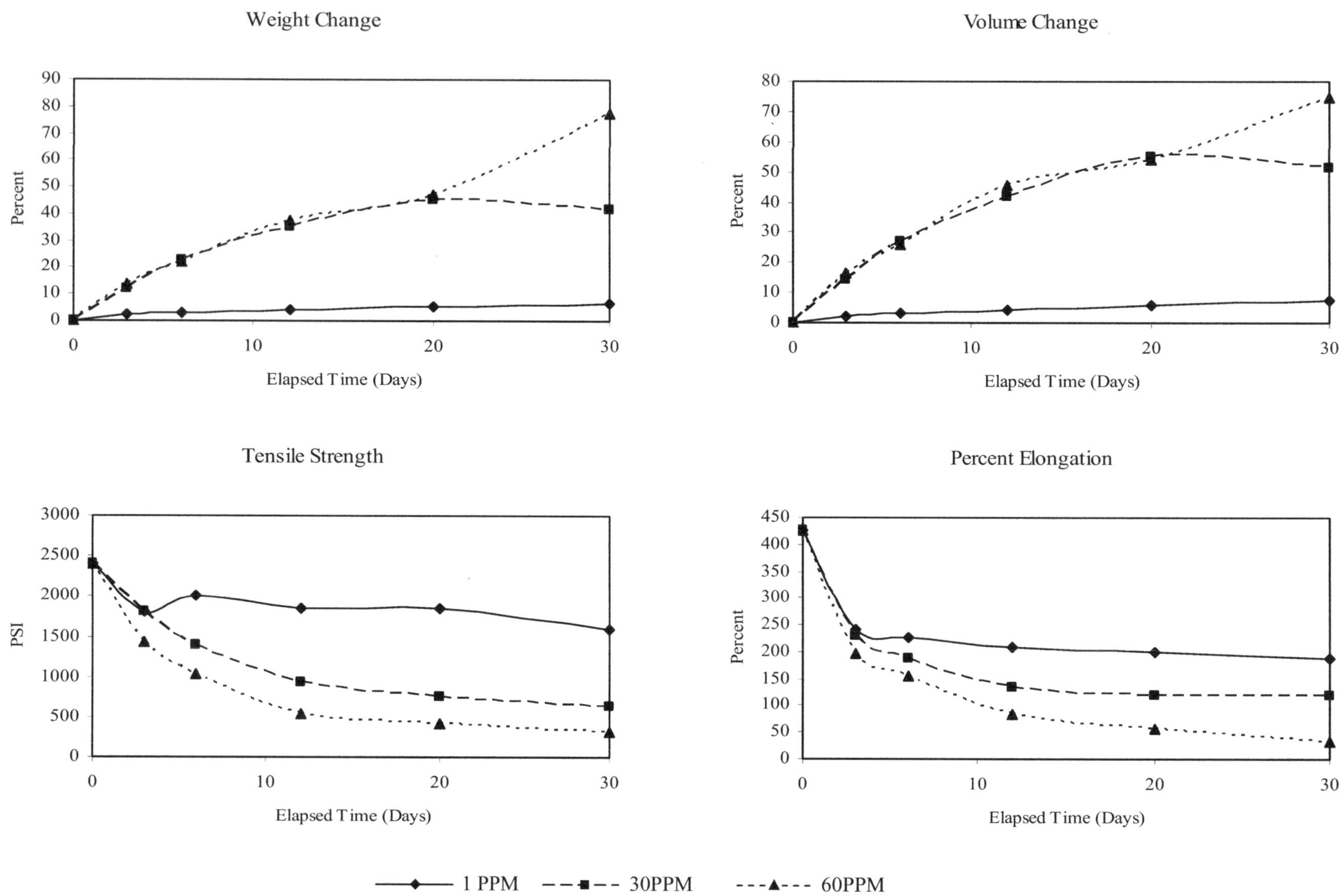

Figure A.67 Concentration of chloramine effect on elastomeric properties at 70°C and pH 8.3 of SBR

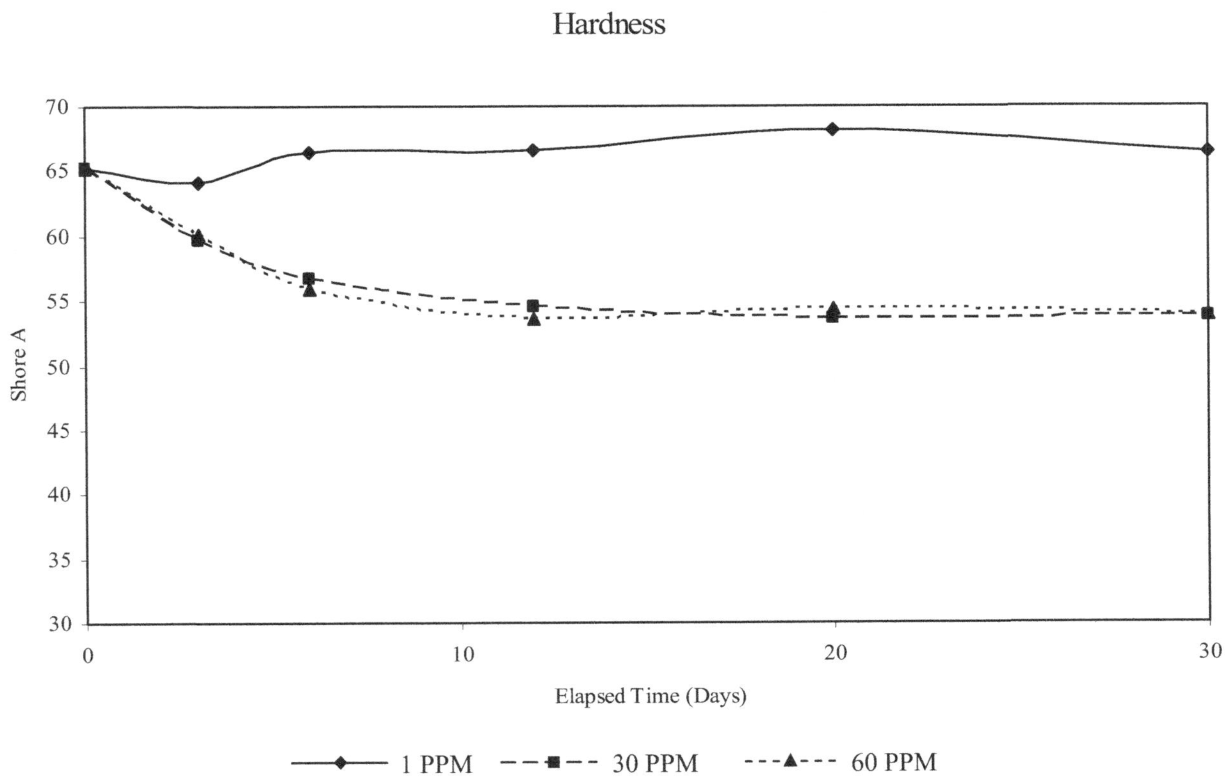

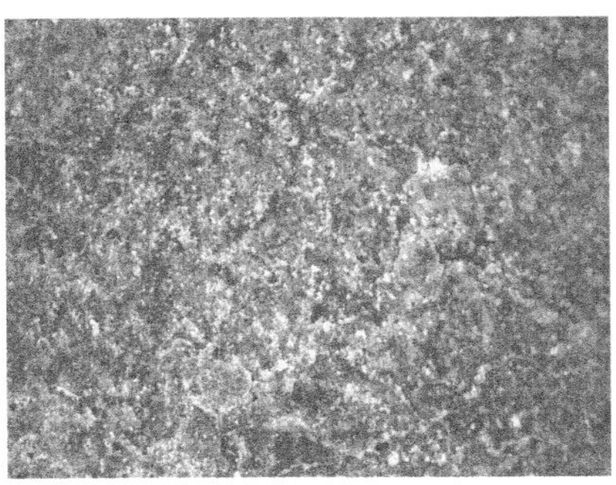

1 PPM at 30 days

30 PPM at 30 days

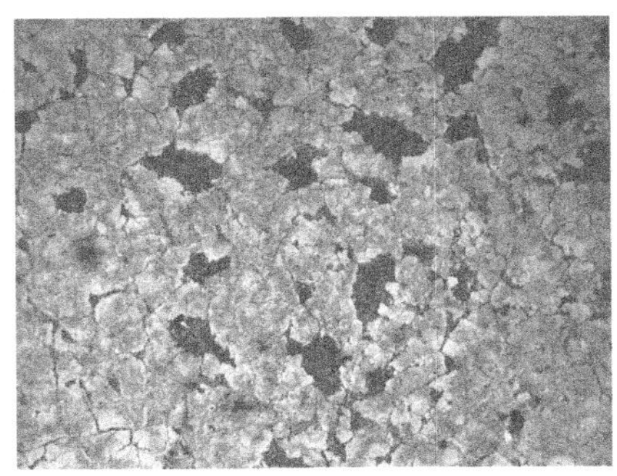

60 PPM at 30 days

Figure A.68 Concentration of chloramine effect on hardness at 70°C and pH 8.3 of SBR with 40× magnified pictographs

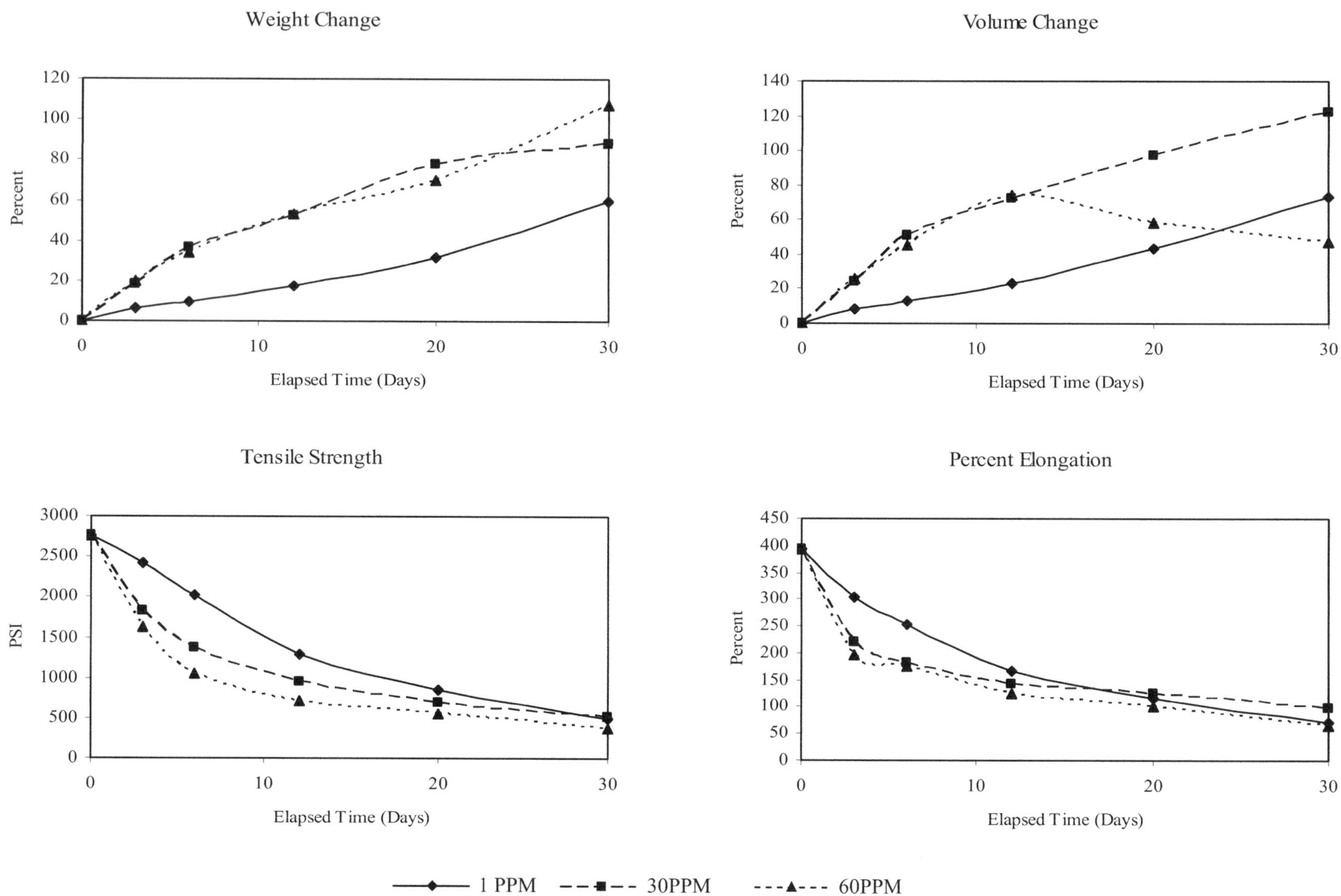

Figure A.69 Concentration of chloramine effect on elastomeric properties at 70°C and pH 8.3 of neoprene

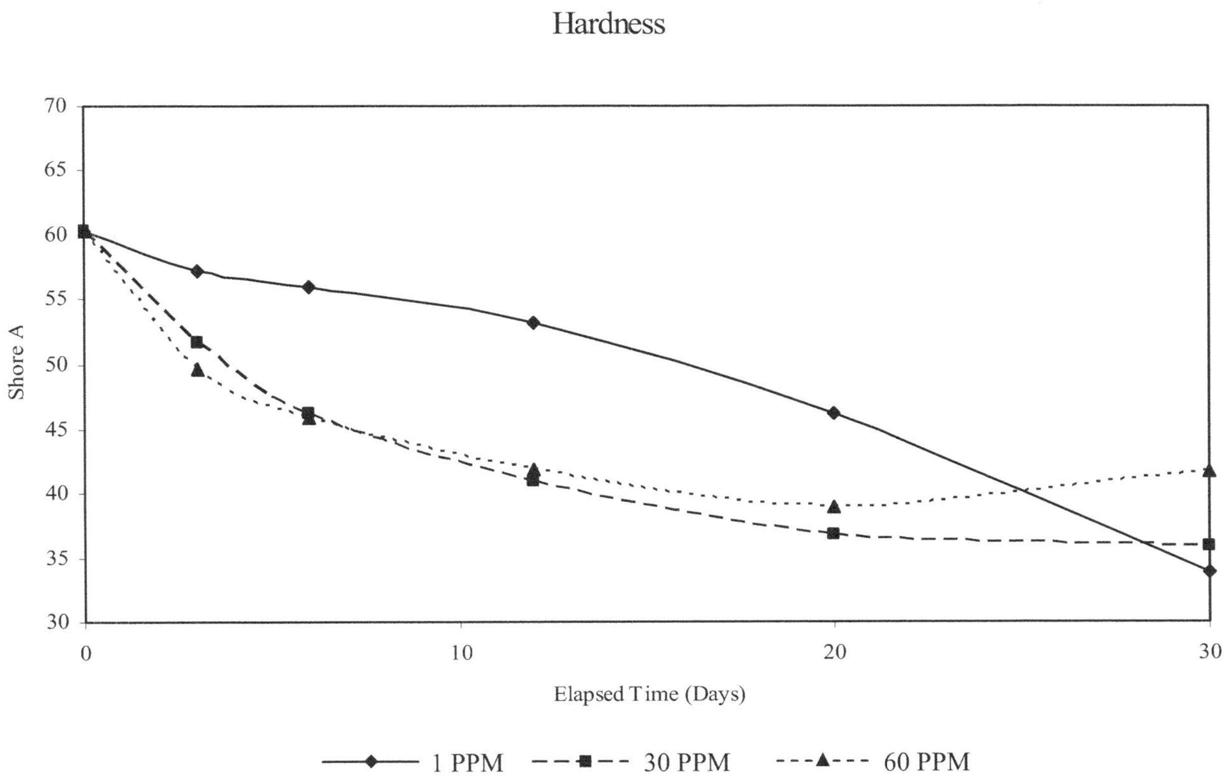

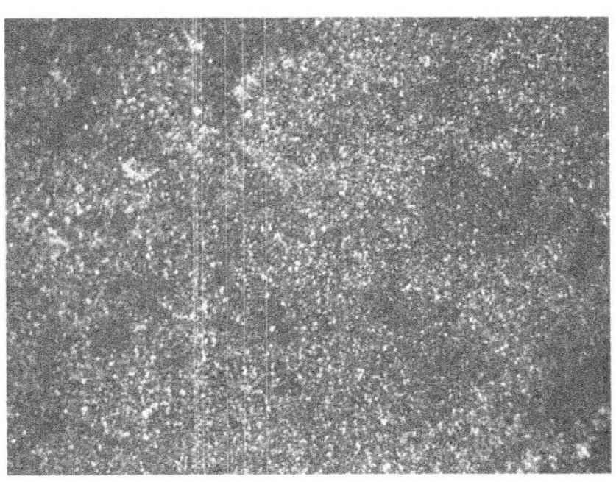

1 PPM at 30 days

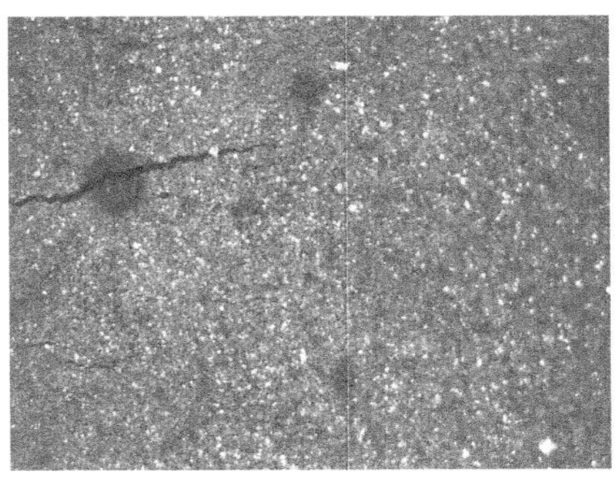

30 PPM at 30 days

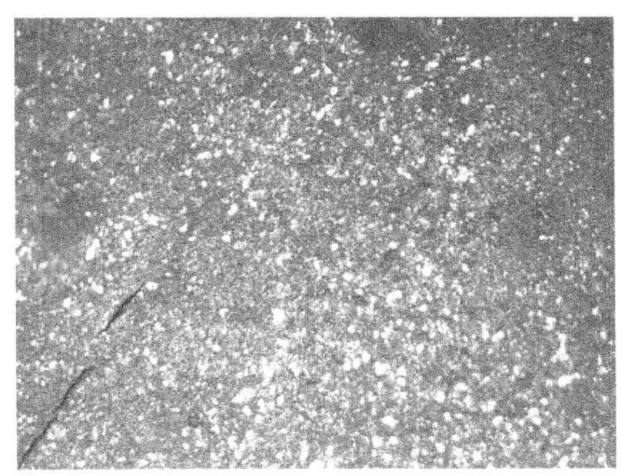

60 PPM at 30 days

Figure A.70 Concentration of chloramine effect on hardness at 70°C and pH 8.3 of neoprene with 40× magnified pictographs

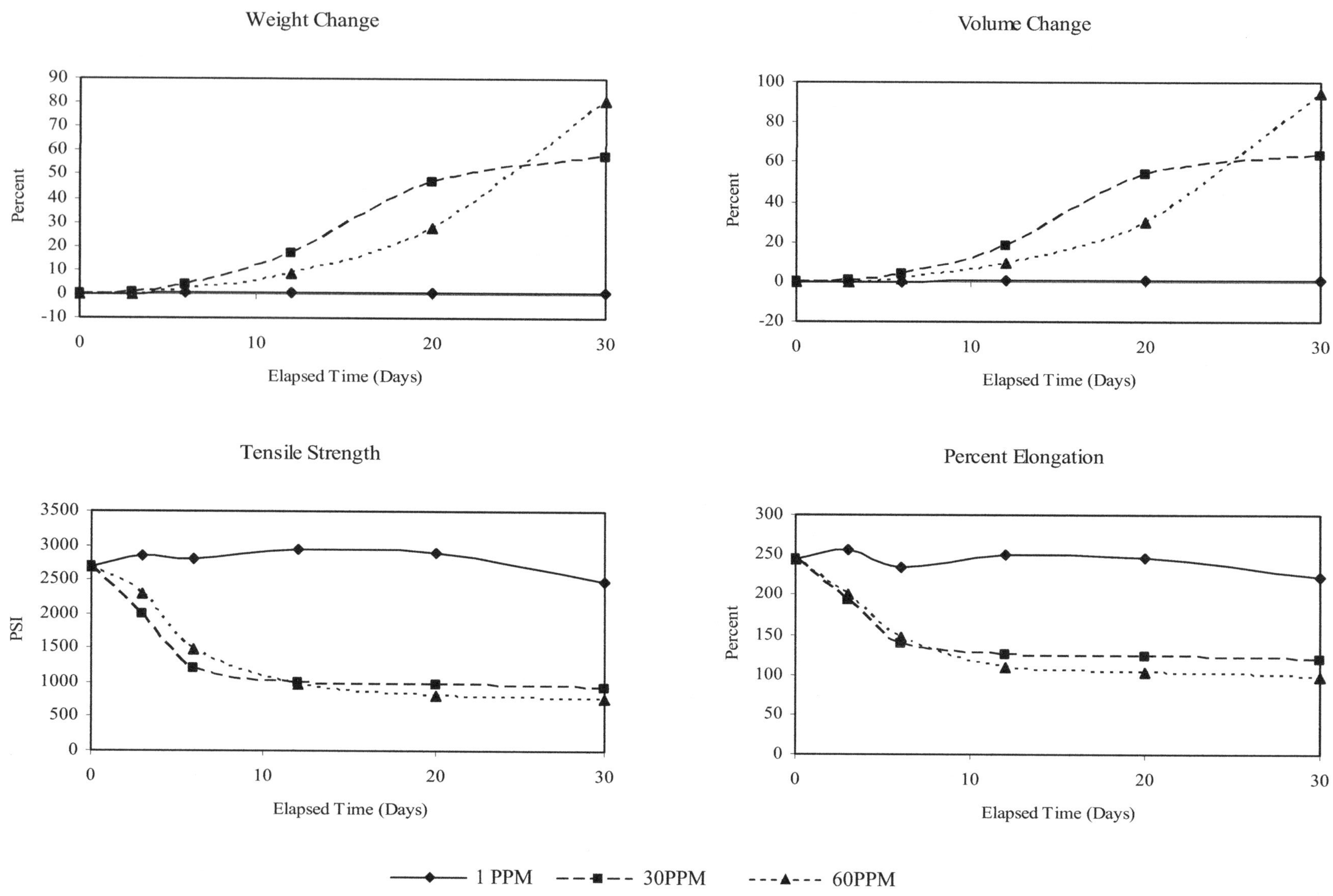

Figure A.71 Concentration of chloramine effect on elastomeric properties at 70°C and pH 8.3 of EPDM-P

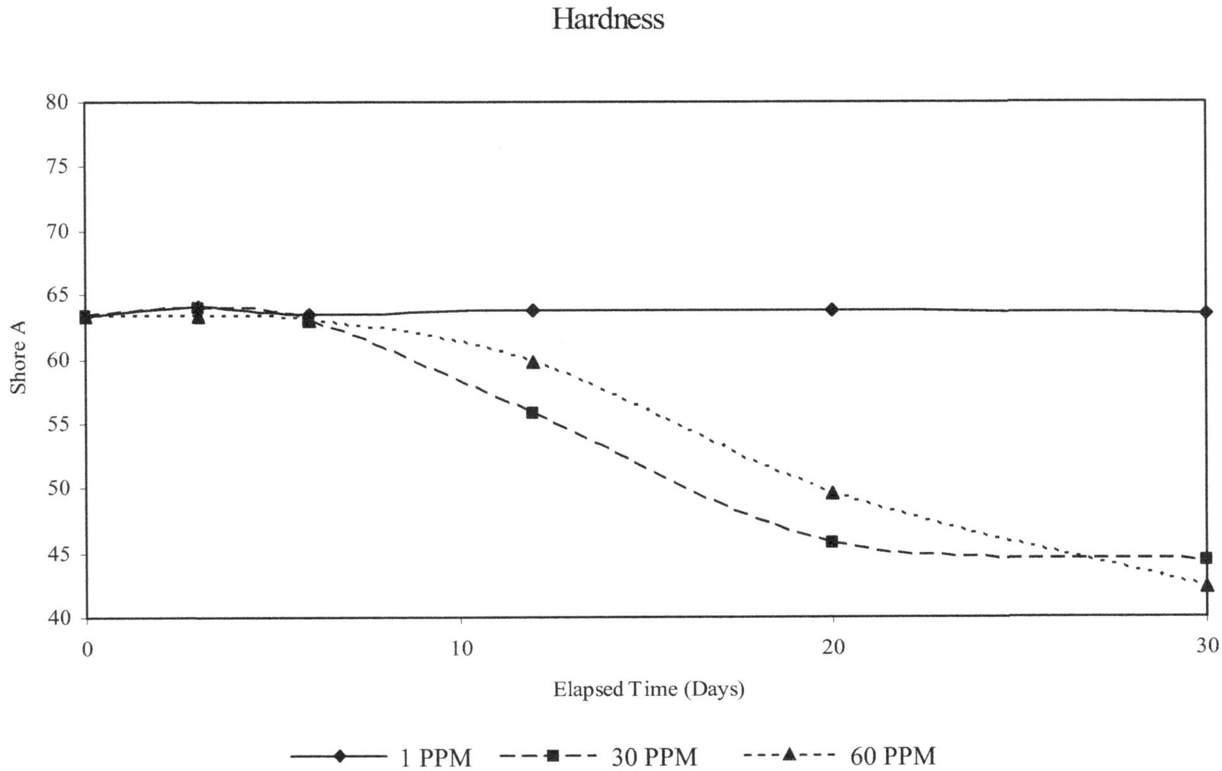

1 PPM at 30 days

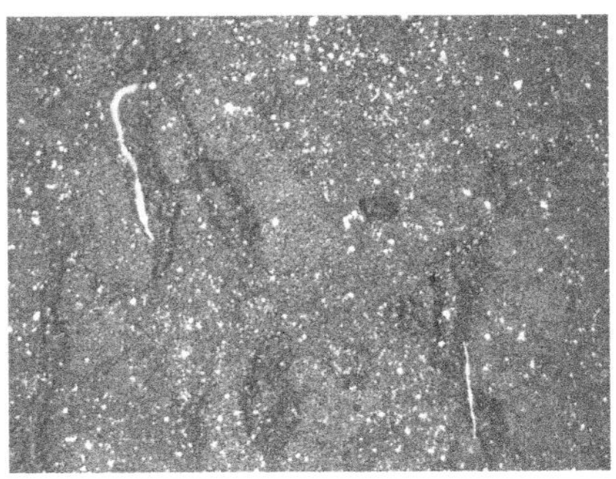

30 PPM at 30 days

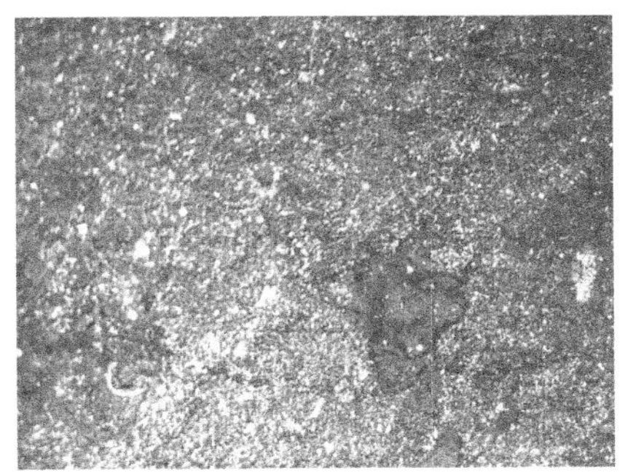

60 PPM at 30 days

Figure A.72 Concentration of chloramine effect on hardness at 70°C and pH 8.3 of EPDM-P with 40× magnified pictographs

PH EFFECTS

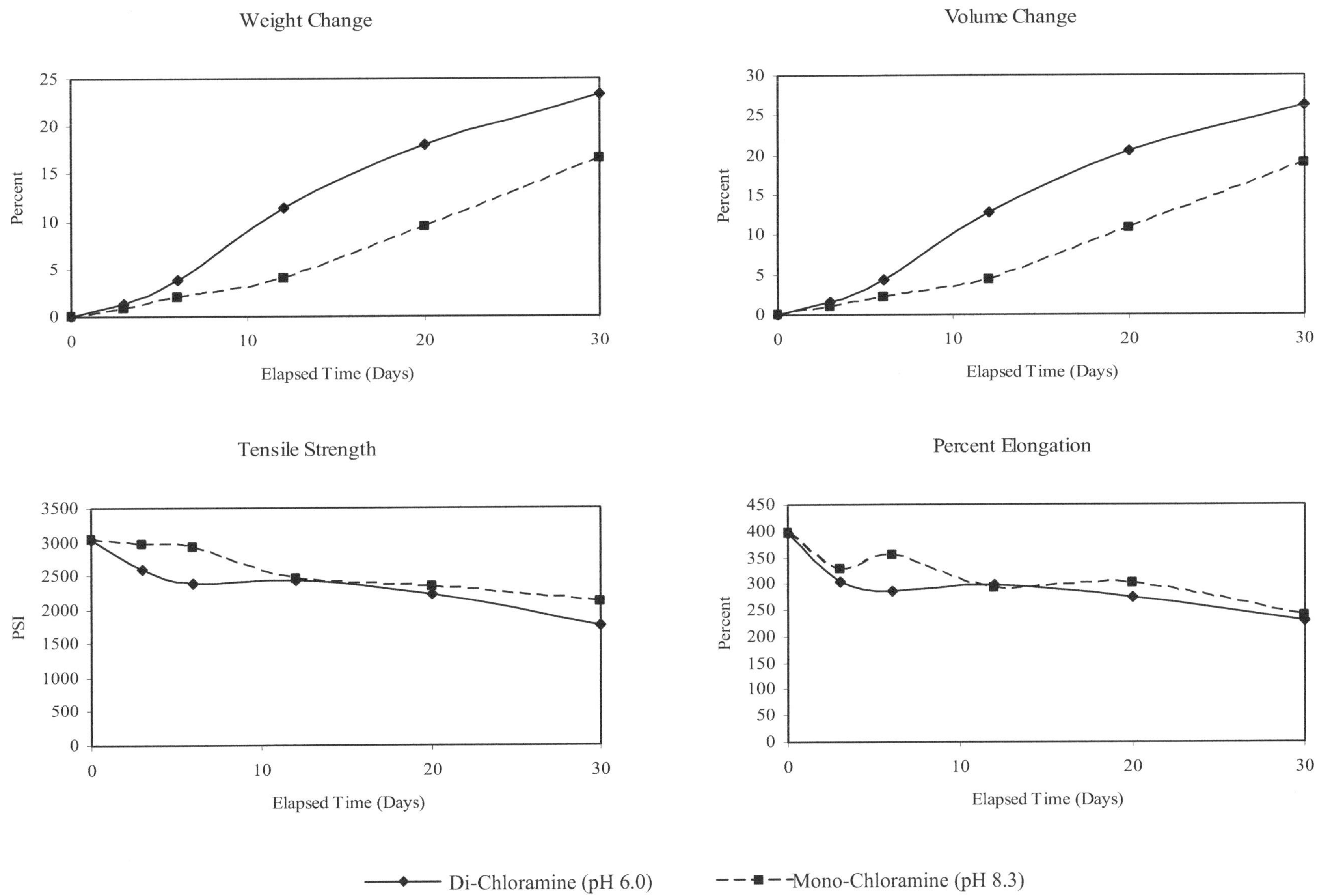

Figure A.73 pH effect on elastomeric properties at 23°C and 60 ppm chloramine of natural rubber

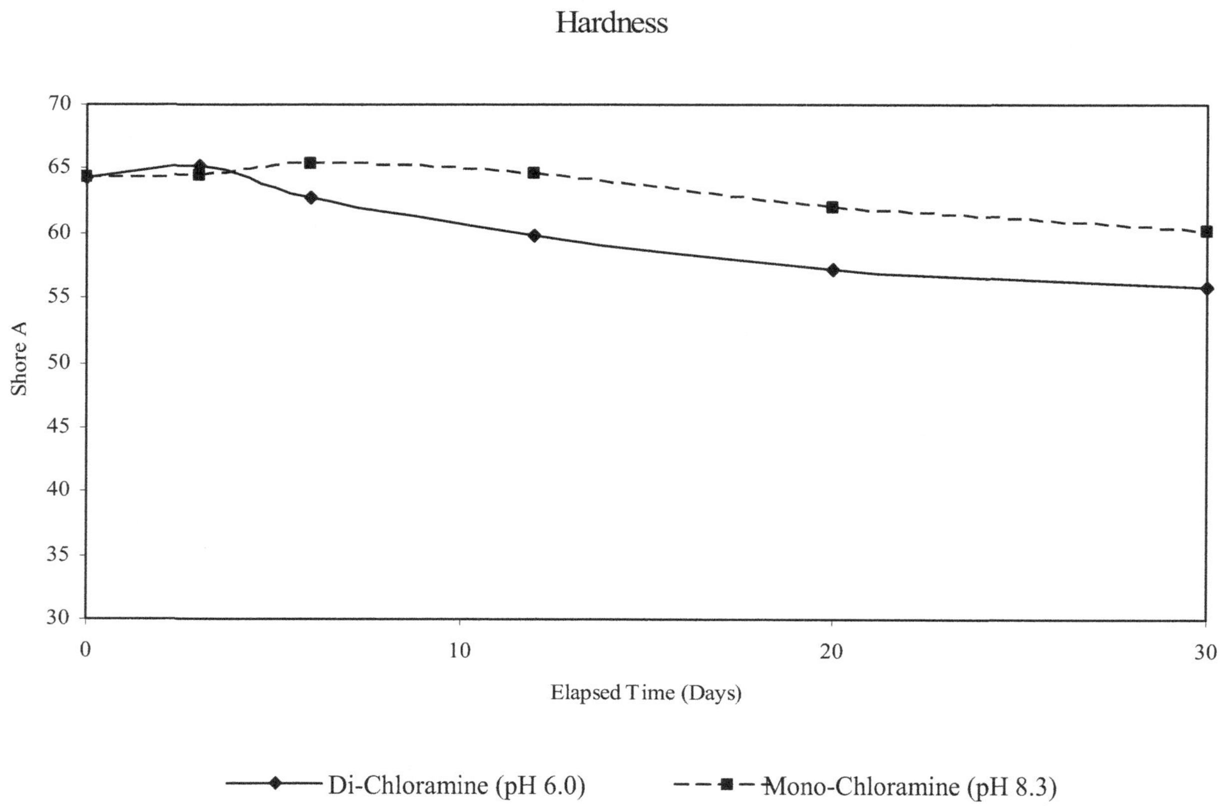

Di-Chloramine at 30 days

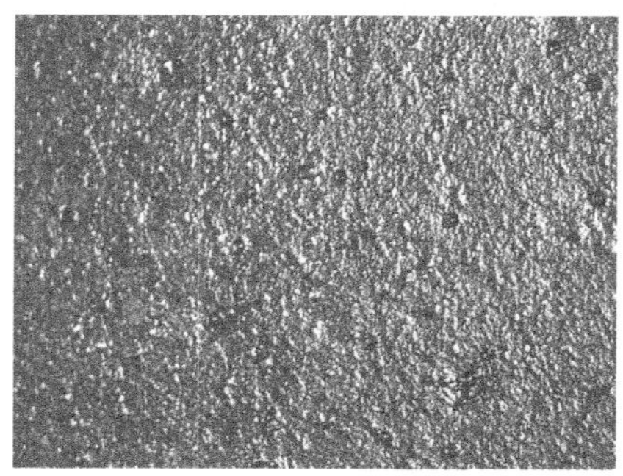

Mono-Chloramine at 30 days

Figure A.74 pH effect on hardness at 23°C and 60 ppm chloramine of natural rubber with 40× magnified pictographs

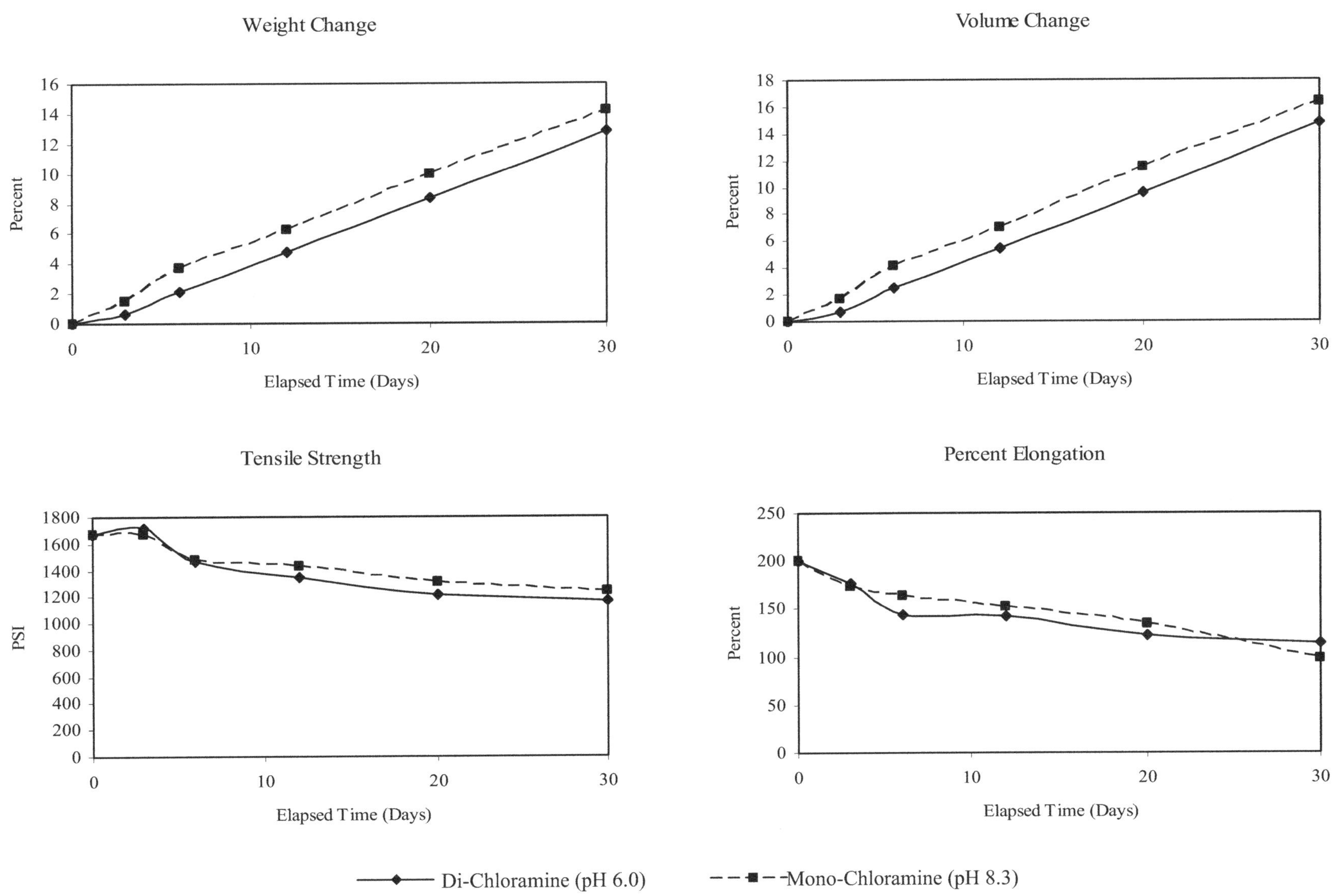

Figure A.75 pH effect on elastomeric properties at 23°C and 60 ppm chloramine of EPDM-S

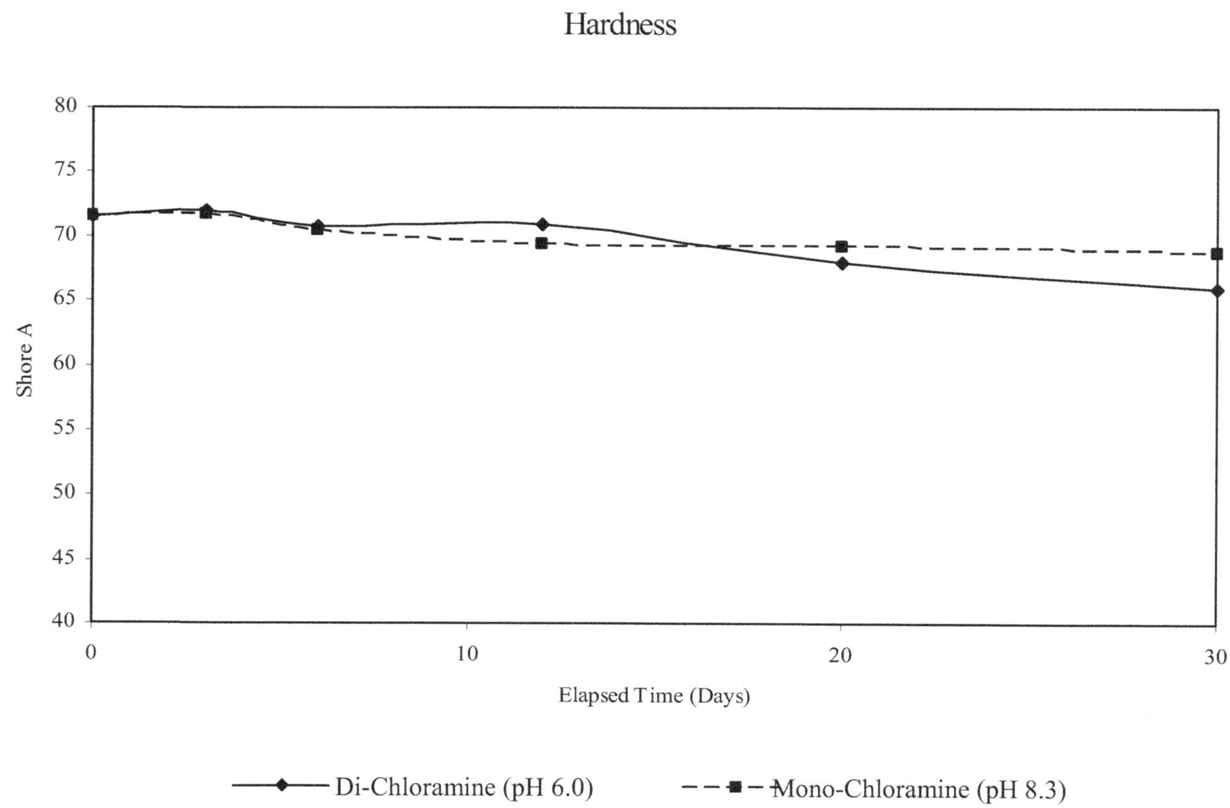

Di-Chloramine at 30 days

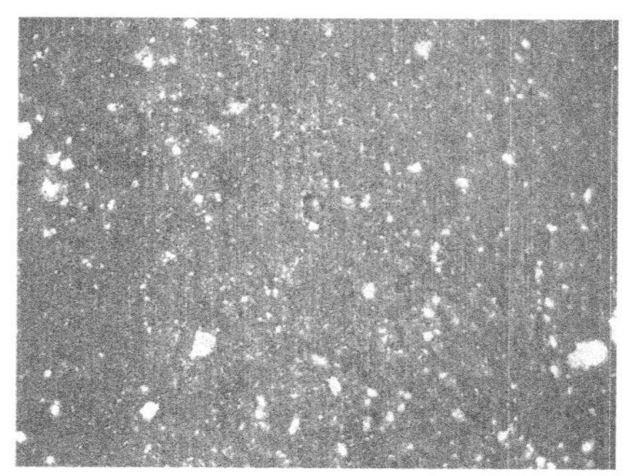

Mono-Chloramine at 30 days

Figure A.76 pH effect on hardness at 23°C and 60 ppm chloramine of EPDM-S with 40× magnified pictographs

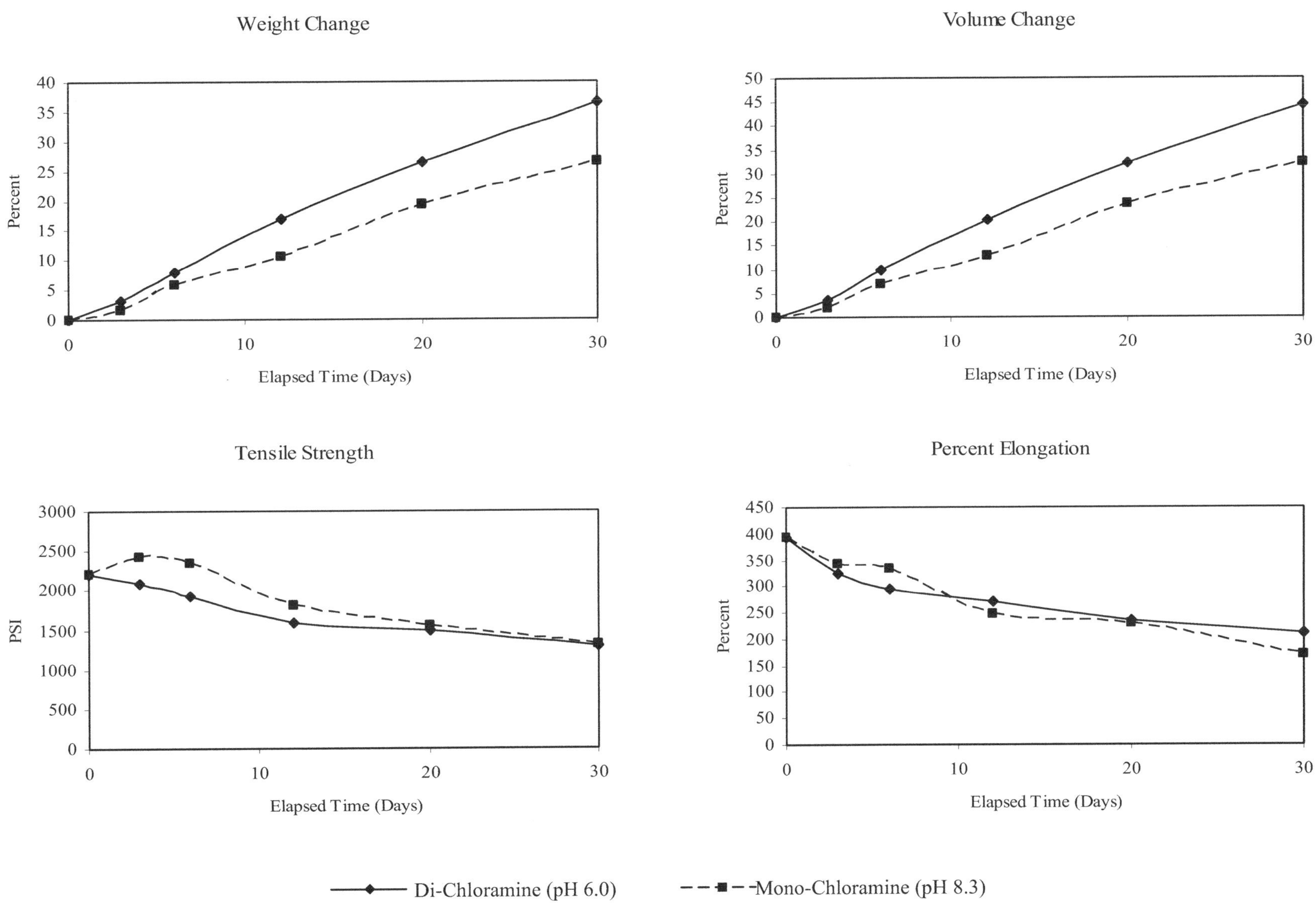

Figure A.77 pH effect on elastomeric properties at 23°C and 60 PPM chloramine of nitrile (NBR)

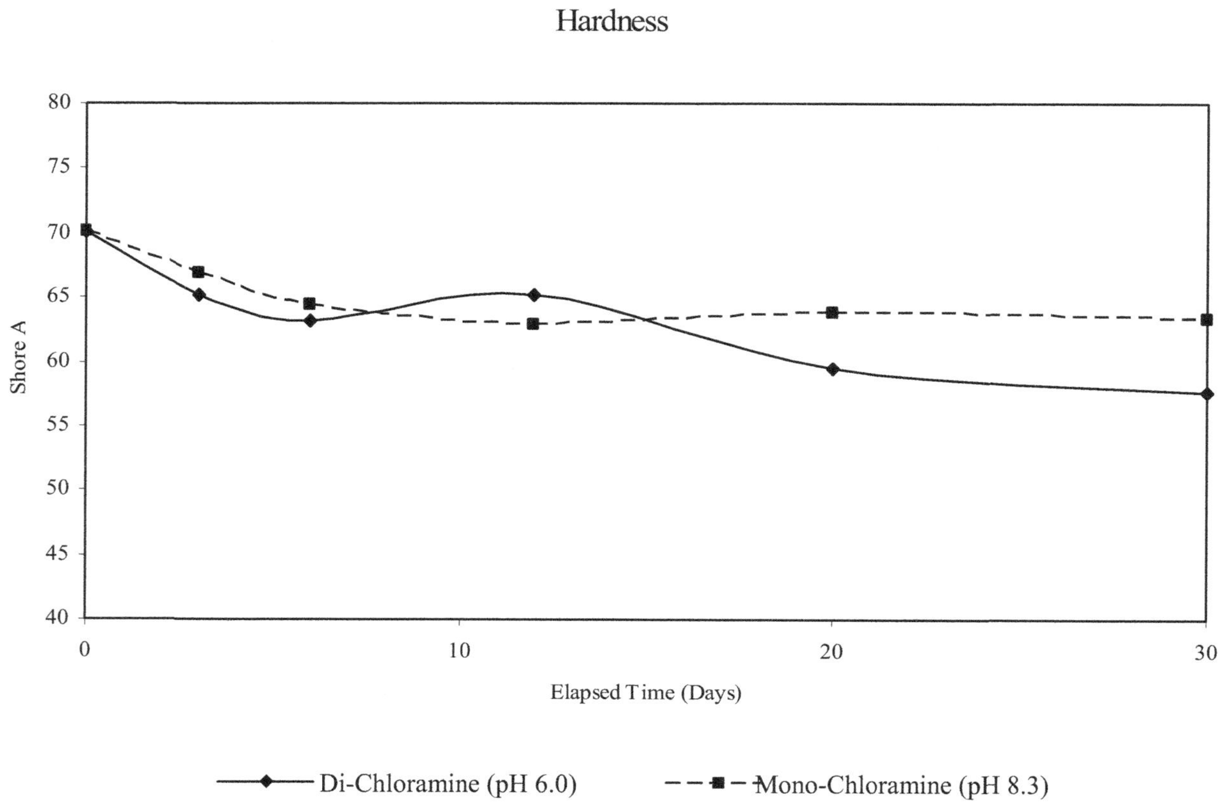

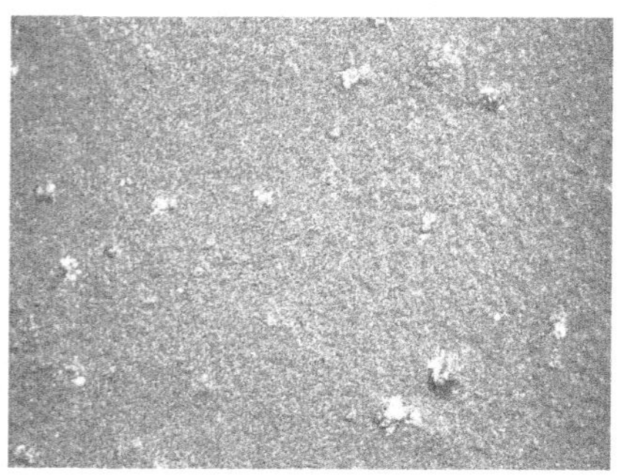

Di-Chloramine at 30 days

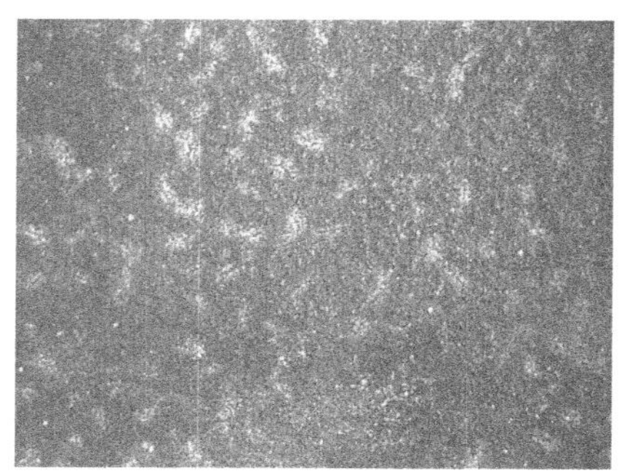

Mono-Chloramine at 30 days

Figure A.78 pH effect on hardness at 23°C and 60 ppm chloramine of nitrile (NBR) with 40× magnified pictographs

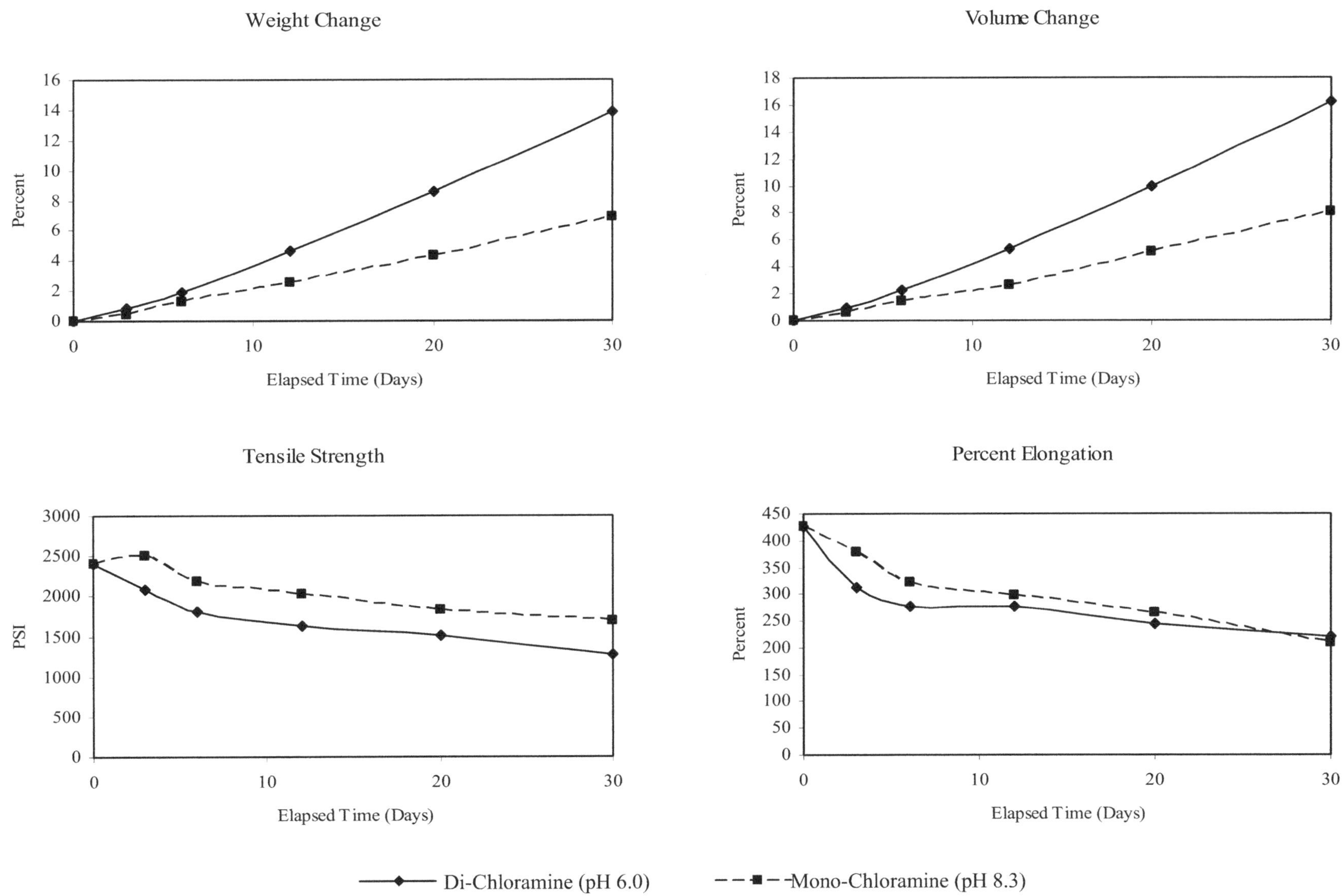

Figure A.79 pH effect on elastomeric properties at 23°C and 60 ppm chloramine of SBR

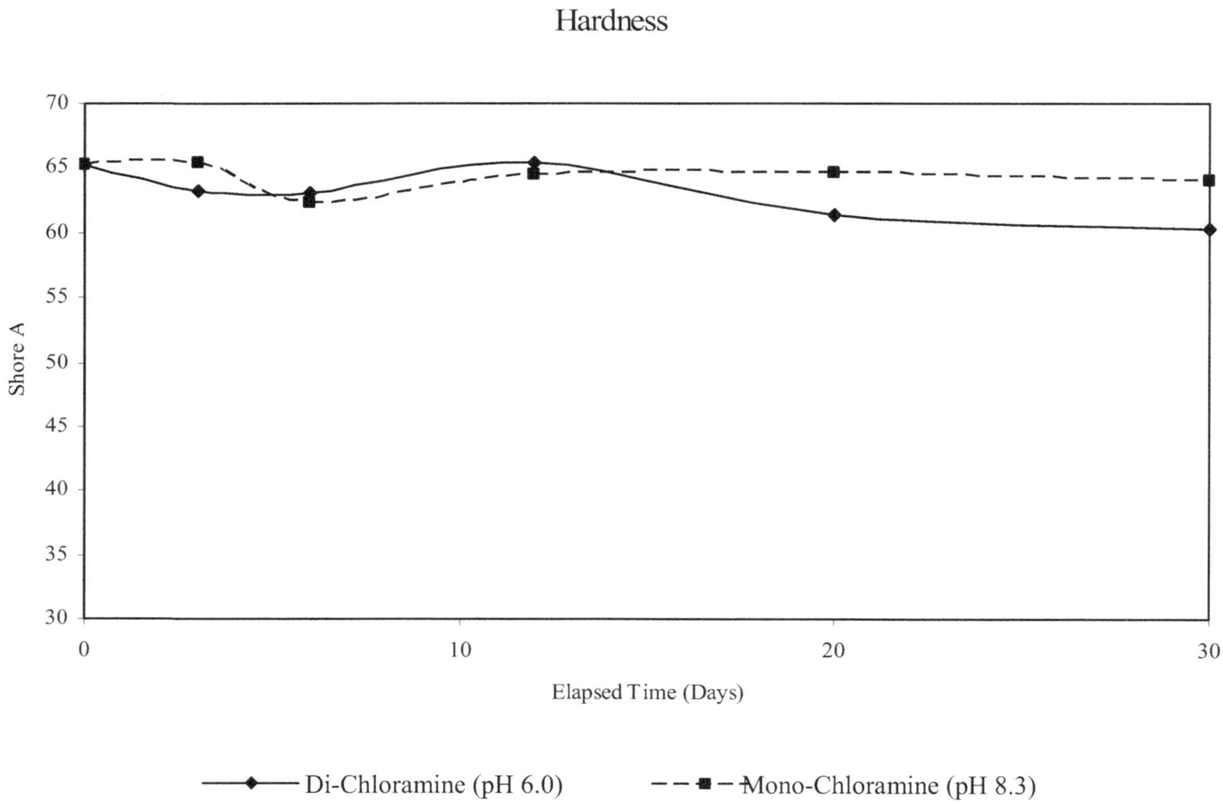

Di-Chloramine at 30 days

Mono-Chloramine at 30 days

Figure A.80 pH effect on hardness at 23°C and 60 ppm chloramine of SBR with 40× magnified pictographs

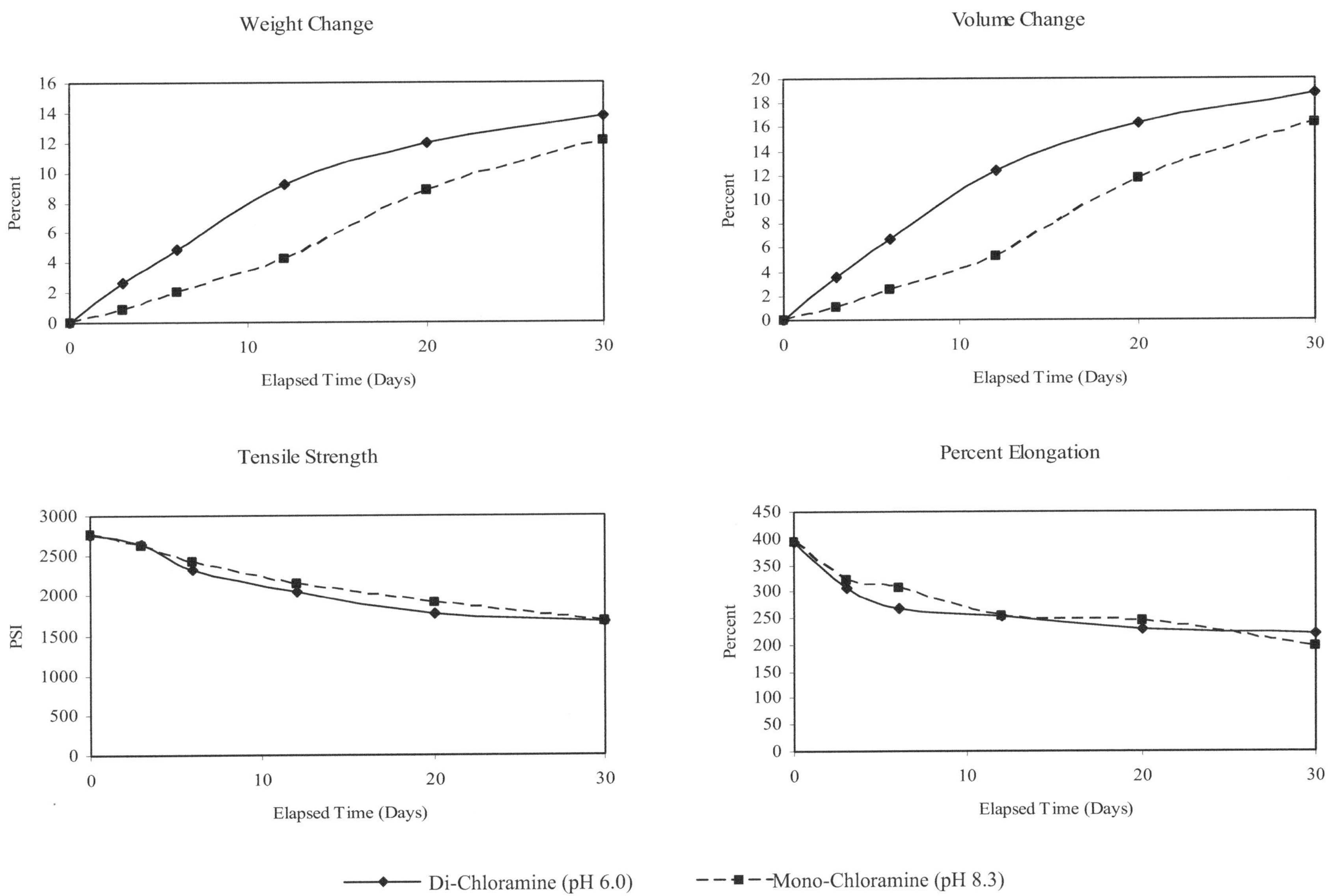

Figure A.81 pH effect on elastomeric properties at 23°C and 60 ppm chloramine of neoprene

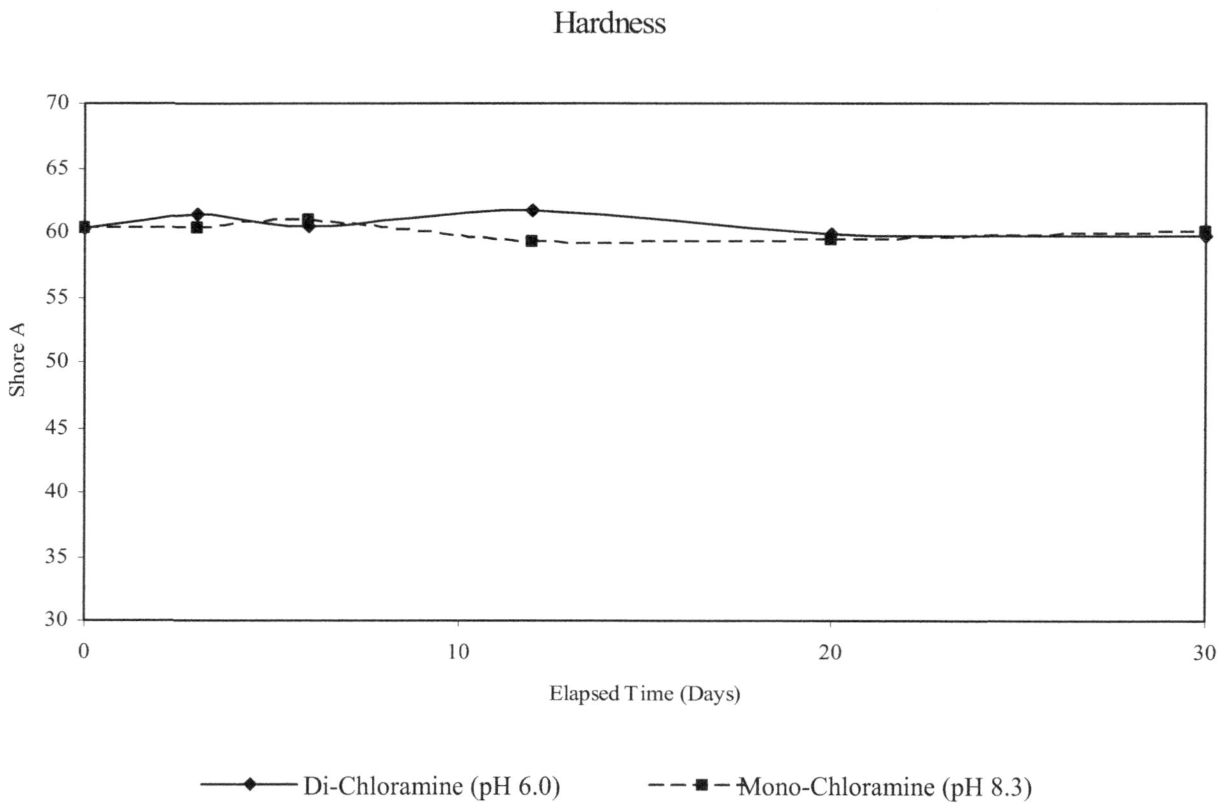

Di-Chloramine at 30 days

Mono-Chloramine at 30 days

Figure A.82 pH effect on hardness at 23°C and 60 ppm chloramine of neoprene with 40× magnified pictographs

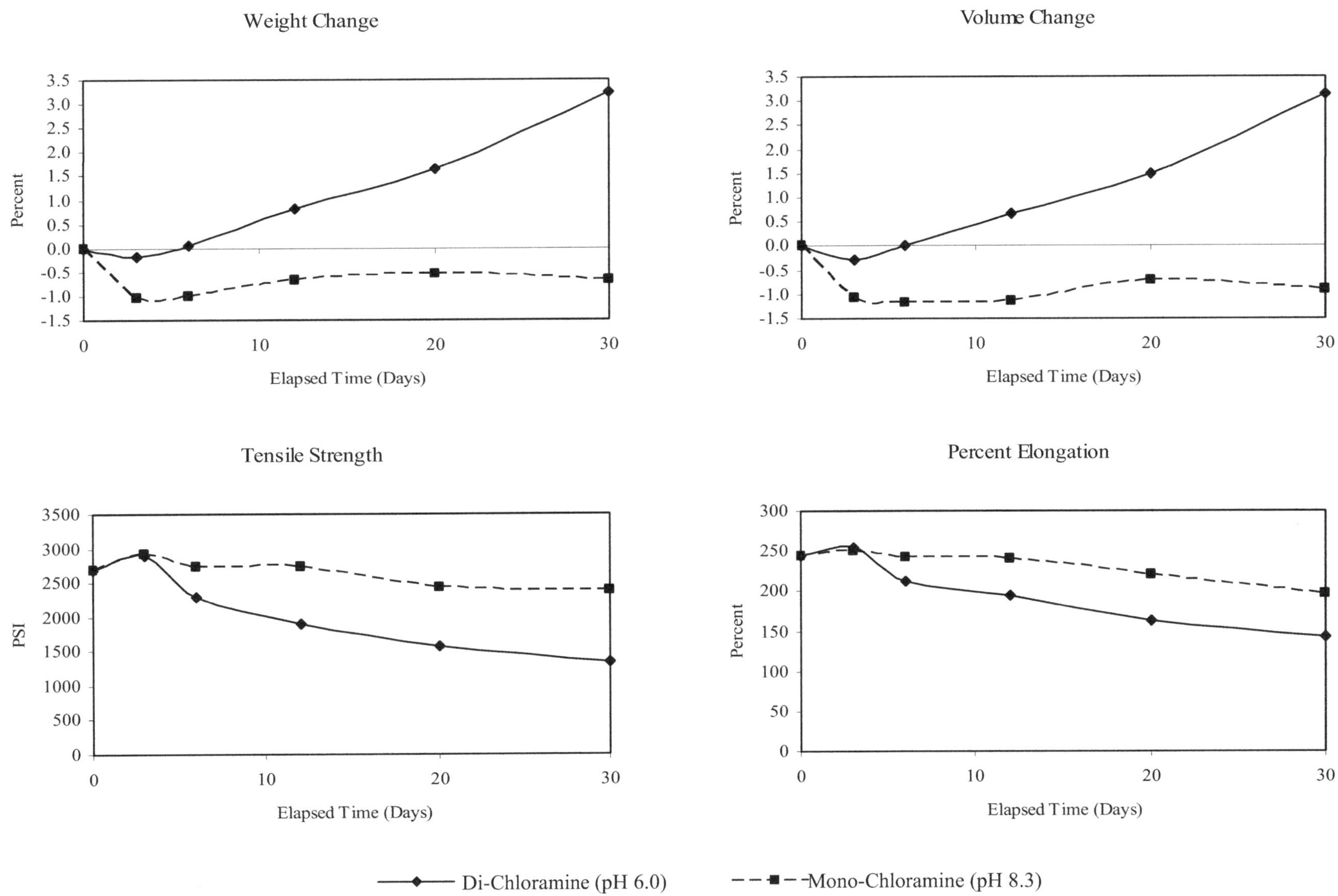

Figure A.83 pH effect on elastomeric properties at 23°C and 60 ppm chloramine of EPDM-P

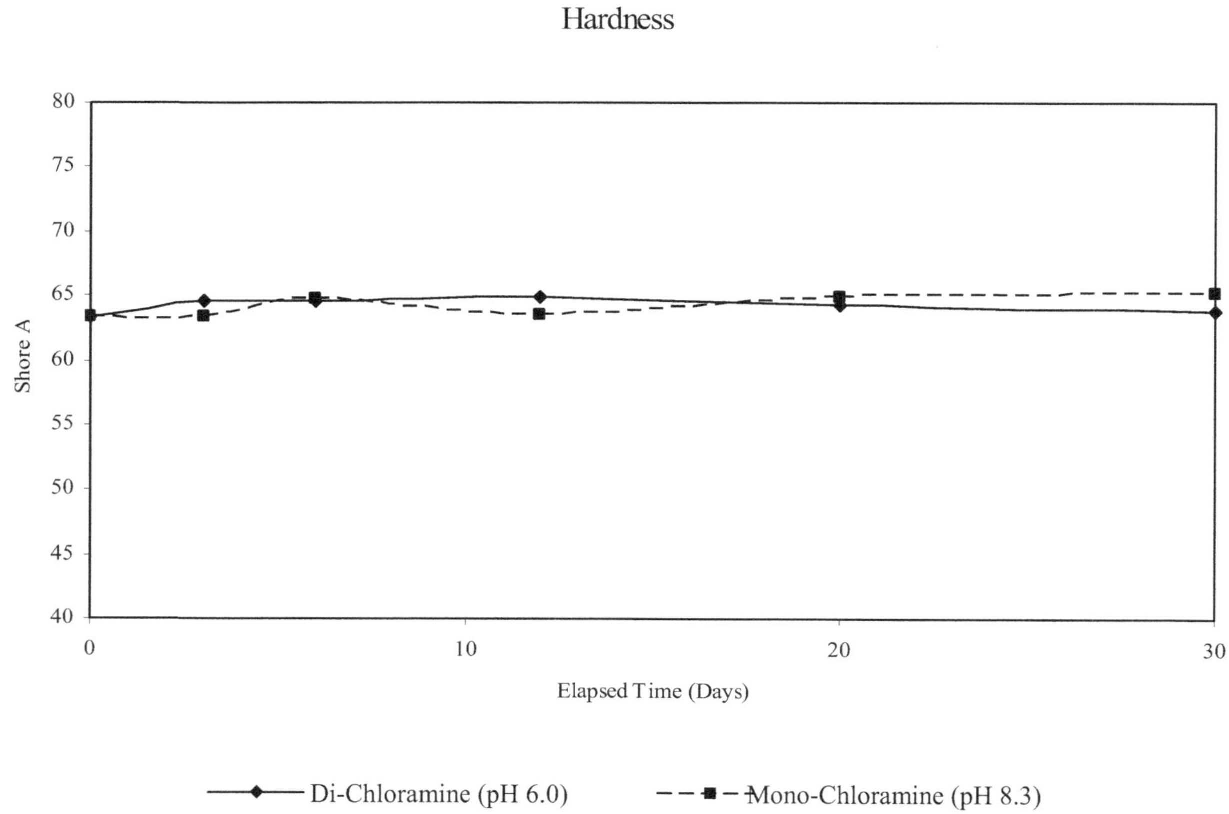

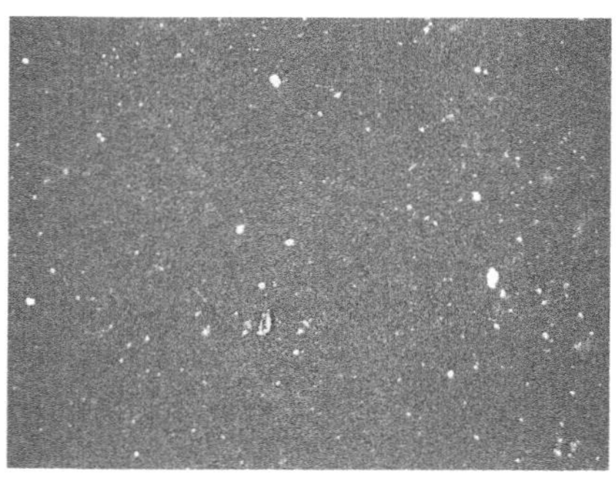

Di-Chloramine at 30 days

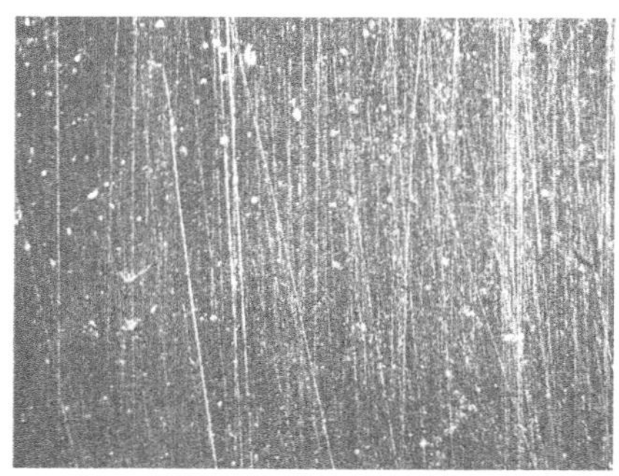

Mono-Chloramine at 30 days

Figure A.84 pH effect on hardness at 23°C and 60 ppm chloramine of EPDM-P with 40× magnified pictographs

LOOP STRAIN EFFECTS

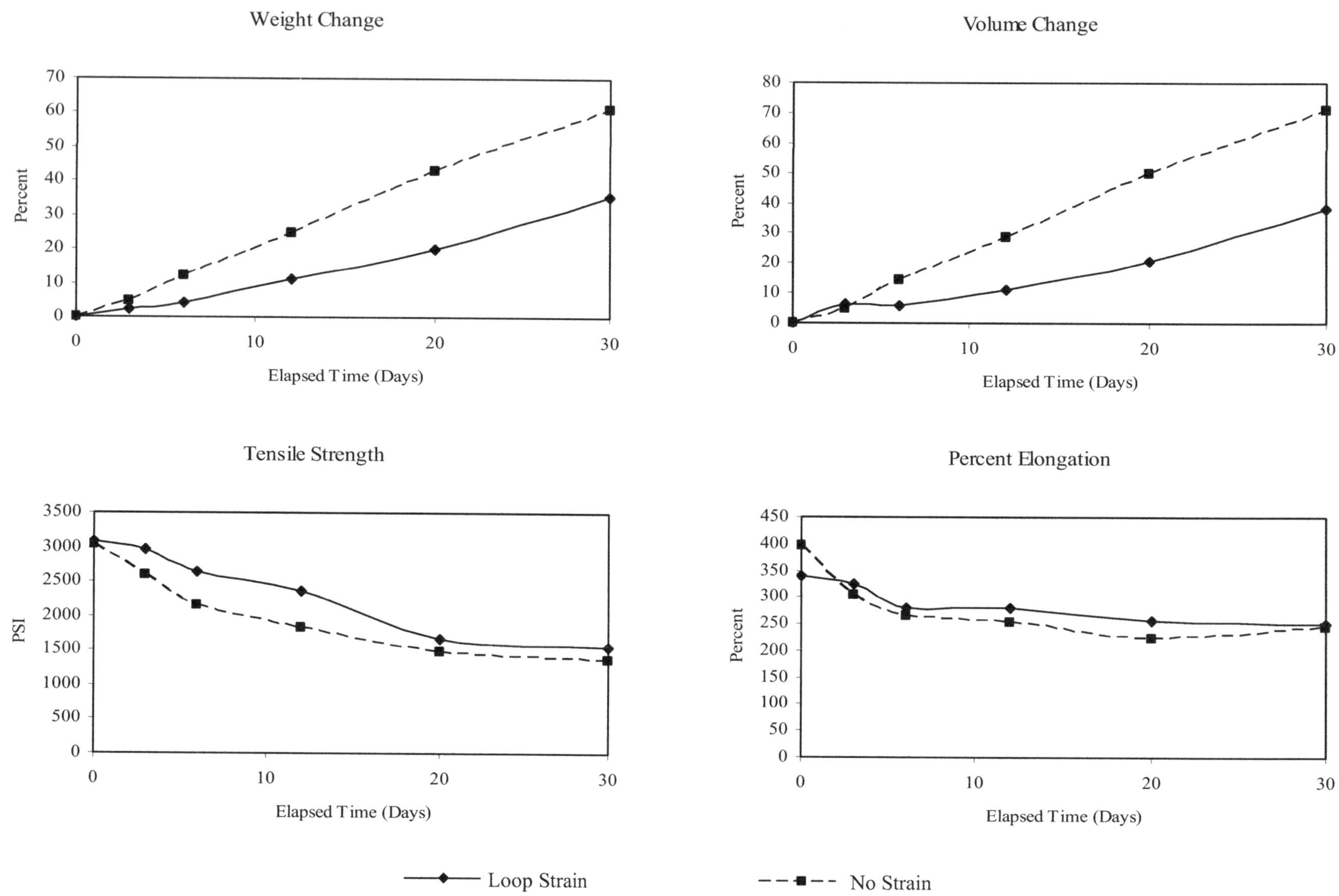

Figure A.85 Loop strain effects vs. no strain on elastomeric properties at 45°C, 30 ppm chloramine, and pH 8.3 of natural rubber

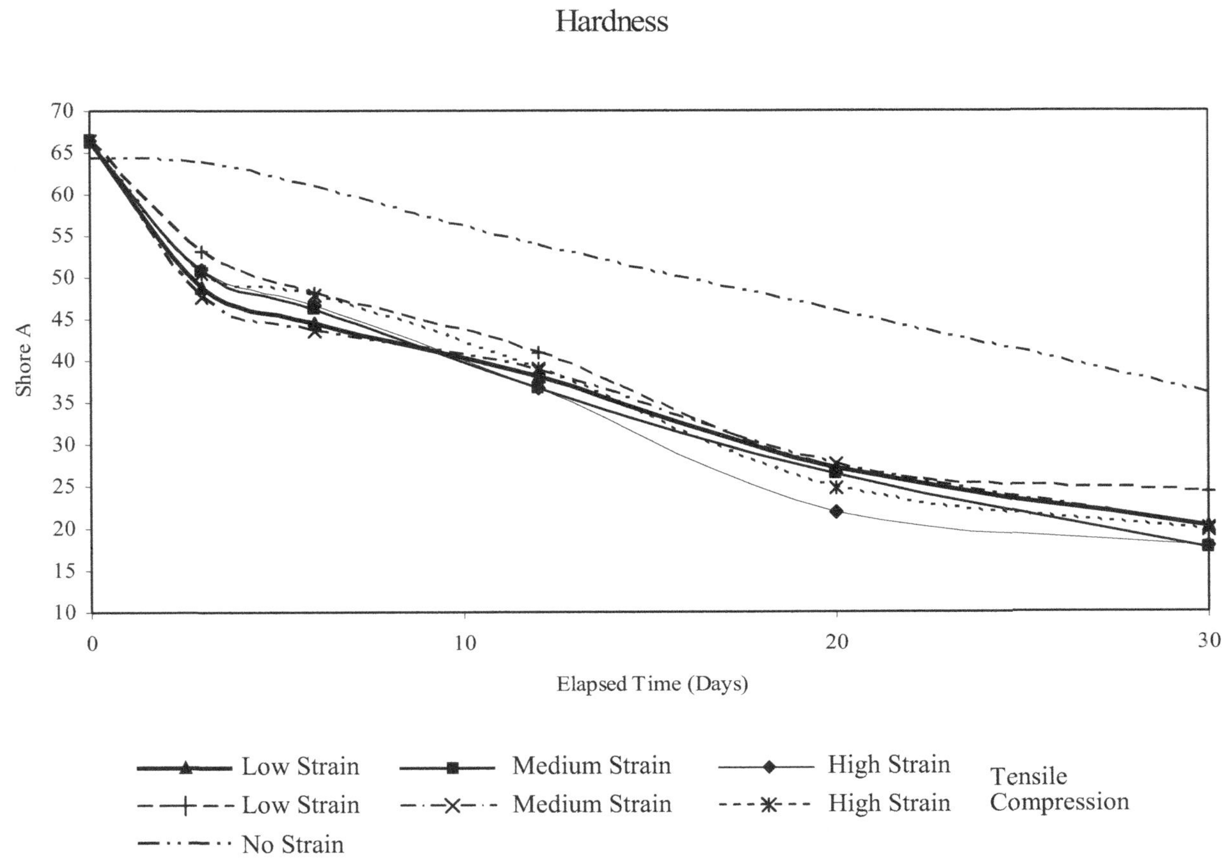

Tensile at 30 days

Compression at 30 days

No Strain at 30 days

Figure A.86 Loop strain effects vs. no strain on hardness at 45°C, 30 ppm chloramine, and pH 8.3 of natural rubber with 40× magnified pictographs

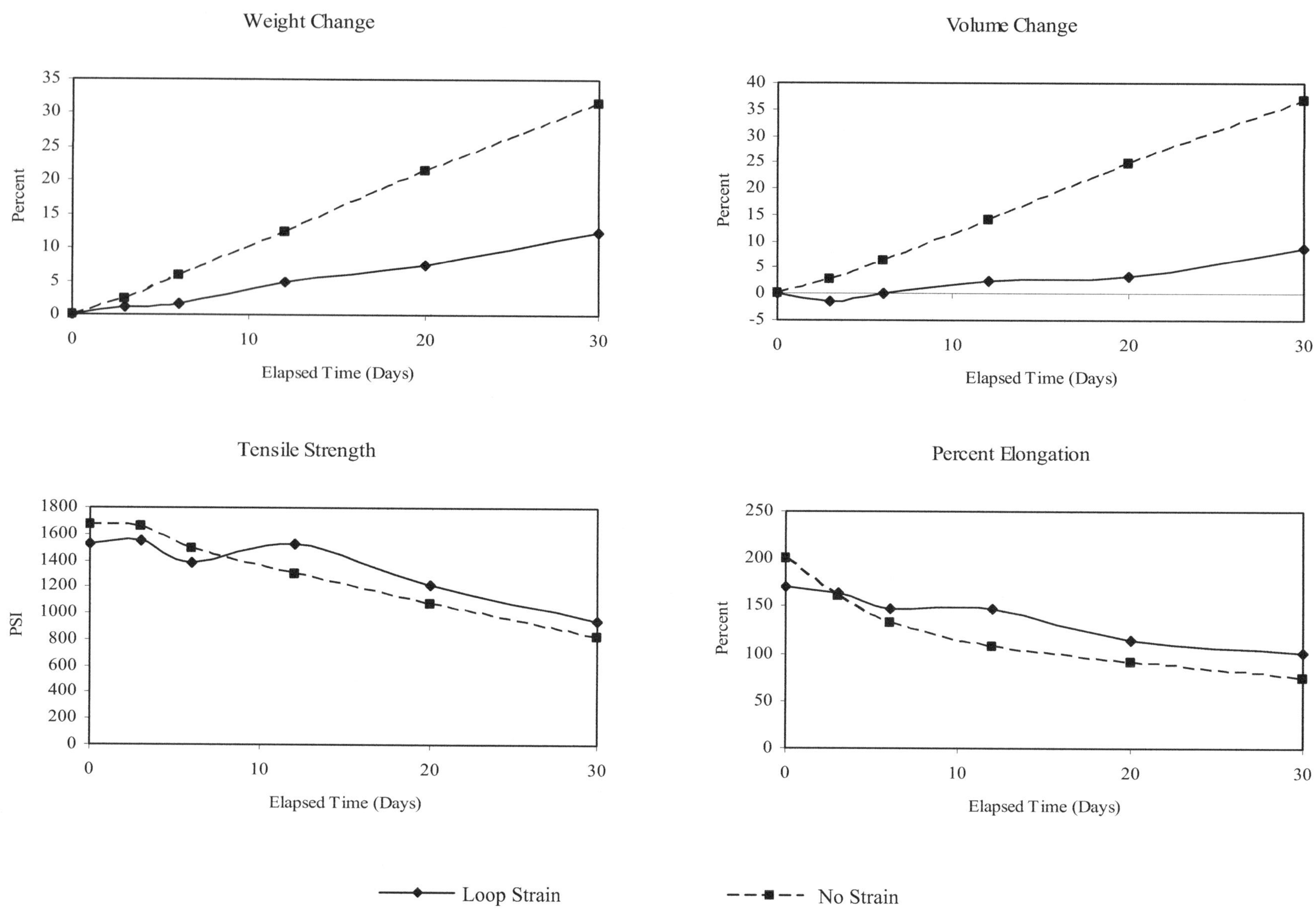

Figure A.87 Loop strain effects vs. no strain on elastomeric properties at 45°C, 30 ppm chloramine, and pH 8.3 of EPDM-S

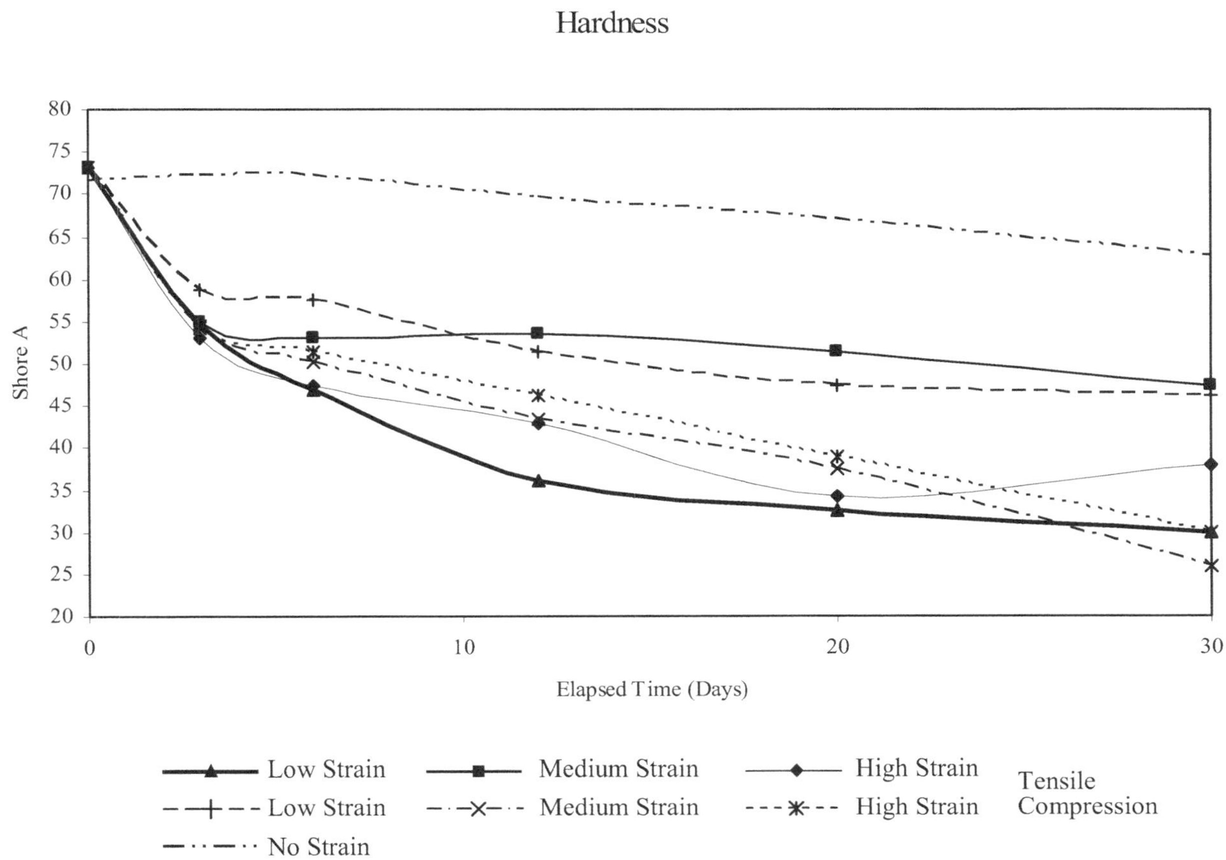

Tensile at 30 days

Compression at 30 days

No Strain at 30 days

Figure A.88 Loop strain effects vs. no strain on hardness at 45°C, 30 ppm chloramine, and pH 8.3 of EPDM-S with 40× magnified pictographs

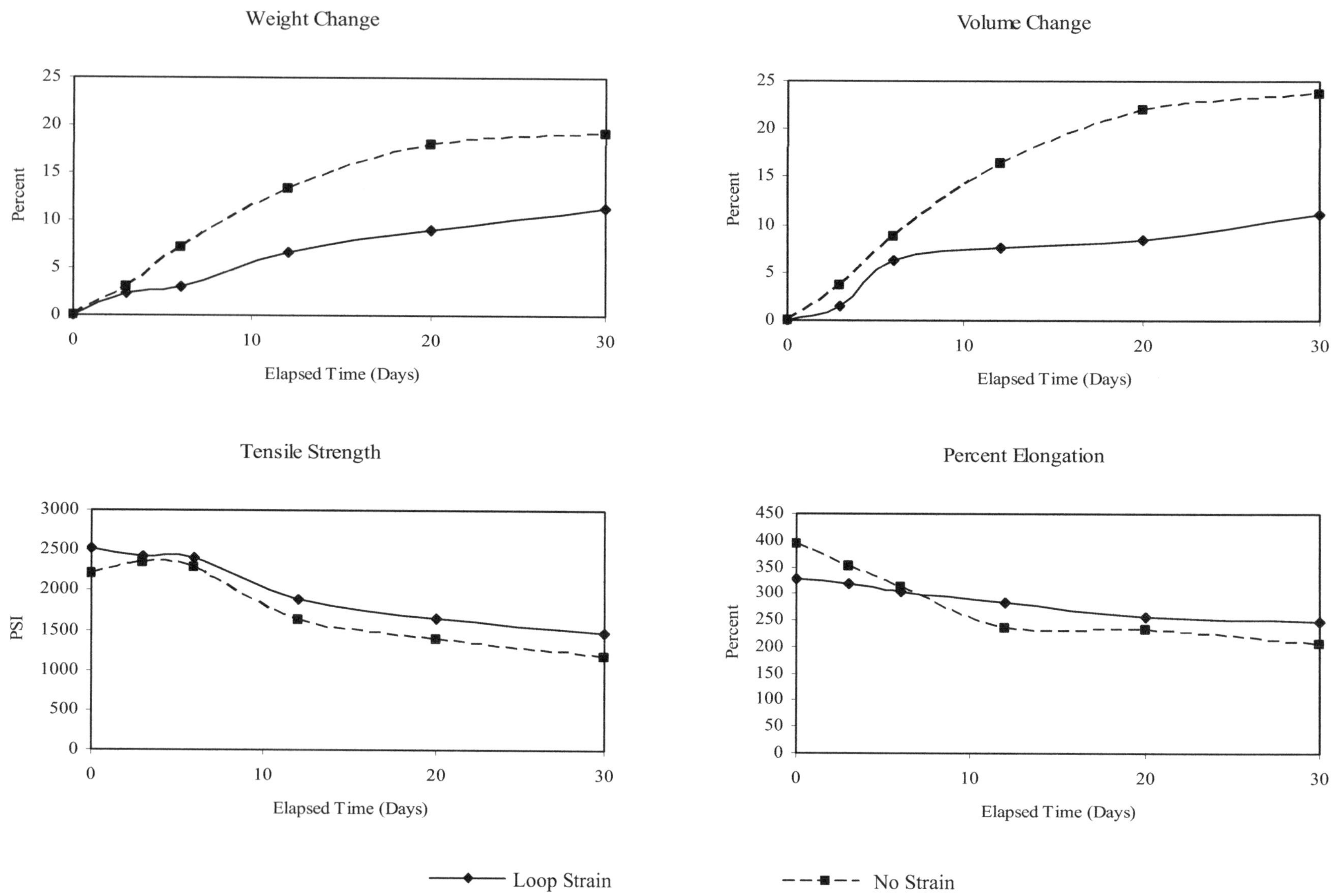

Figure A.89 Loop strain effects vs. no strain on elastomeric properties at 45°C, 30 ppm chloramine, and pH 8.3 of nitrile (NBR)

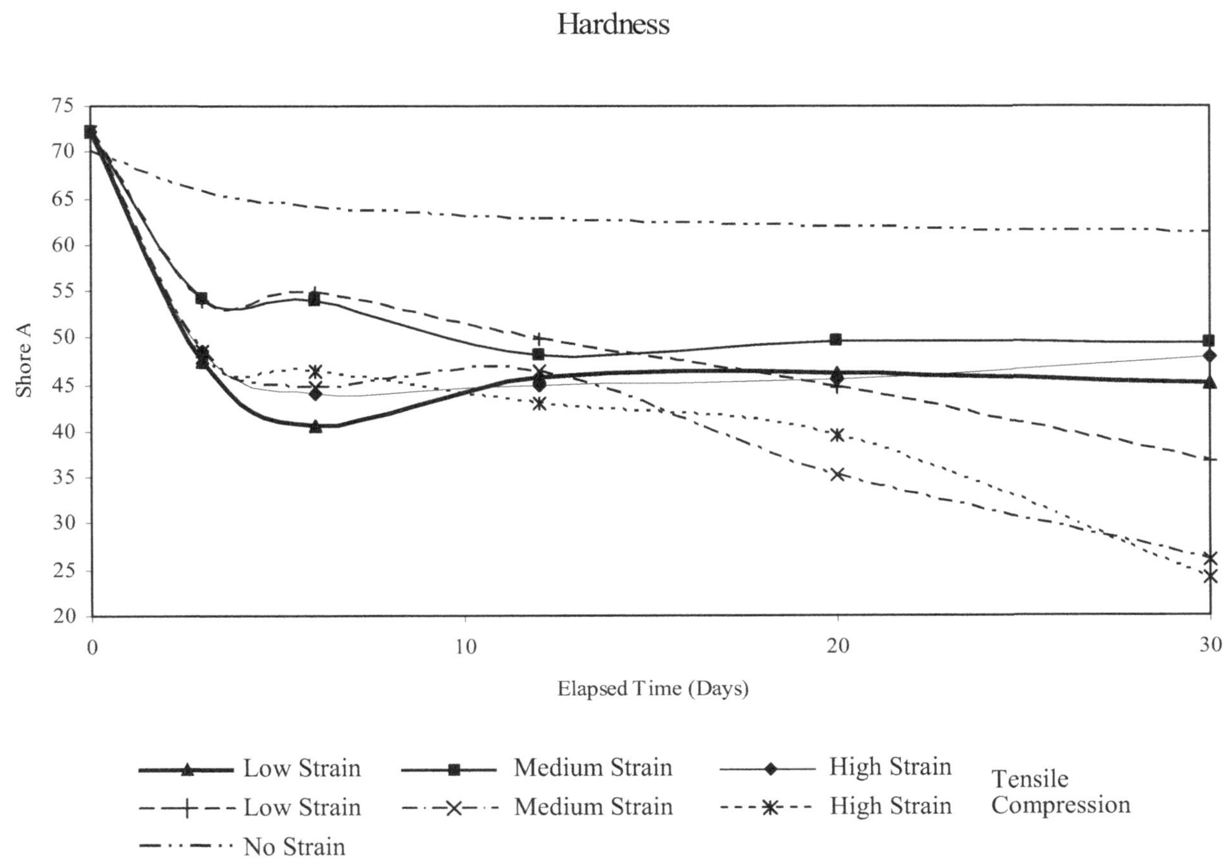

Tensile at 30 days

Compression at 30 days

No Strain at 30 days

Figure A.90 Loop strain effects vs. no strain on hardness at 45°C, 30 ppm chloramine, and pH 8.3 of nitrile (NBR) with 40× magnified pictographs

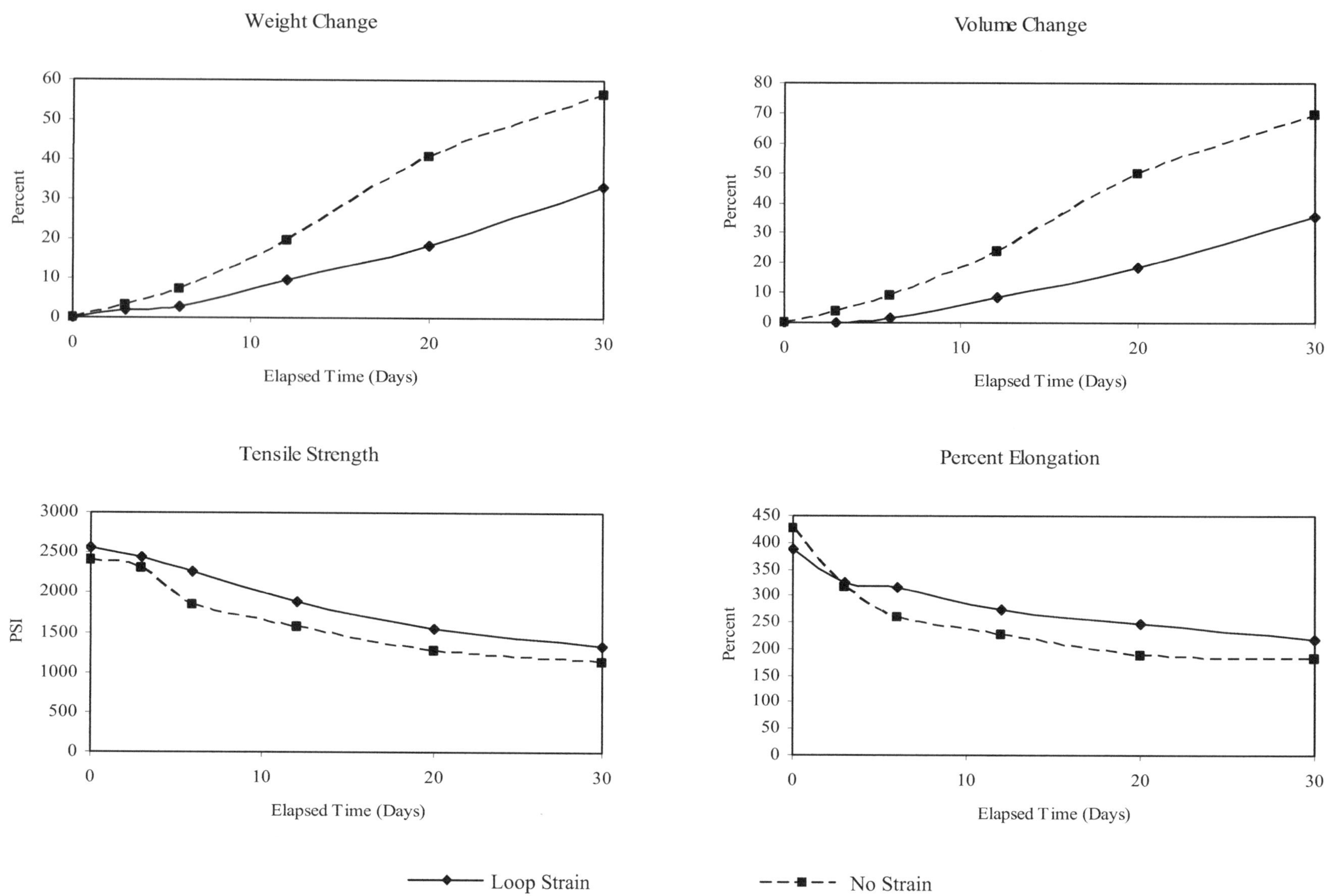

Figure A.91 Loop strain effects vs. no strain on elastomeric properties at 45°C, 30 ppm chloramine, and pH 8.3 of SBR

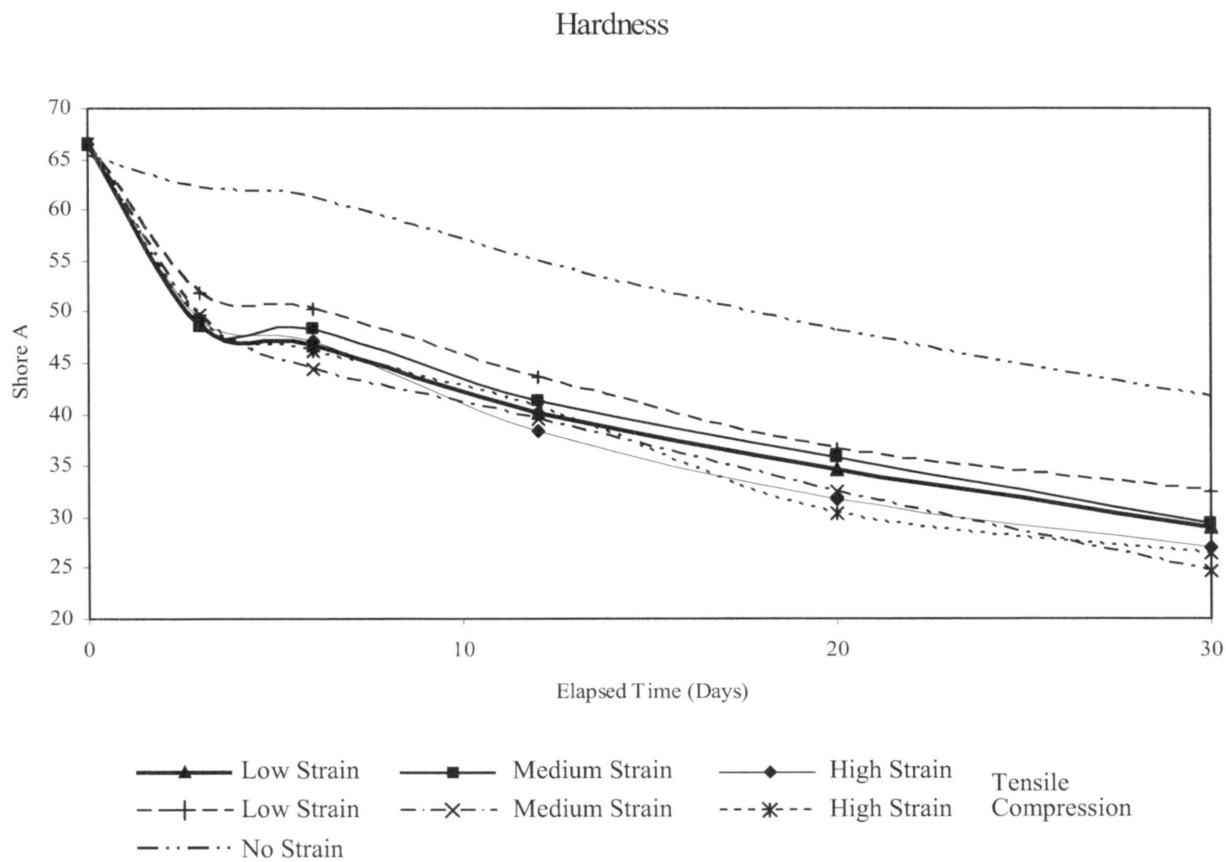

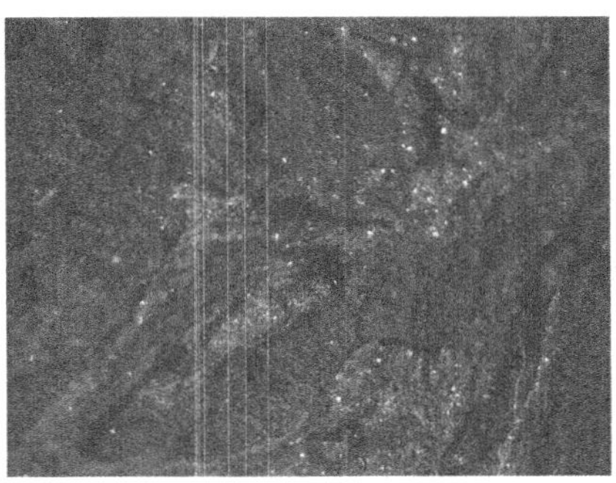

Tensile at 30 days

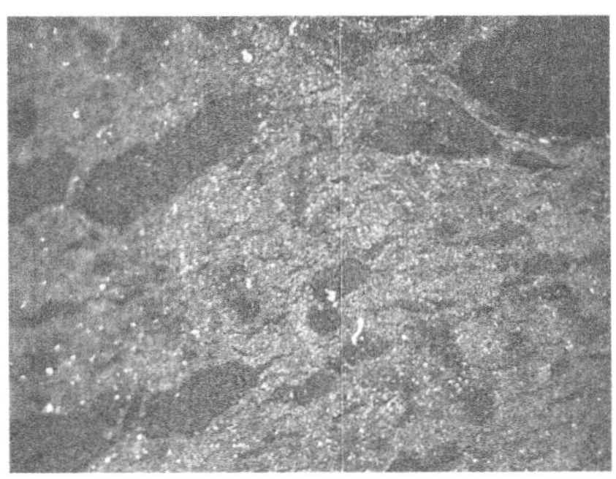

Compression at 30 days

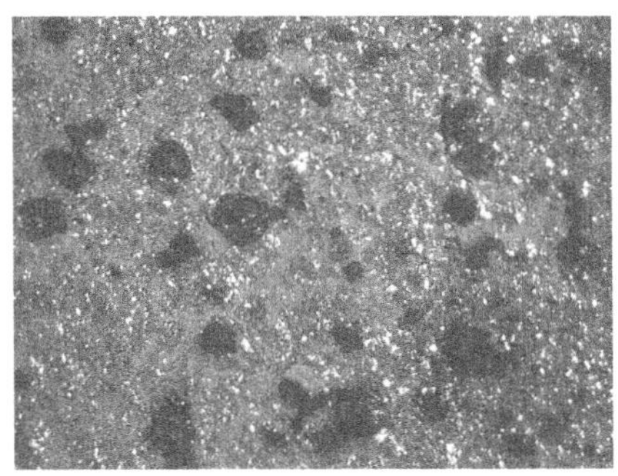

No Strain at 30 days

Figure A.92 Loop strain effects vs. no strain on hardness at 45°C, 30 ppm chloramine, and pH 8.3 of SBR with 40× magnified pictographs

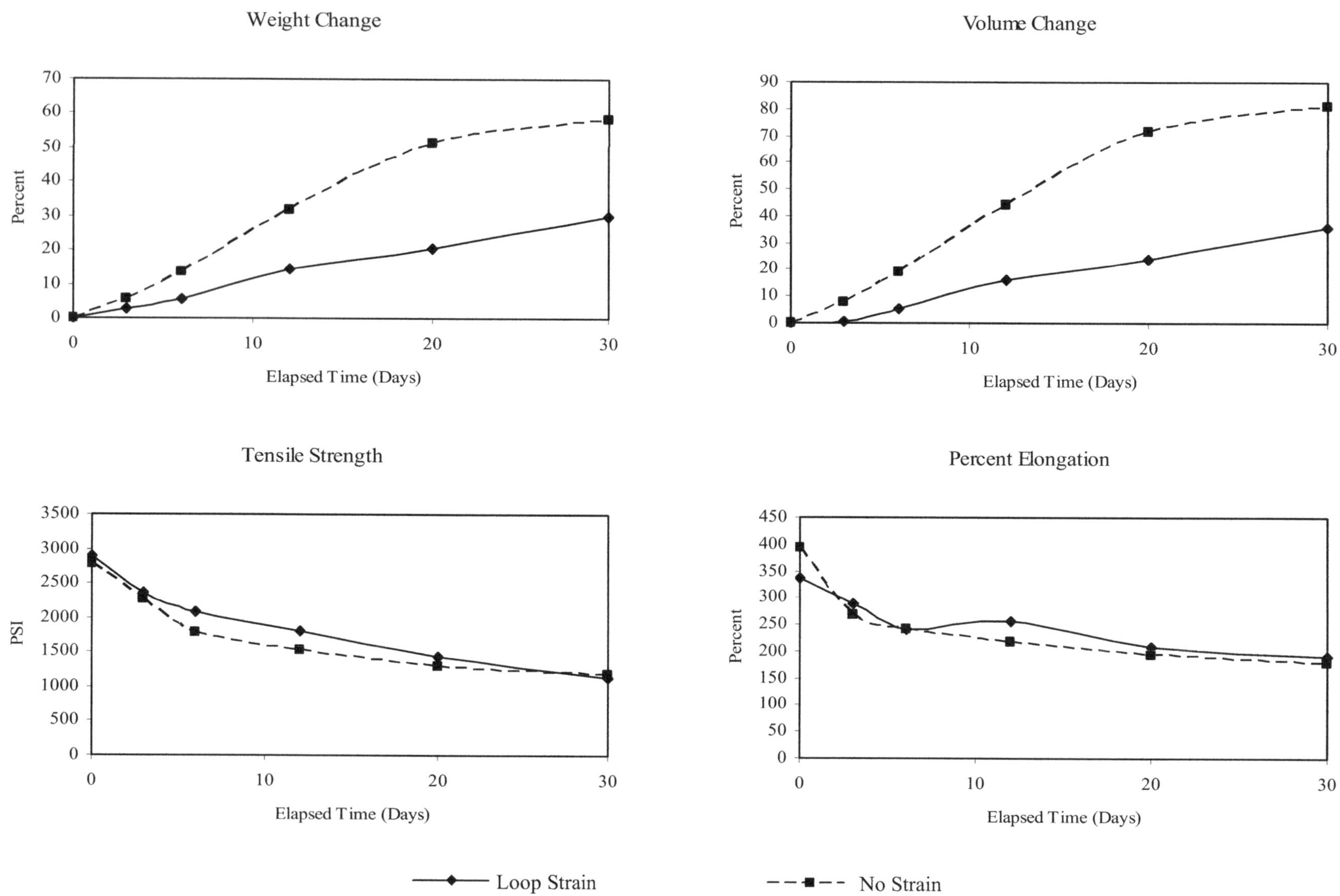

Figure A.93 Loop strain effects vs. no strain on elastomeric properties at 45°C, 30 ppm chloramine, and pH 8.3 of neoprene

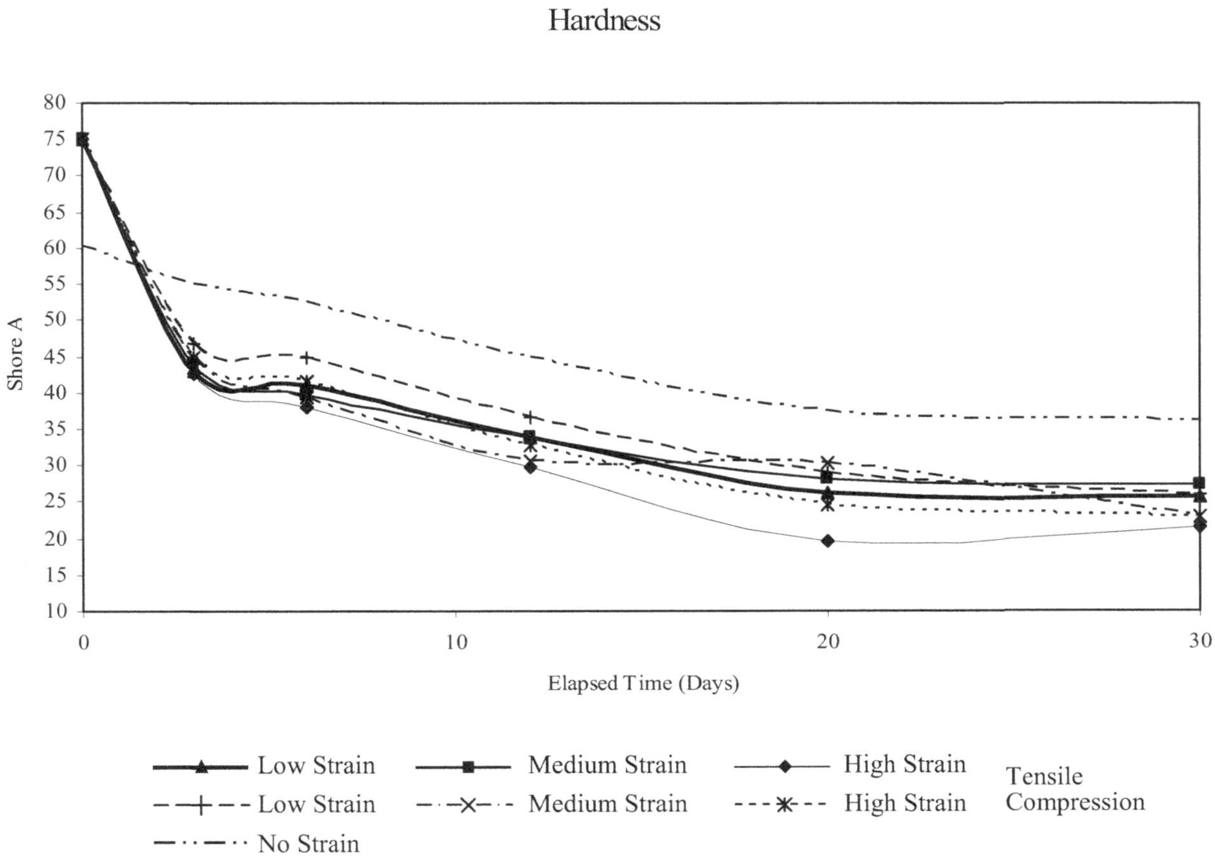

Tensile at 30 days

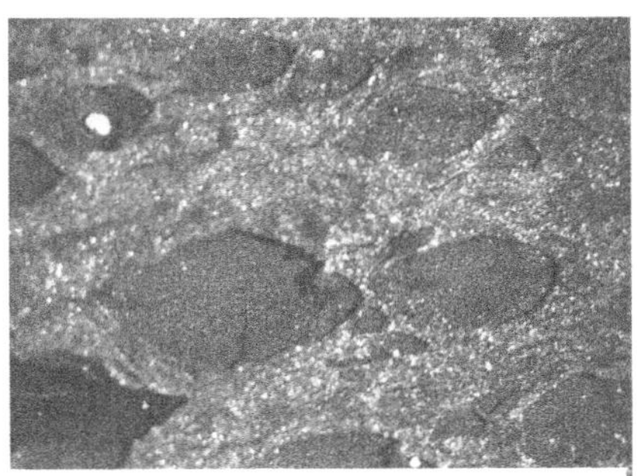

Compression at 30 days

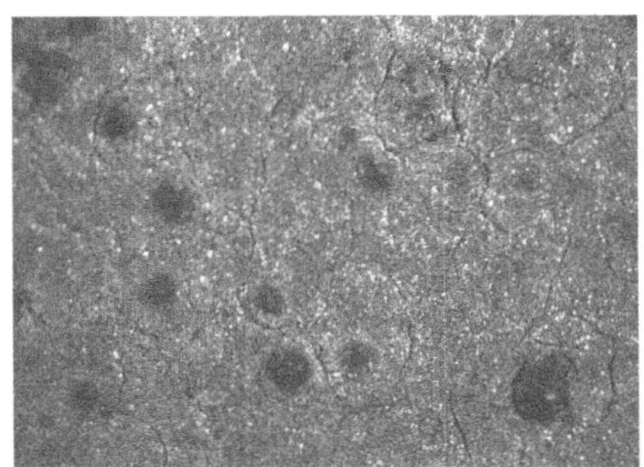

No Strain at 30 days

Figure A.94 Loop strain effects vs. no strain on hardness at 45°C, 30 ppm chloramine, and pH 8.3 of neoprene with 40× magnified pictographs

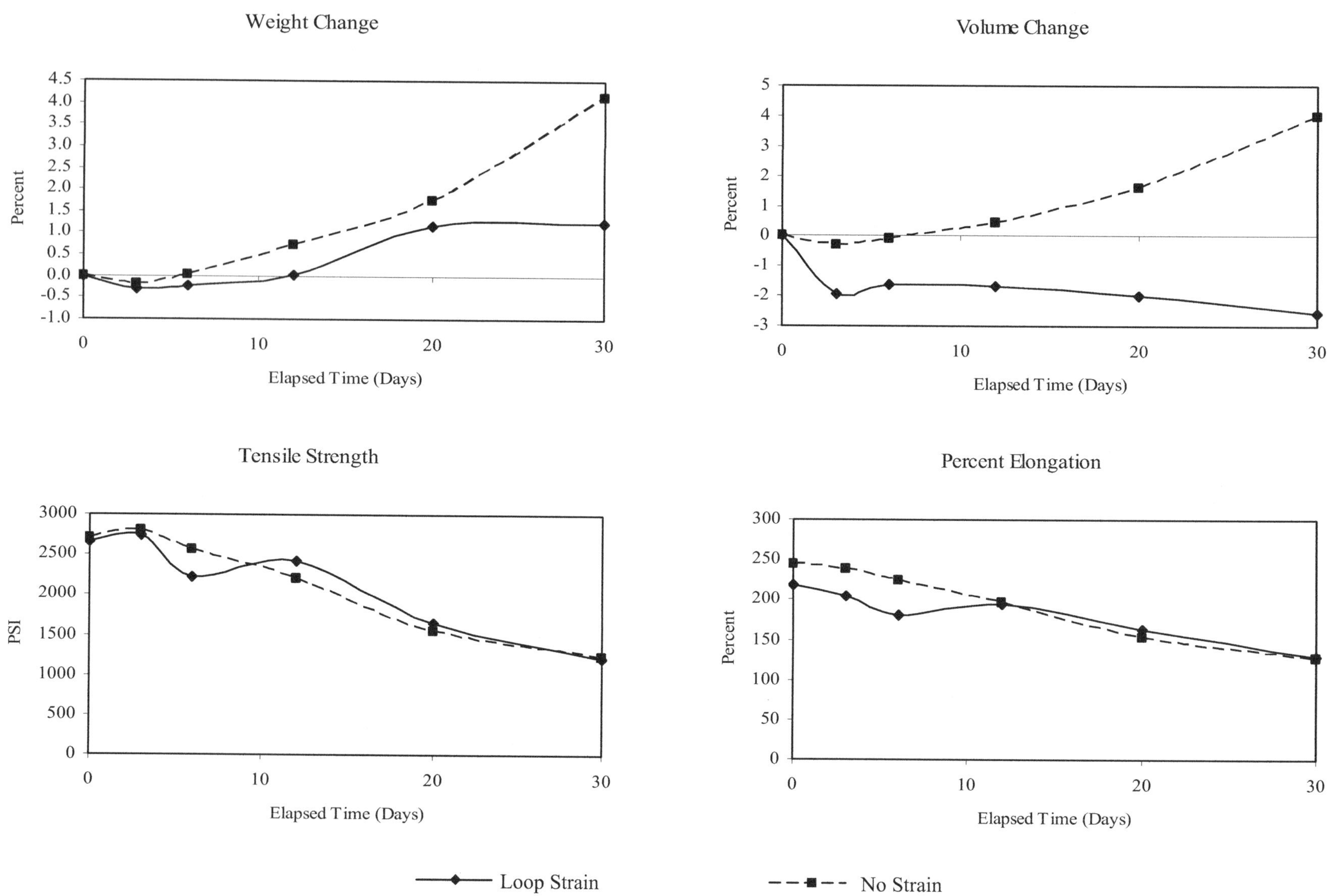

Figure A.95 Loop strain effects vs. no strain on elastomeric properties at 45°C, 30 ppm chloramine, and pH 8.3 of EPDM-P

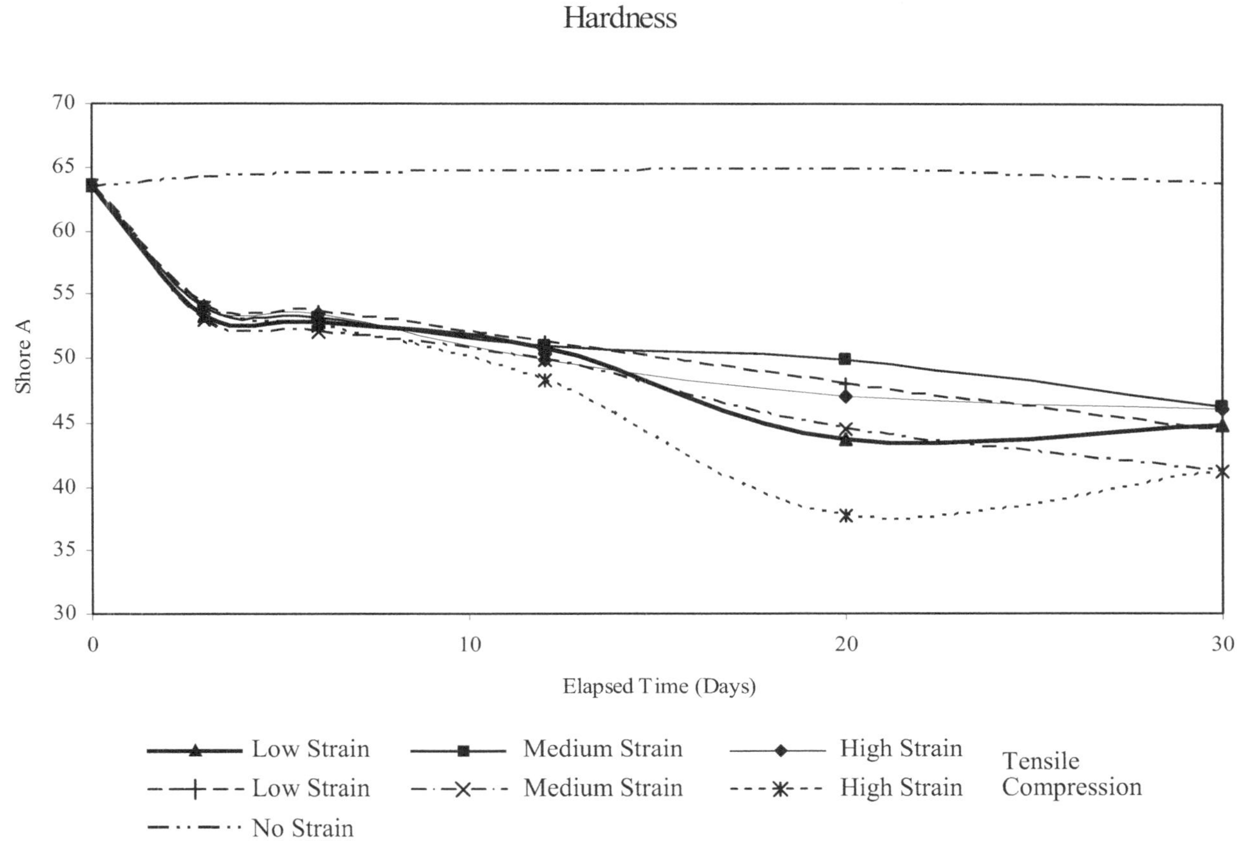

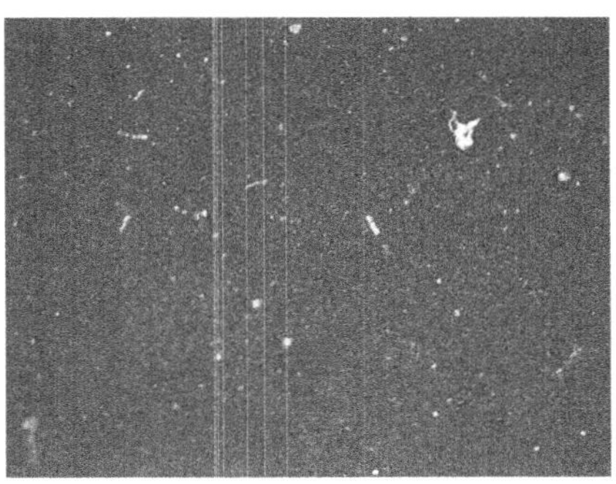

Tensile at 30 days

Compression at 30 days

No Strain at 30 days

Figure A.96 Loop strain effects vs. no strain on hardness at 45°C, 30 ppm chloramine, and pH 8.3 of EPDM-P with 40× magnified pictographs

CHLORINE VERSUS CHLORAMINES EFFECTS

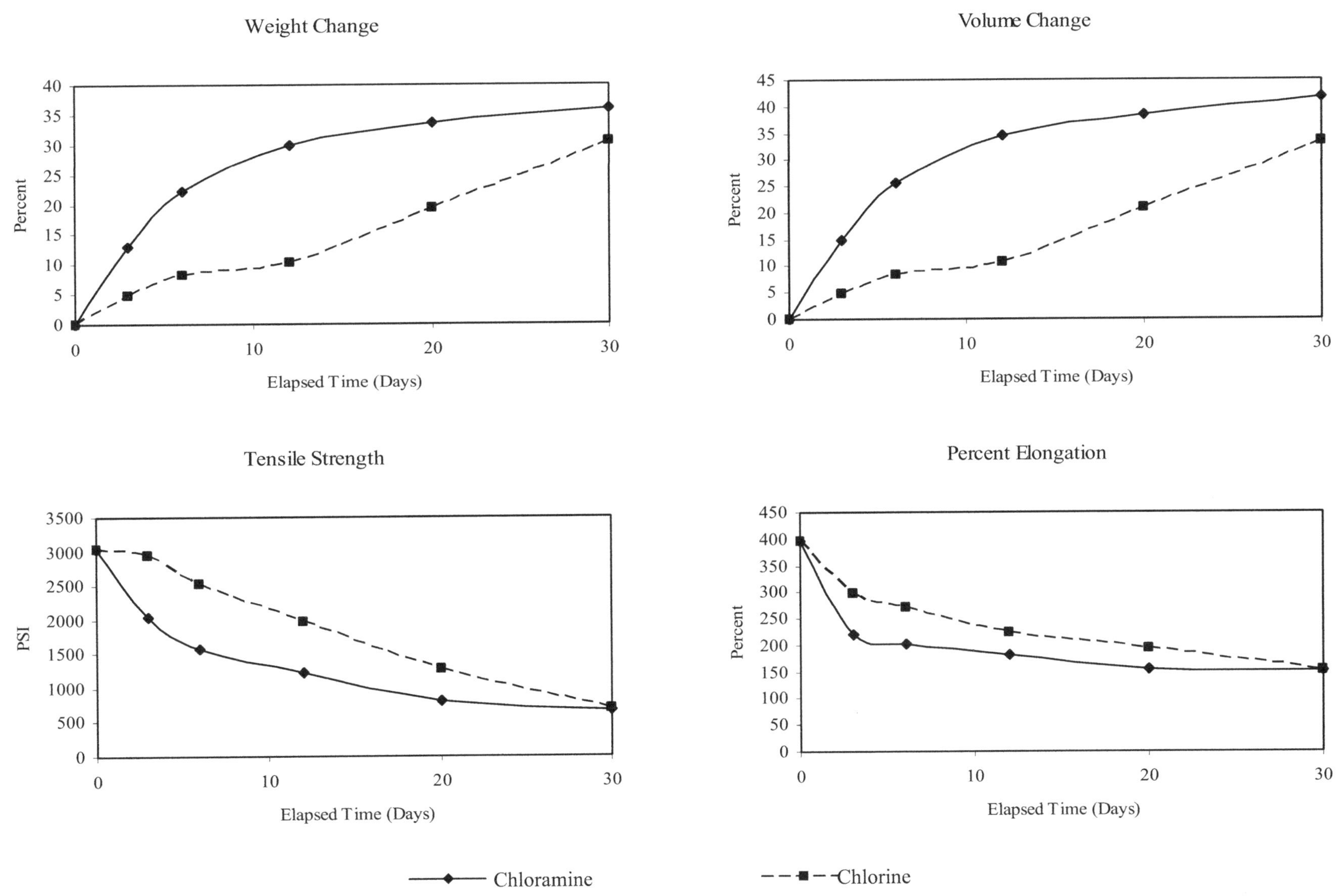

Figure A.97 Chloramine vs. chlorine effect on elastomeric properties at 70°C, 30 ppm, and pH 8.3 of natural rubber

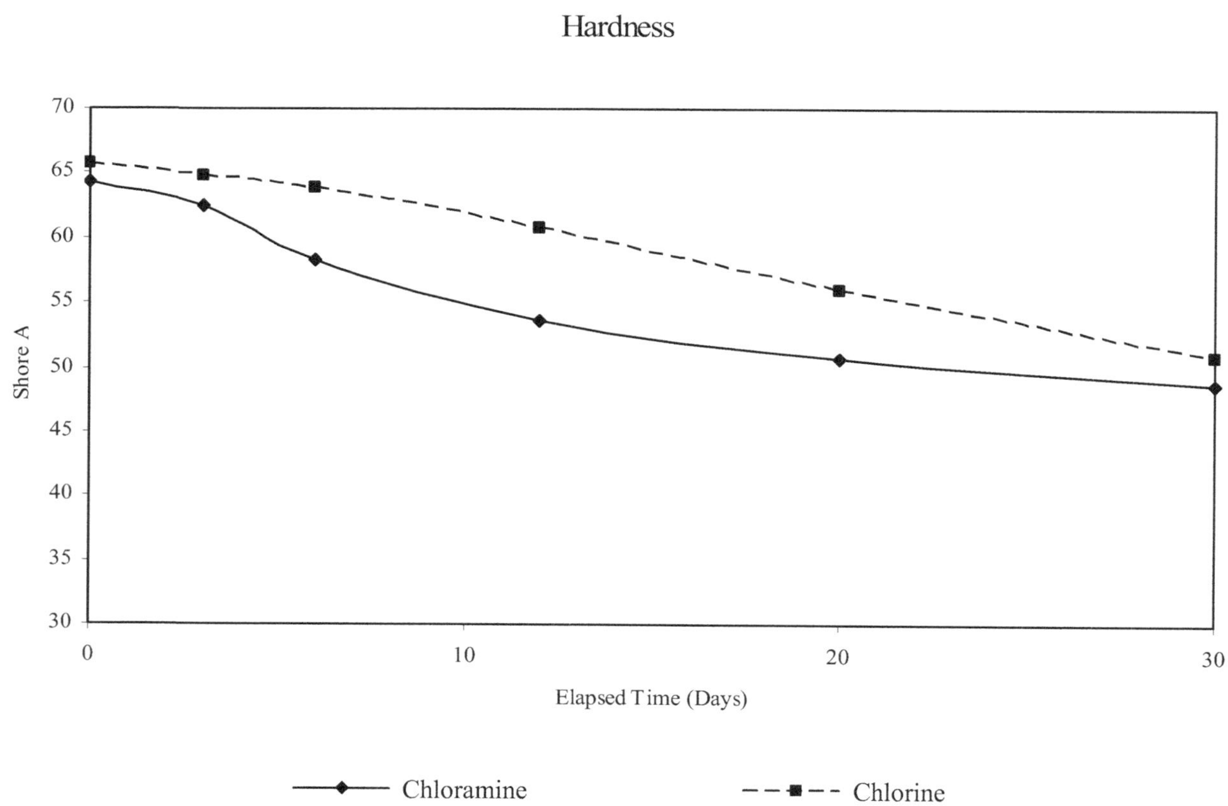

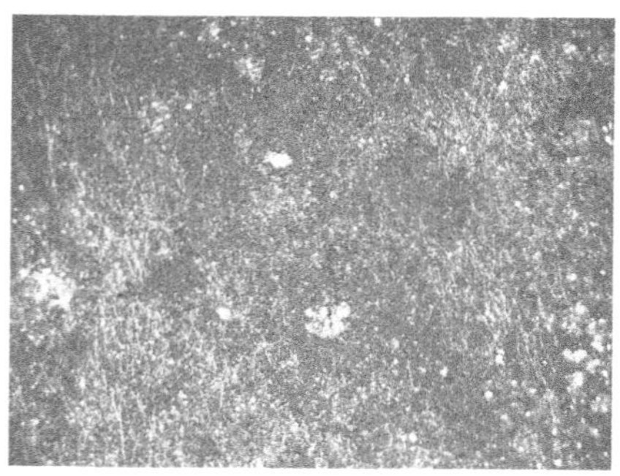

Chloramine at 30 days

Chlorine at 30 days

Figure A.98 Chloramine vs. chlorine effect on hardness at 70°C, 30 ppm, and pH 8.3 of natural rubber with 40× magnified pictographs

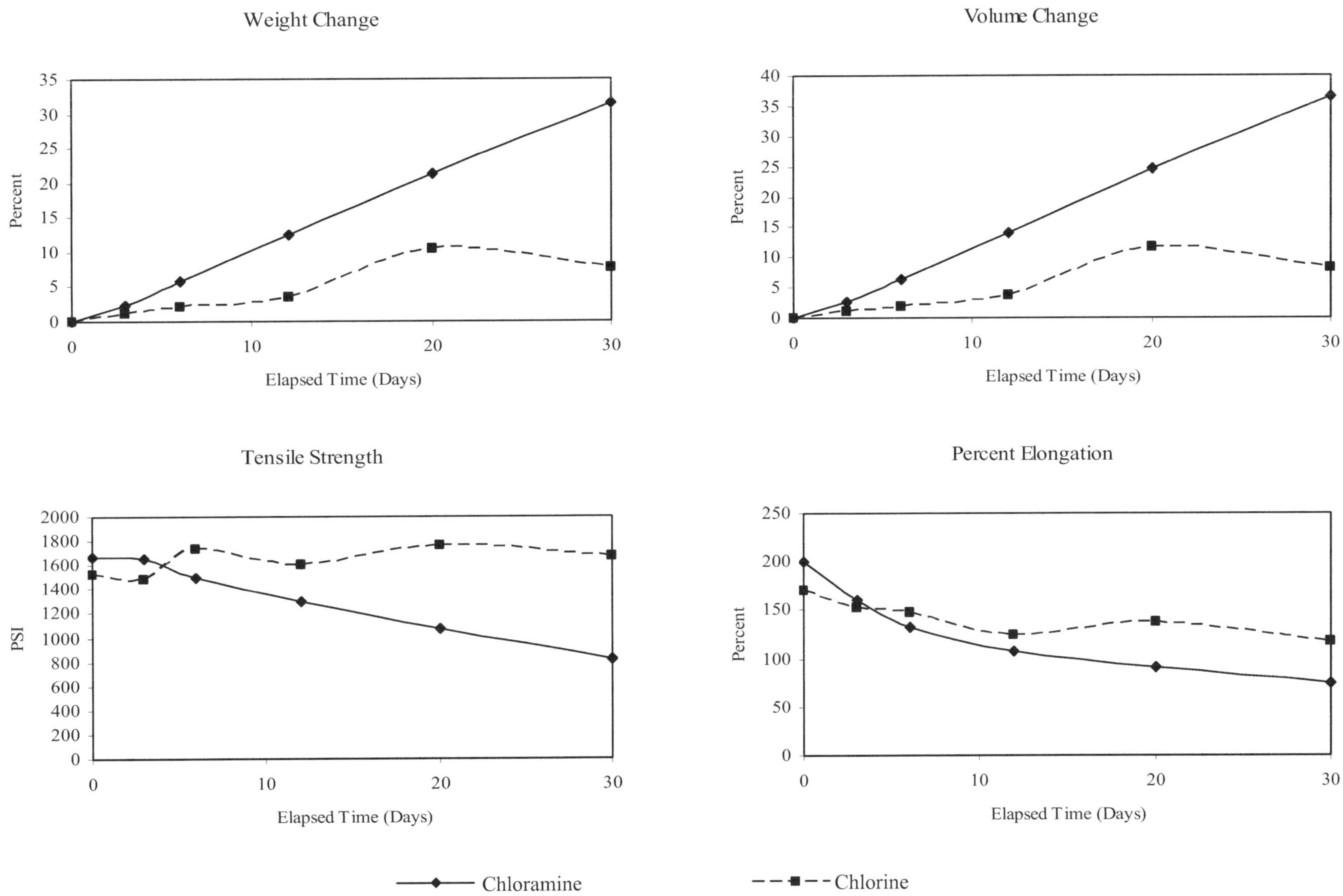

Figure A.99 Chloramine vs. chlorine effect on elastomeric properties at 70°C, 30 ppm, and pH 8.3 of EPDM-S

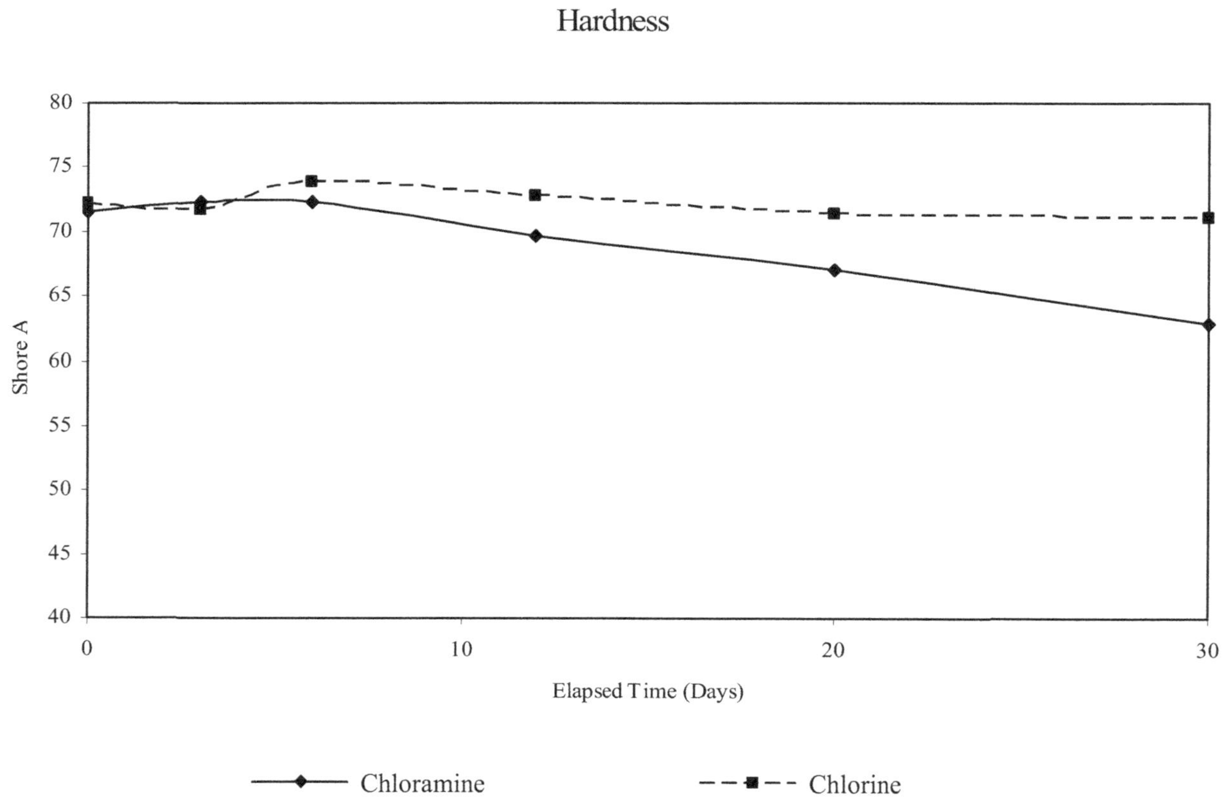

Chloramine at 30 days

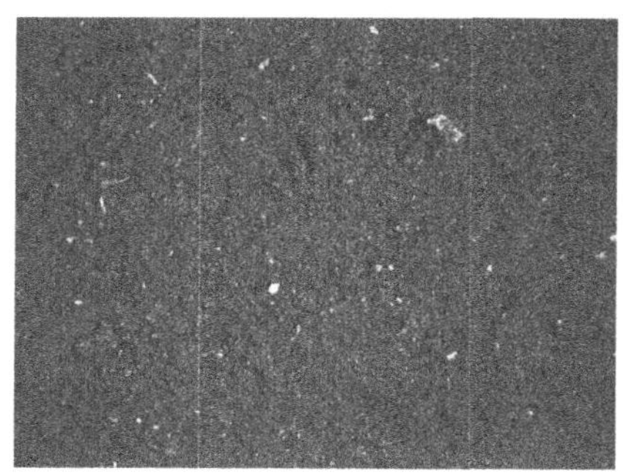

Chlorine at 30 days

Figure A.100 Chloramine vs. chlorine effect on hardness at 70°C, 30 ppm, and pH 8.3 of EPDM-S with 40× magnified pictographs

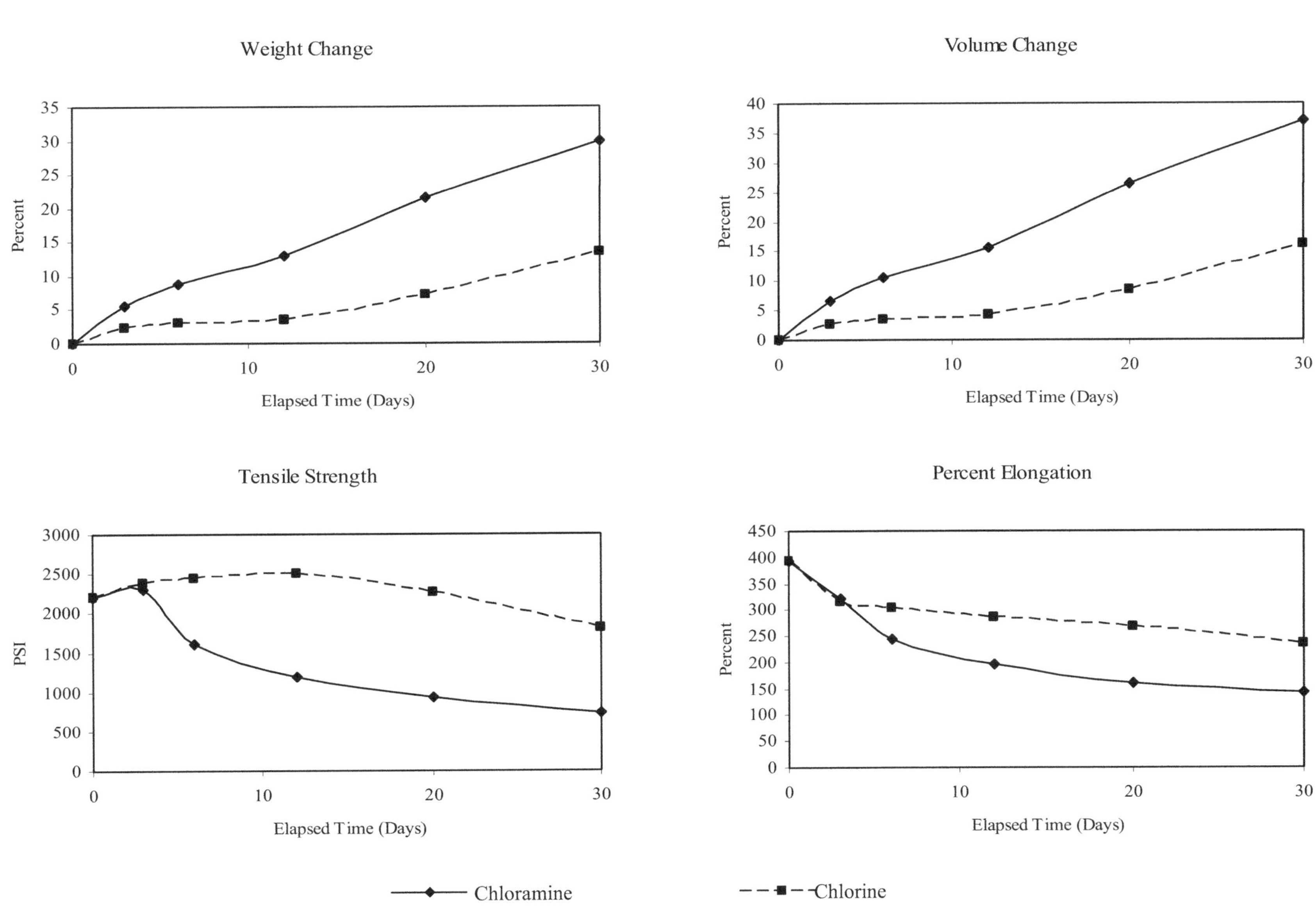

Figure A.101 Chloramine vs. chlorine effect on elastomeric properties at 70°C, 30 ppm, and pH 8.3 of nitrile (NBR)

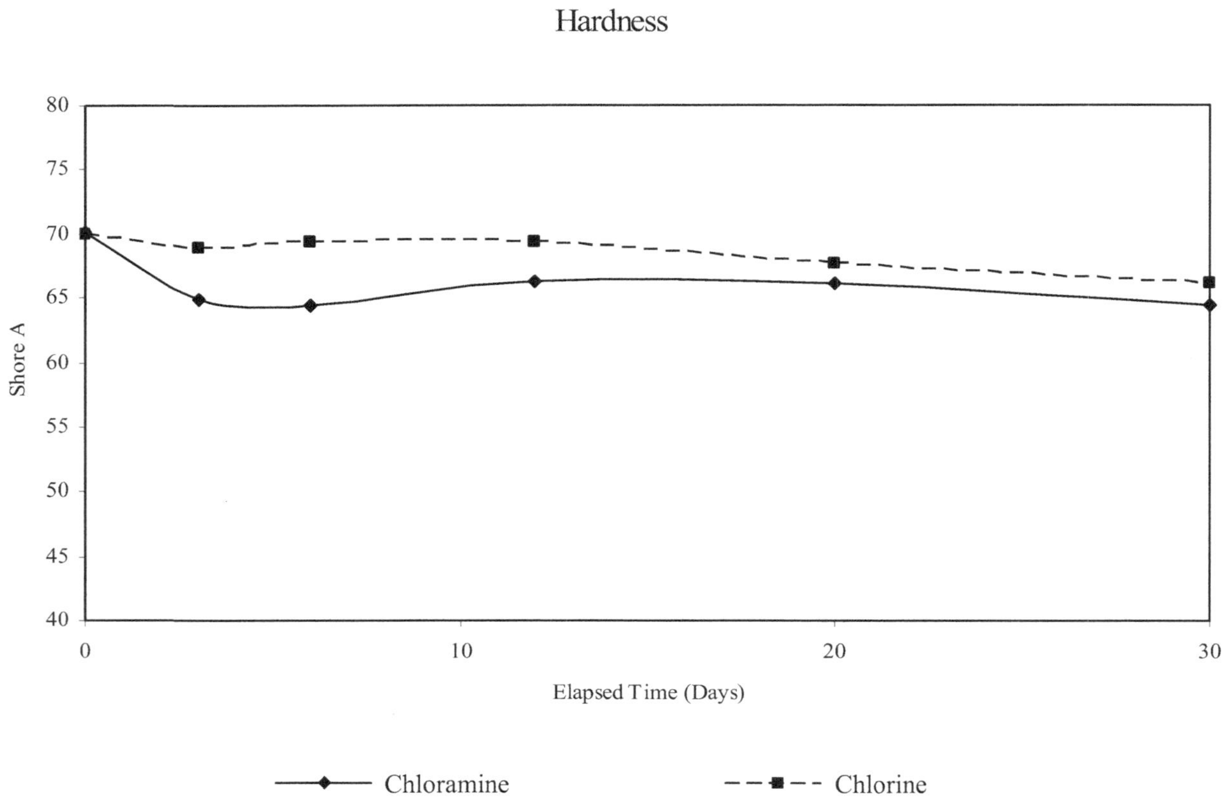

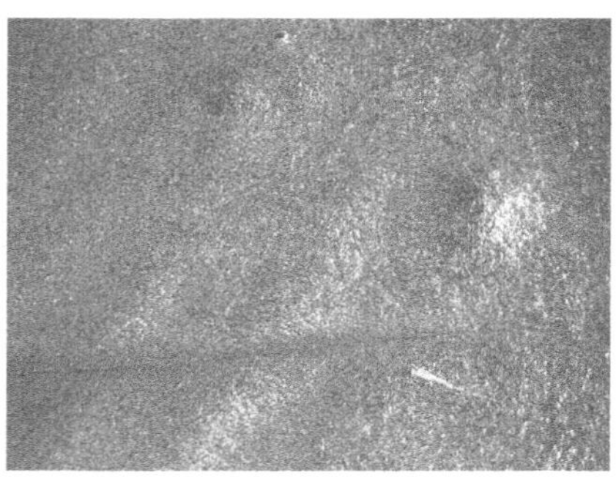

Chloramine at 30 days

Chlorine at 30 days

Figure A.102 Chloramine vs. chlorine effect on hardness at 70°C, 30 ppm, and pH 8.3 of nitrile (NBR) with 40× magnified pictographs

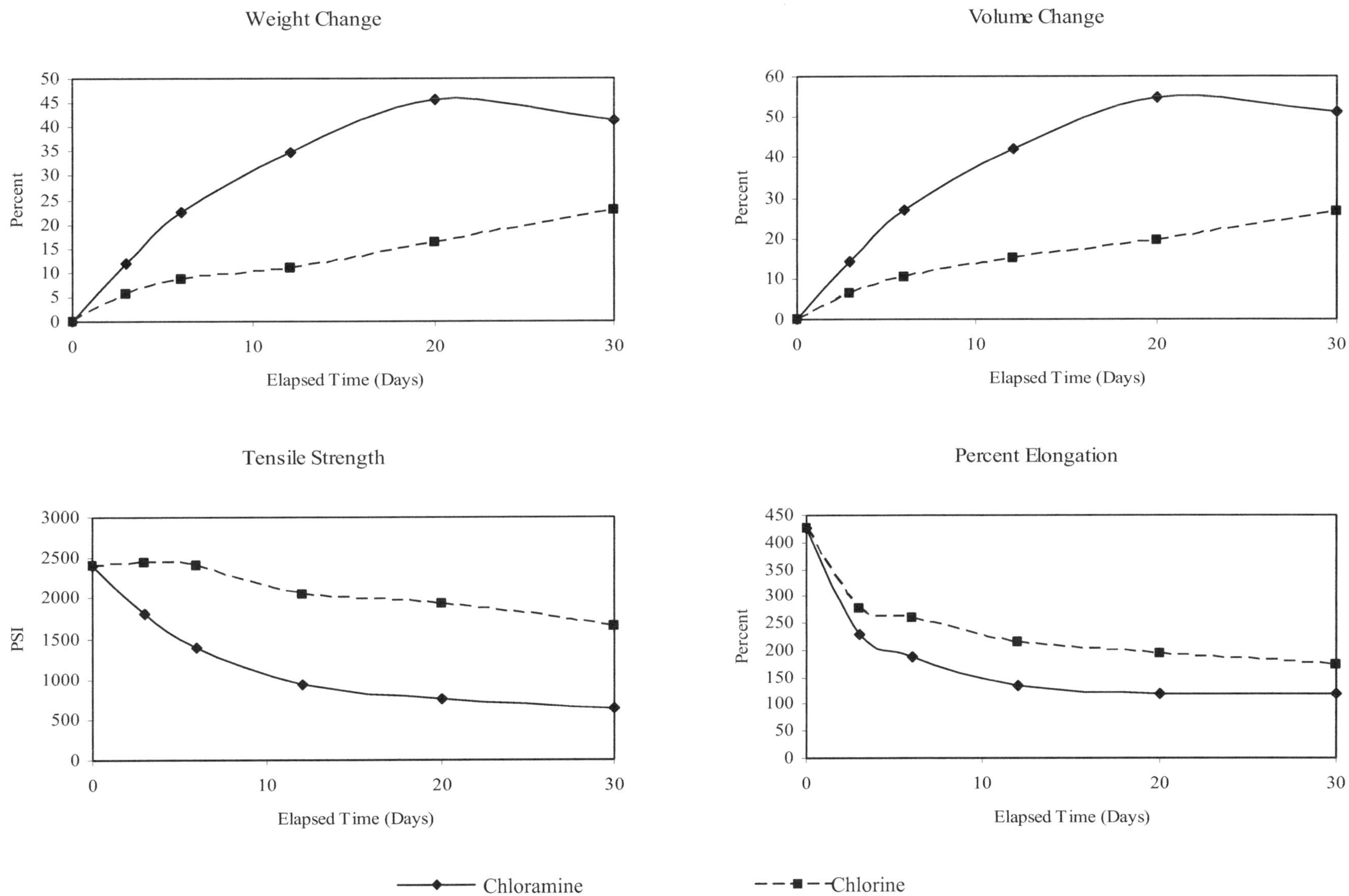

Figure A.103 Chloramine vs. chlorine effect on elastomeric properties at 70°C, 30 ppm, and pH 8.3 of SBR

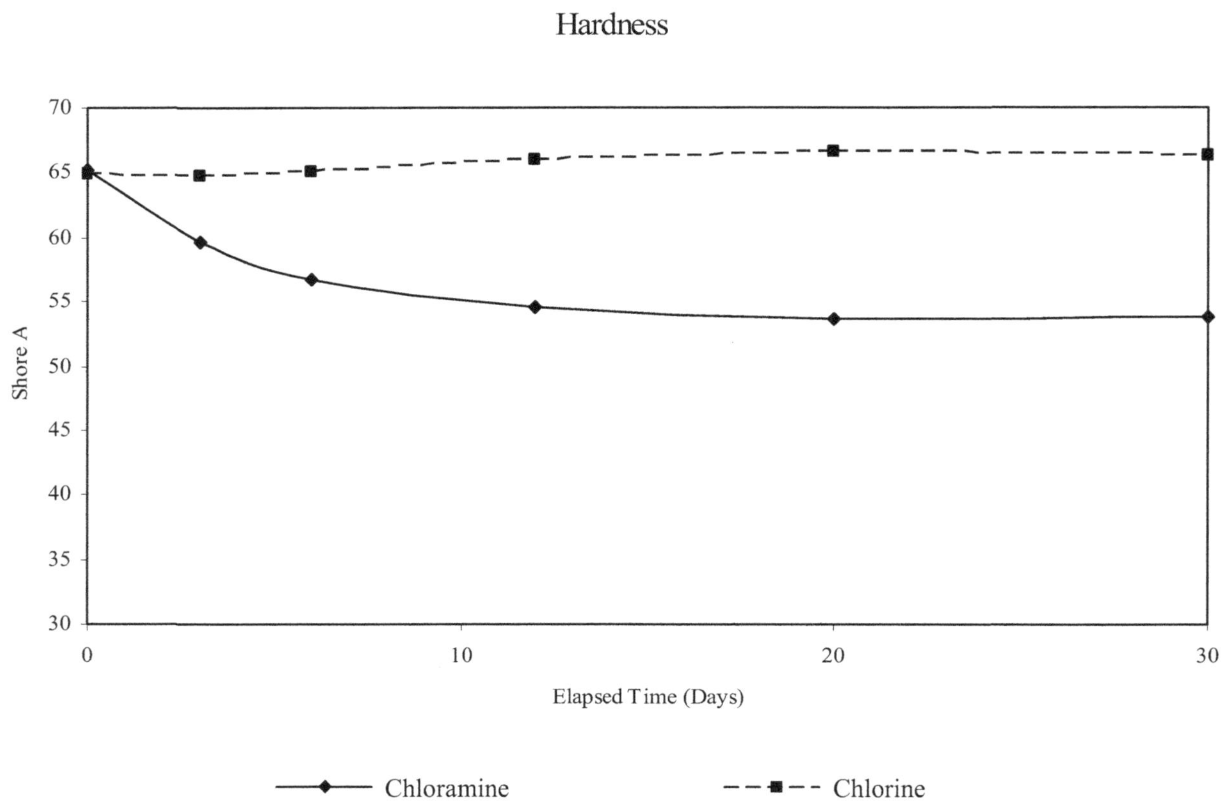

Chloramine at 30 days

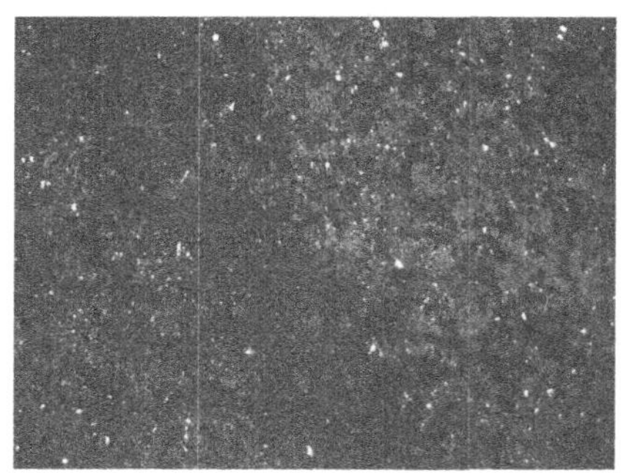

Chlorine at 30 days

Figure A.104 Chloramine vs. chlorine effect on hardness at 70°C, 30 ppm, and pH 8.3 of SBR with 40× magnified pictographs

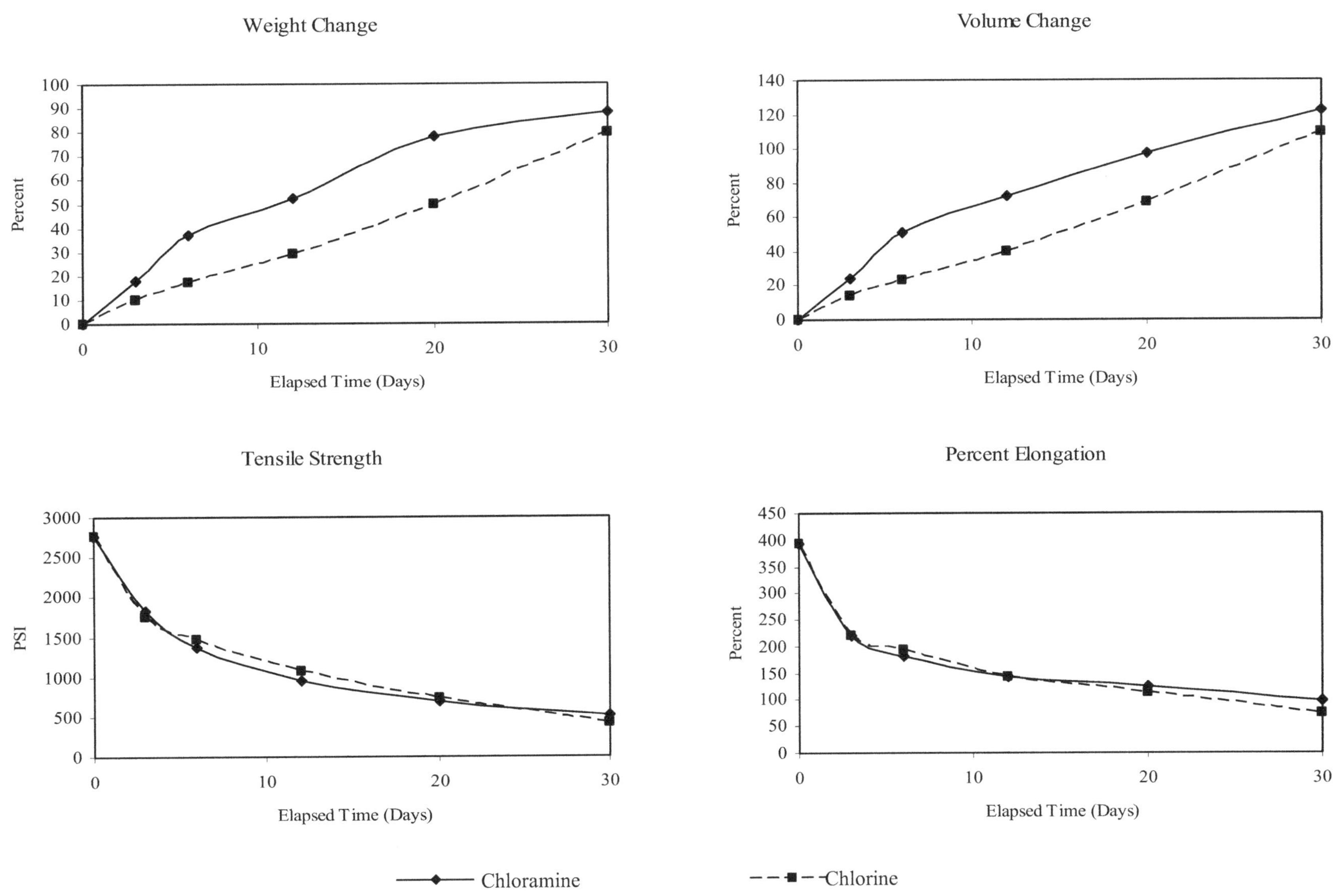

Figure A.105 Chloramine vs. chlorine effect on elastomeric properties at 70°C, 30 ppm, and pH 8.3 of neoprene

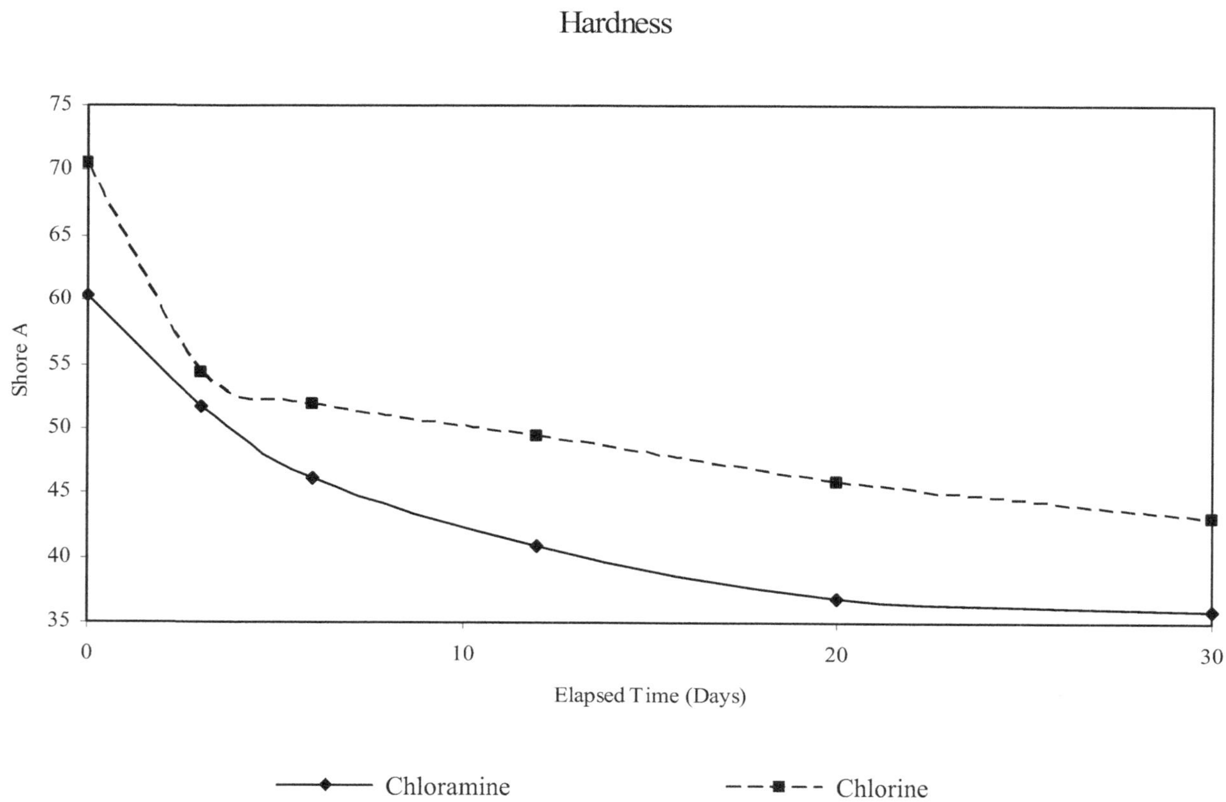

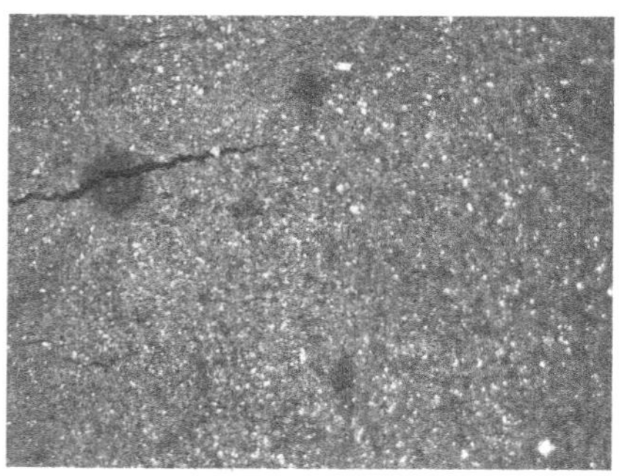

Chloramine at 30 days

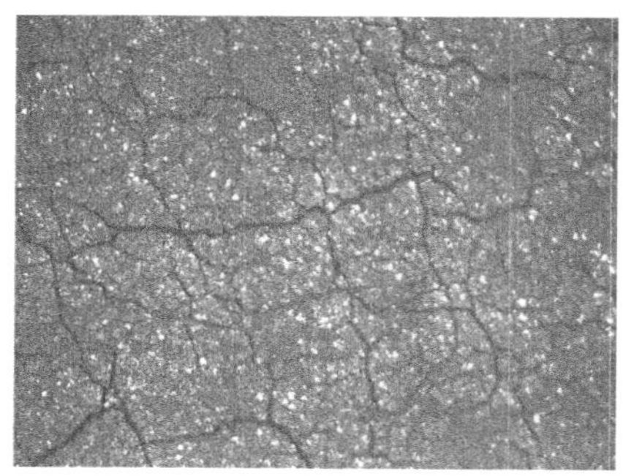

Chlorine at 30 days

Figure A.106 Chloramine vs. chlorine effect on hardness at 70°C, 30 ppm, and pH 8.3 of neoprene with 40× magnified pictographs

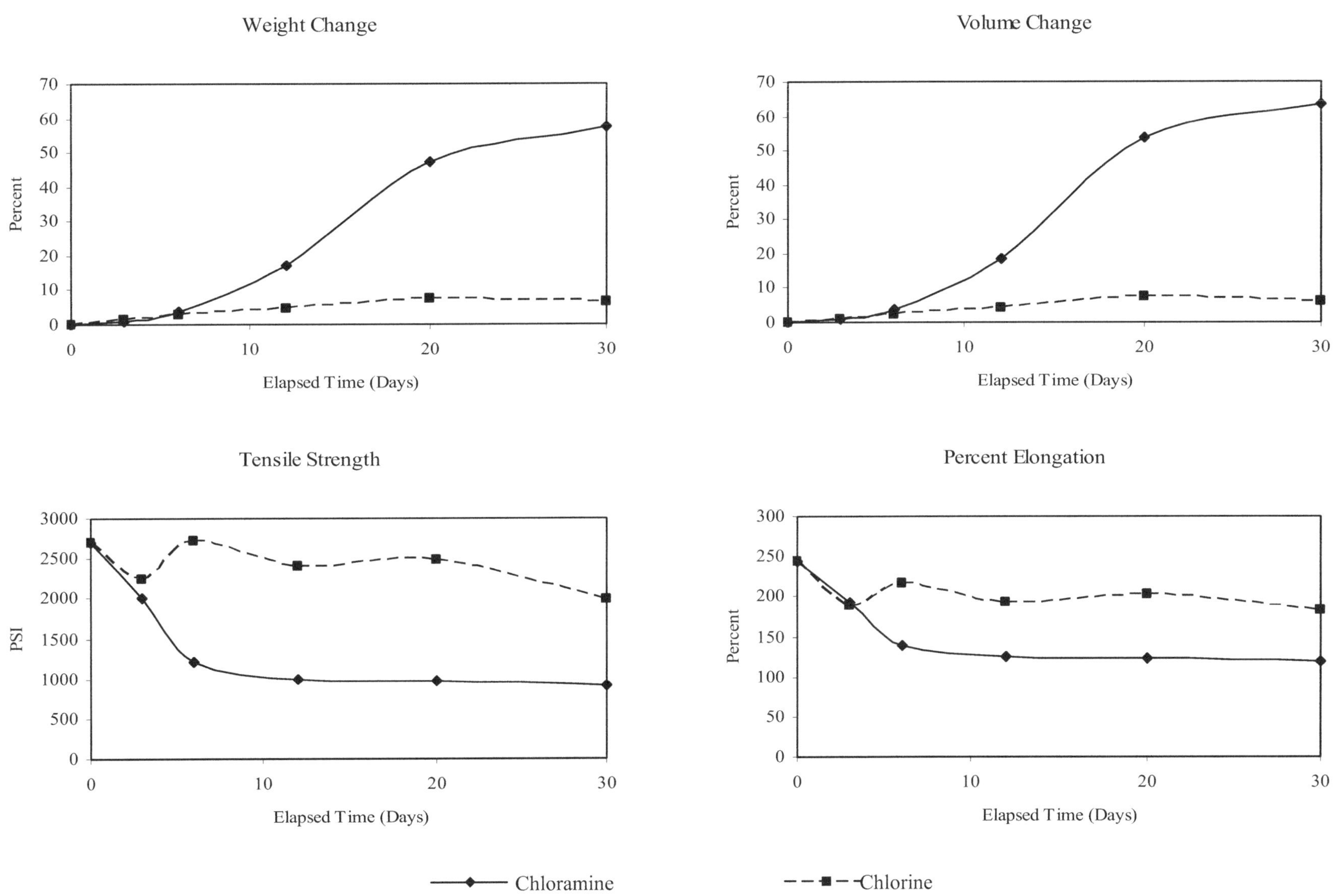

Figure A.107 Chloramine vs. chlorine effect on elastomeric properties at 70°C, 30 ppm, and pH 8.3 of EPDM-P

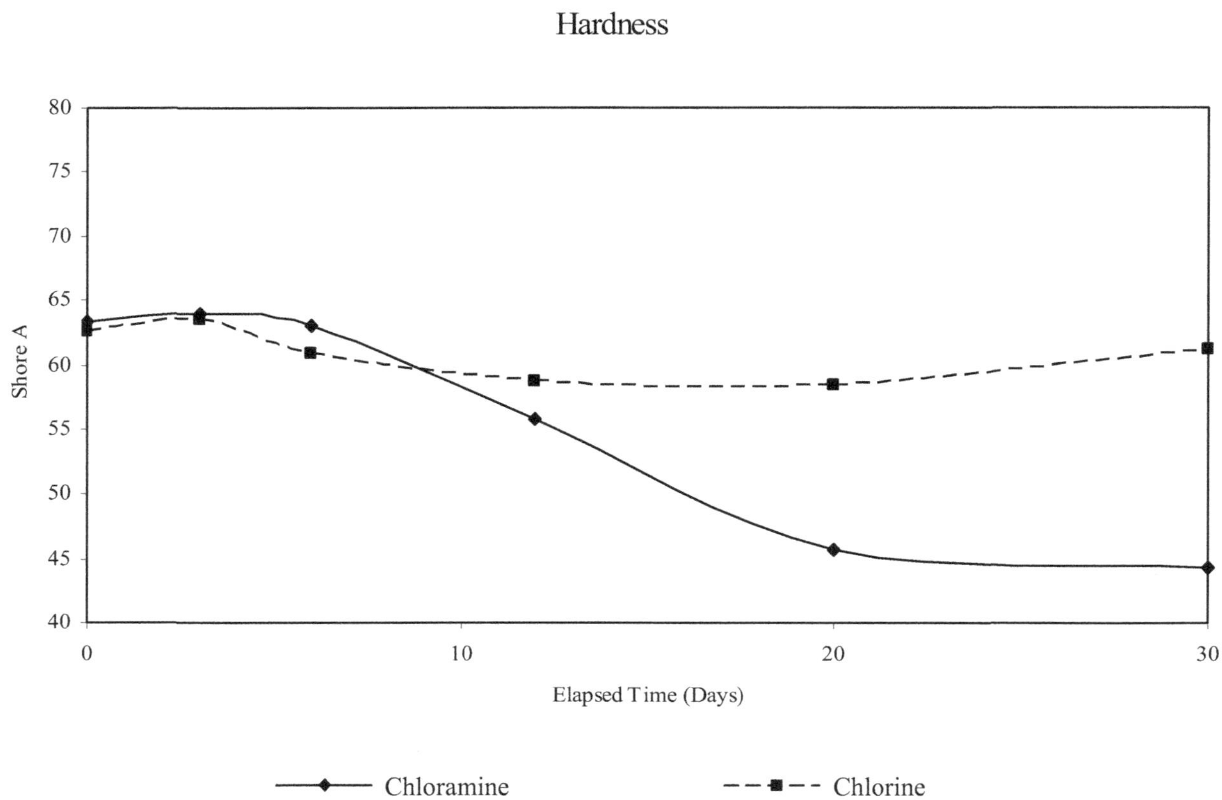

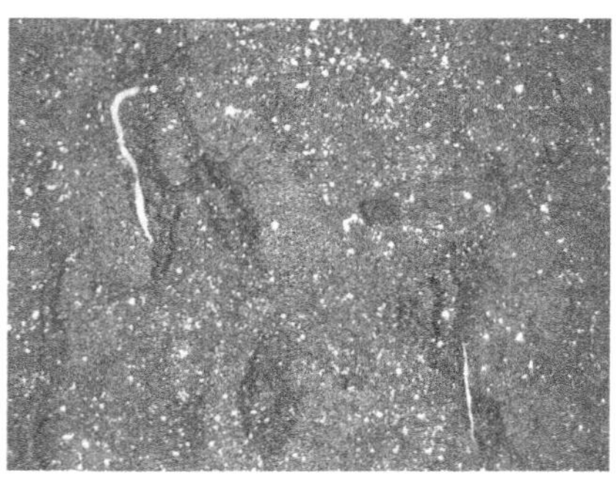

Chloramine at 30 days

Chlorine at 30 days

Figure A.108 Chloramine vs. chlorine effect on hardness at 70°C, 30 ppm, and pH 8.3 of EPDM-P with 40× magnified pictographs

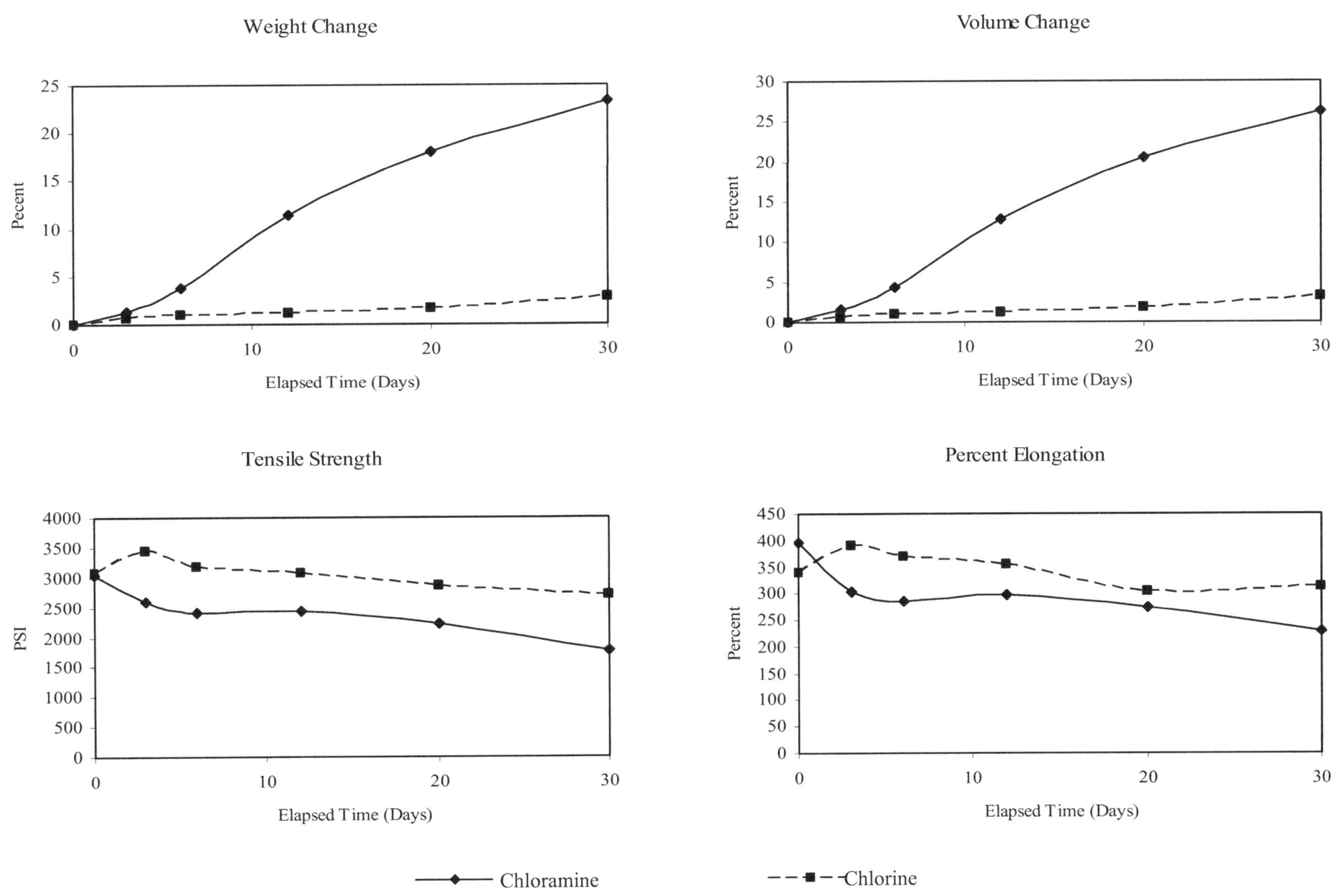

Figure A.109 Chloramine vs. chlorine effect on elastomeric properties at 23°C, 60 ppm, and pH 6.0 of natural rubber

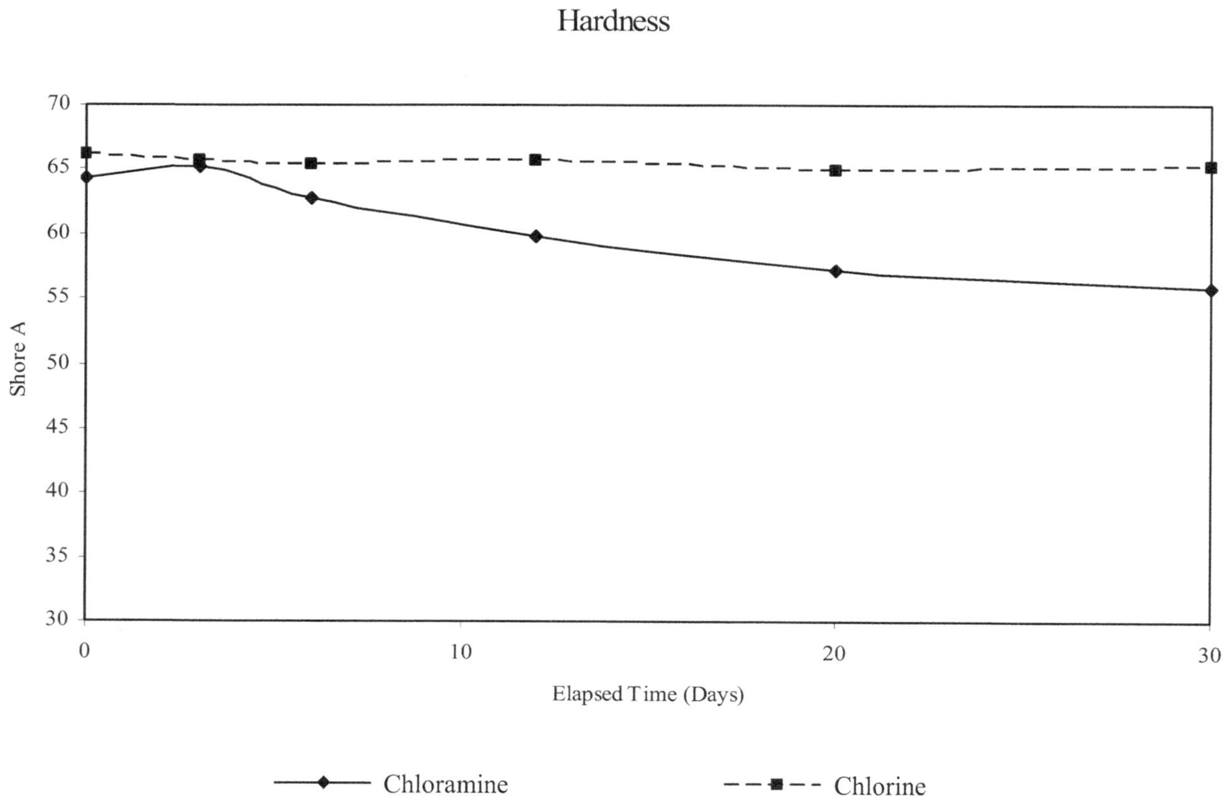

Chloramine at 30 days

Chlorine at 30 days

Figure A.110 Chloramine vs. chlorine effect on hardness at 23°C, 60 ppm, and pH 6.0 of natural rubber with 40× magnified pictographs

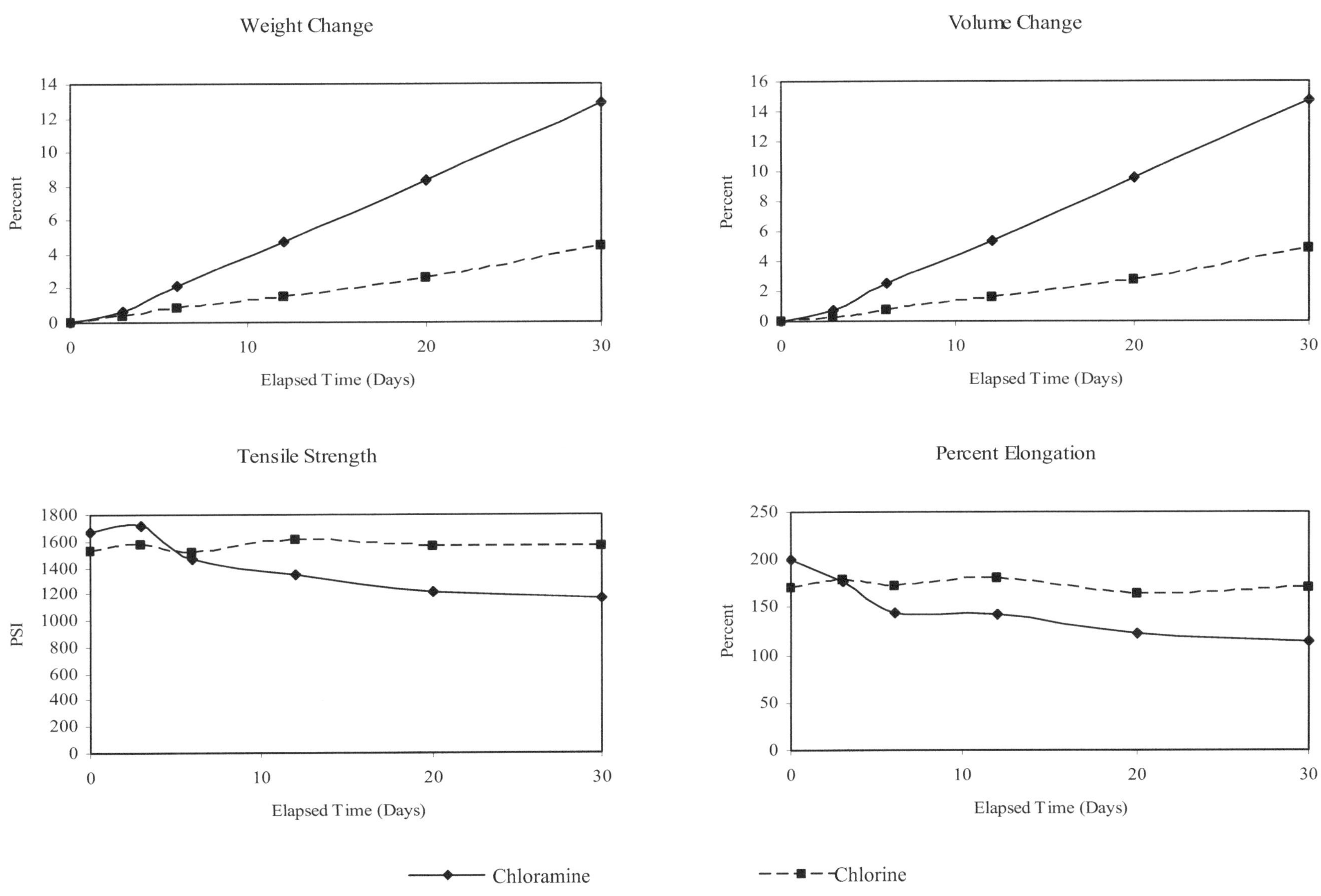

Figure A.111 Chloramine vs. chlorine effect on elastomeric properties at 23°C, 60 ppm, and pH 6.0 of EPDM-S

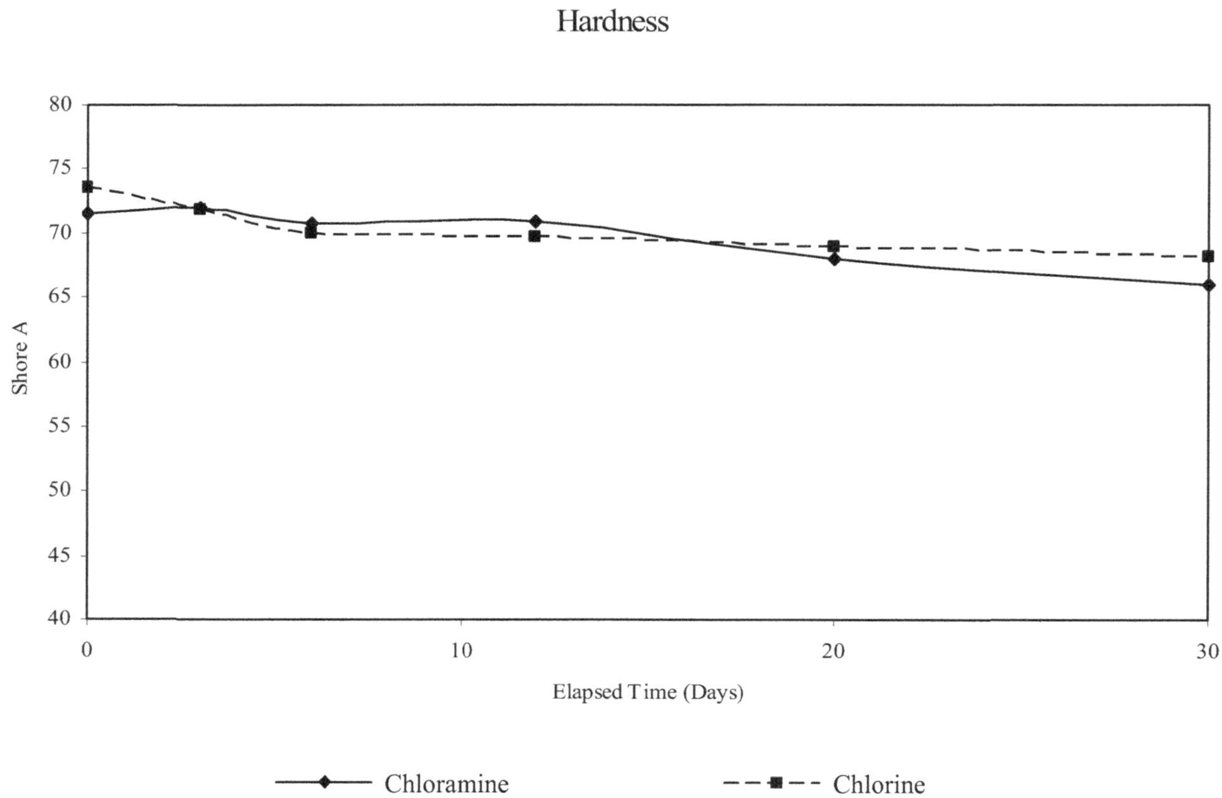

Chloramine at 30 days

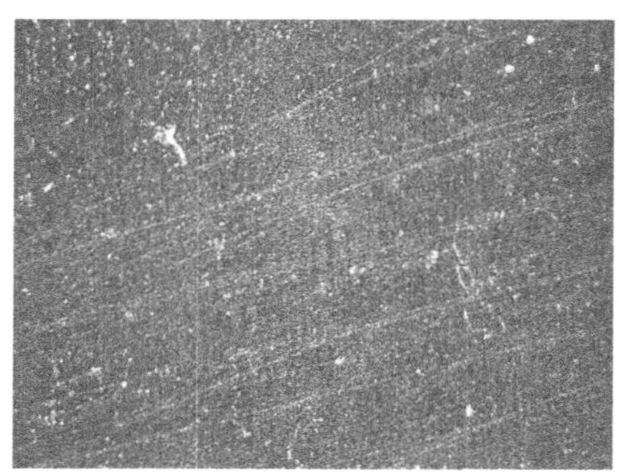

Chlorine at 30 days

Figure A.112 Chloramine vs. chlorine effect on hardness at 23°C, 60 ppm, and pH 6.0 of EPDM-S with 40× magnified pictographs

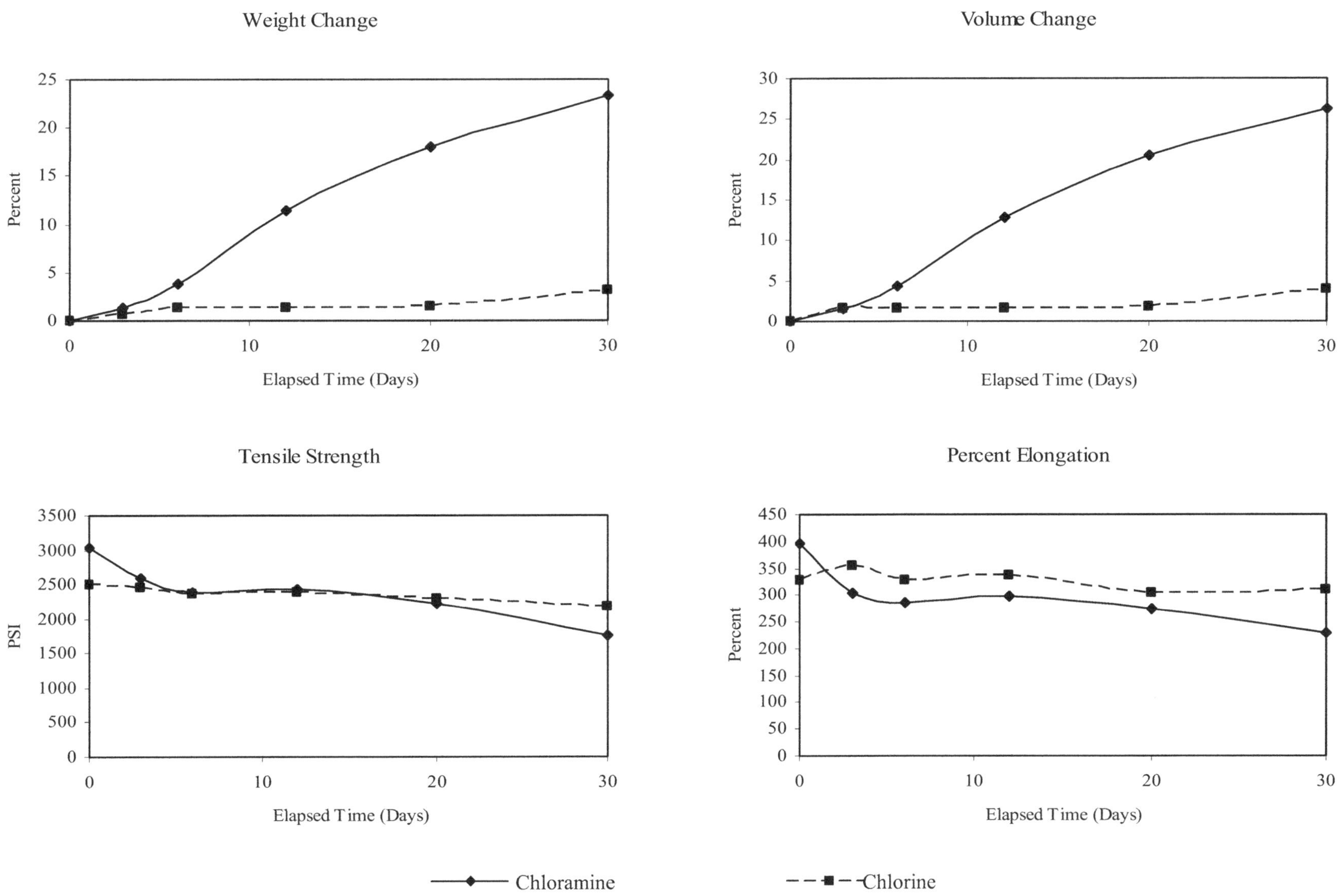

Figure A.113 Chloramine vs. chlorine effect on elastomeric properties at 23°C, 60 ppm, and pH 6.0 of nitrile (NBR)

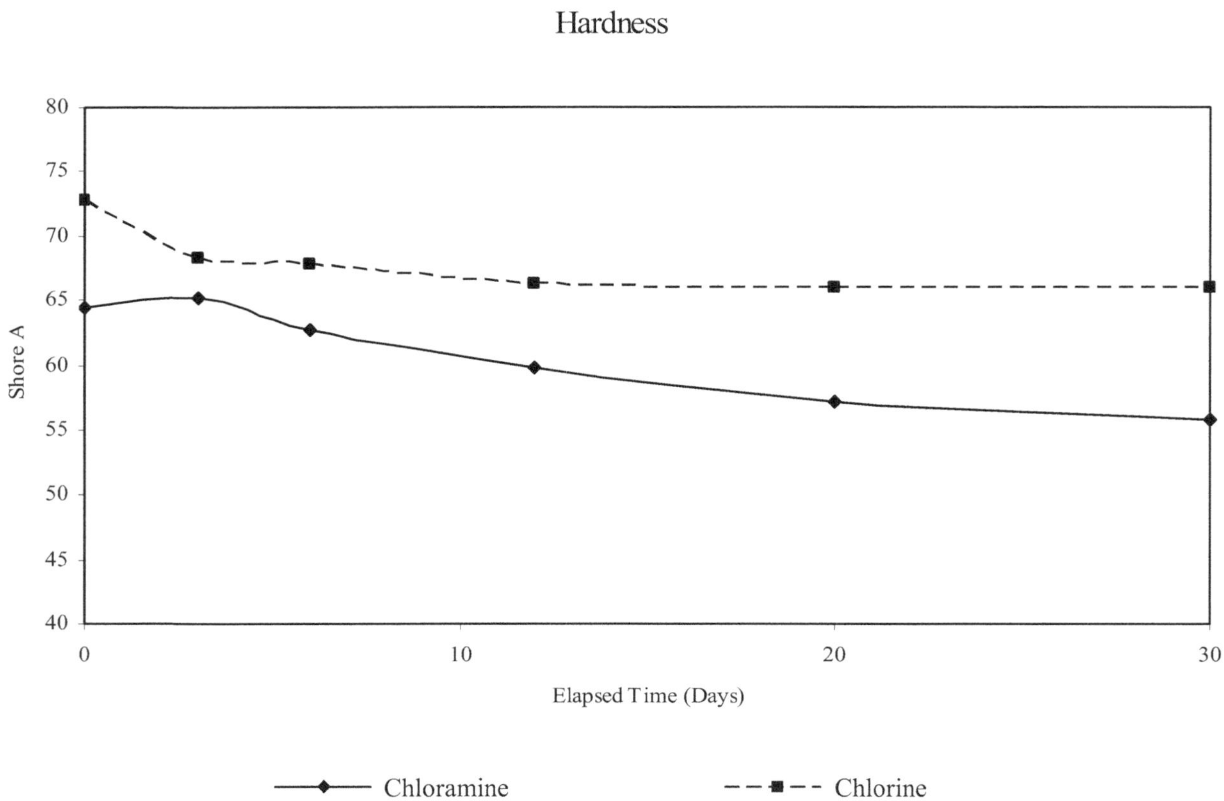

Chloramine at 30 days

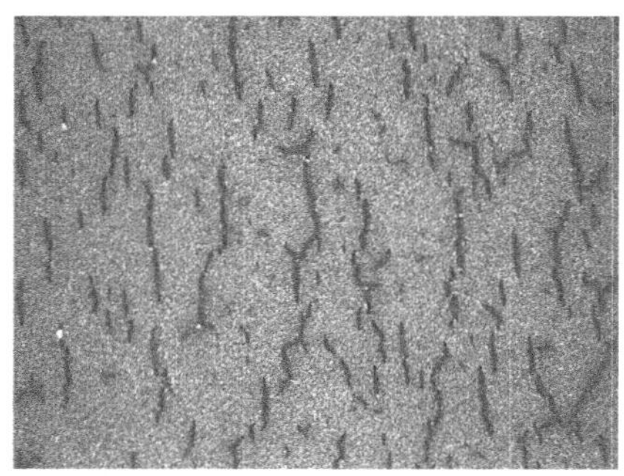

Chlorine at 30 days

Figure A.114 Chloramine vs. chlorine effect on hardness at 23°C, 60 ppm, and pH 6.0 of nitrile (NBR) with 40× magnified pictographs

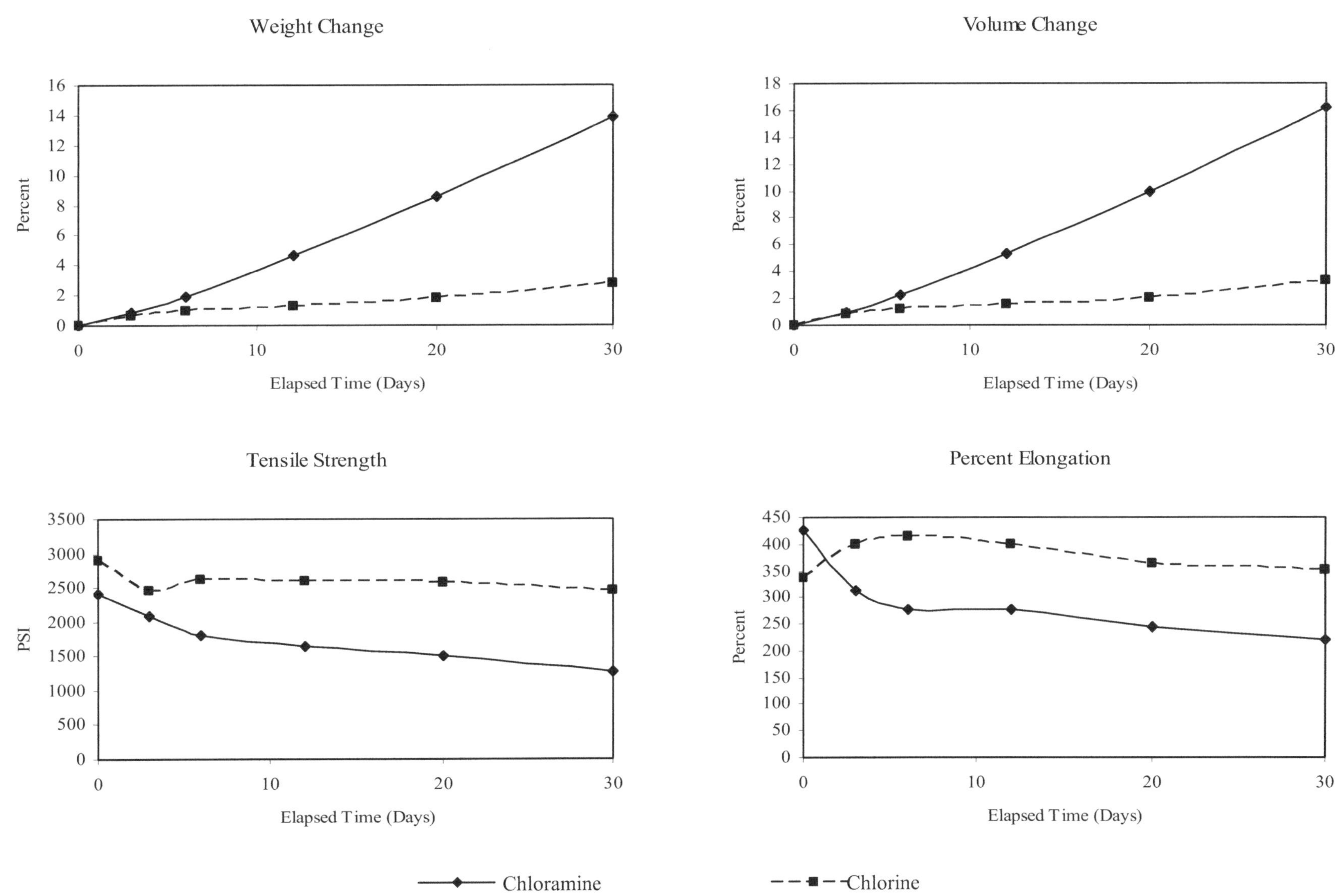

Figure A.115 Chloramine vs. chlorine effect on elastomeric properties at 23°C, 60 ppm, and pH 6.0 of SBR

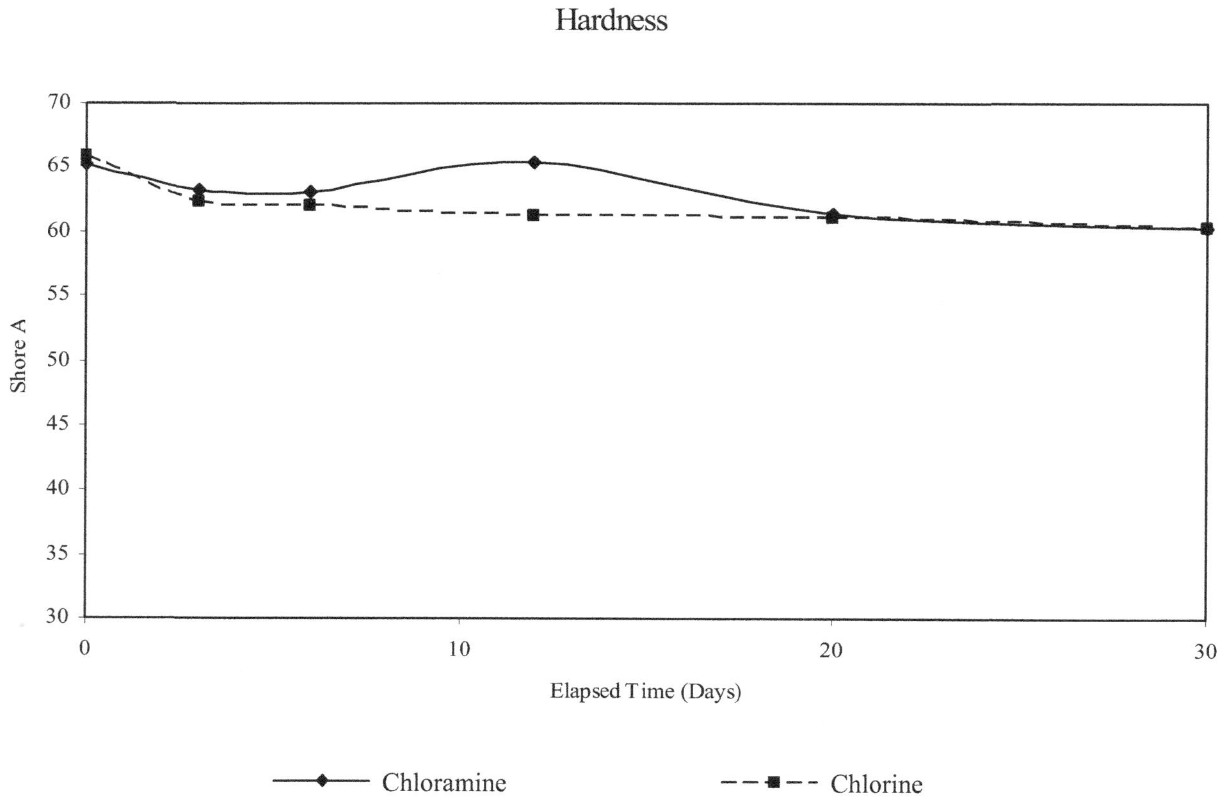

Chloramine at 30 days

Chlorine at 30 days

Figure A.116 Chloramine vs. chlorine effect on hardness at 23°C, 60 ppm, and pH 6.0 of SBR with 40× magnified pictographs

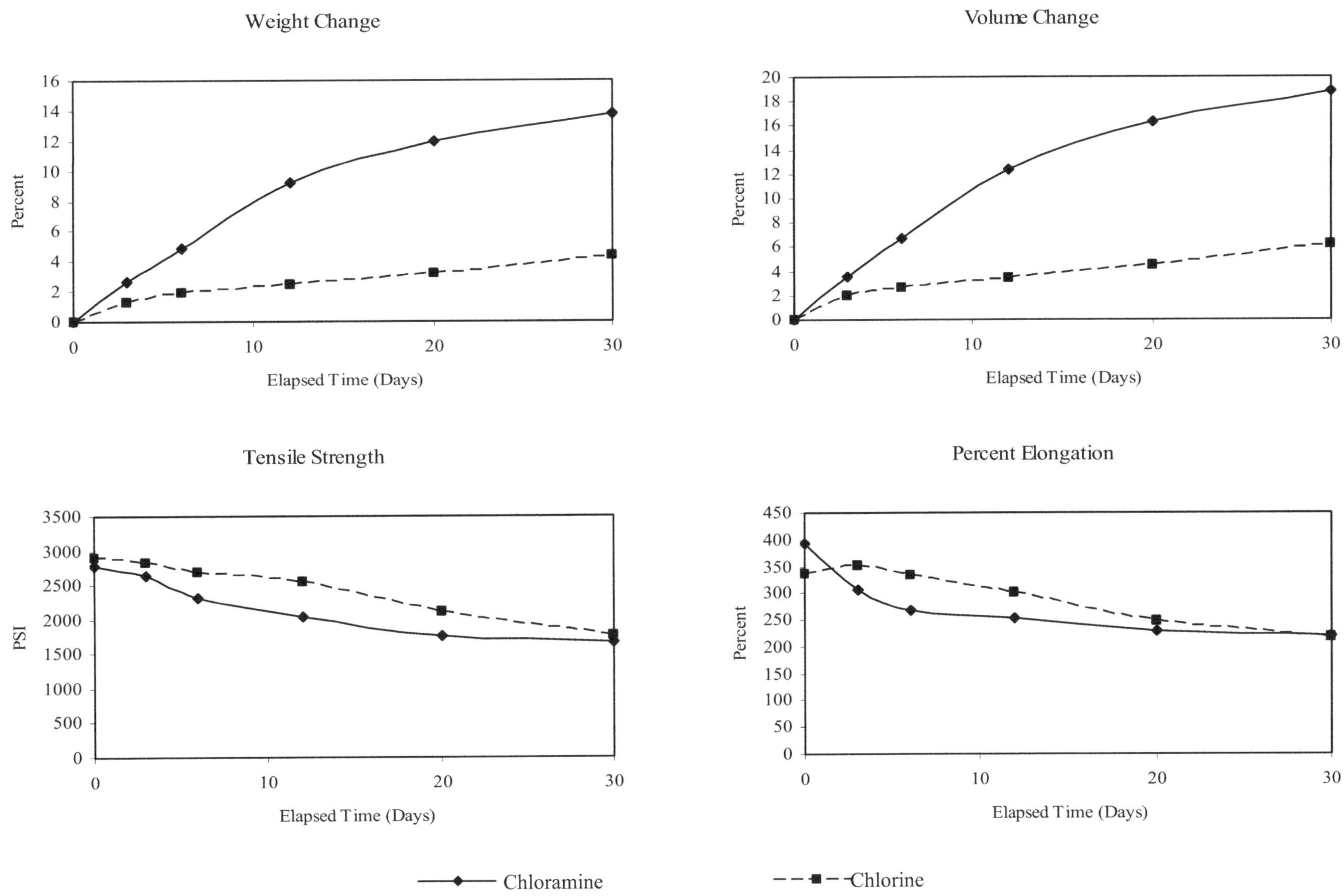

Figure A.117 Chloramine vs. chlorine effect on elastomeric properties at 23°C, 60 ppm, and pH 6.0 of neoprene

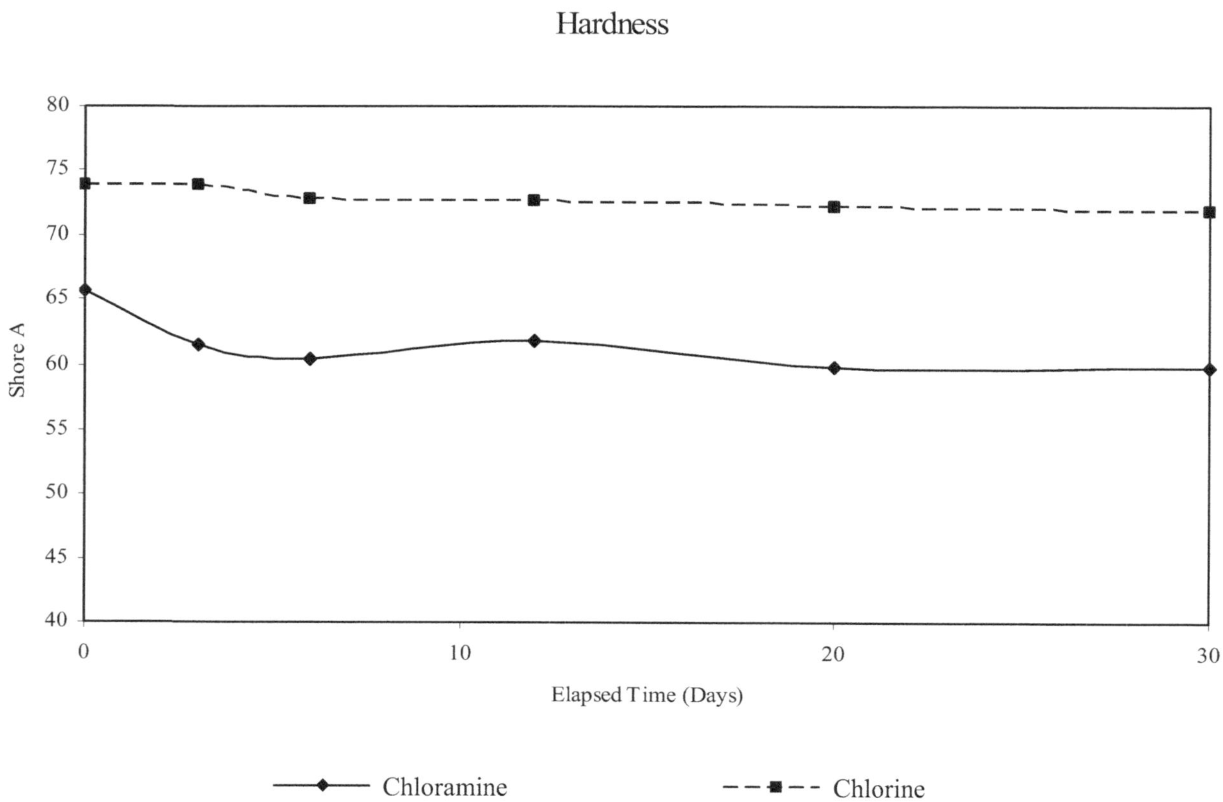

Chloramine at 30 days

Chlorine at 30 days

Figure A.118 Chloramine vs. chlorine effect on hardness at 23°C, 60 ppm, and pH 6.0 of neoprene with 40× magnified pictographs

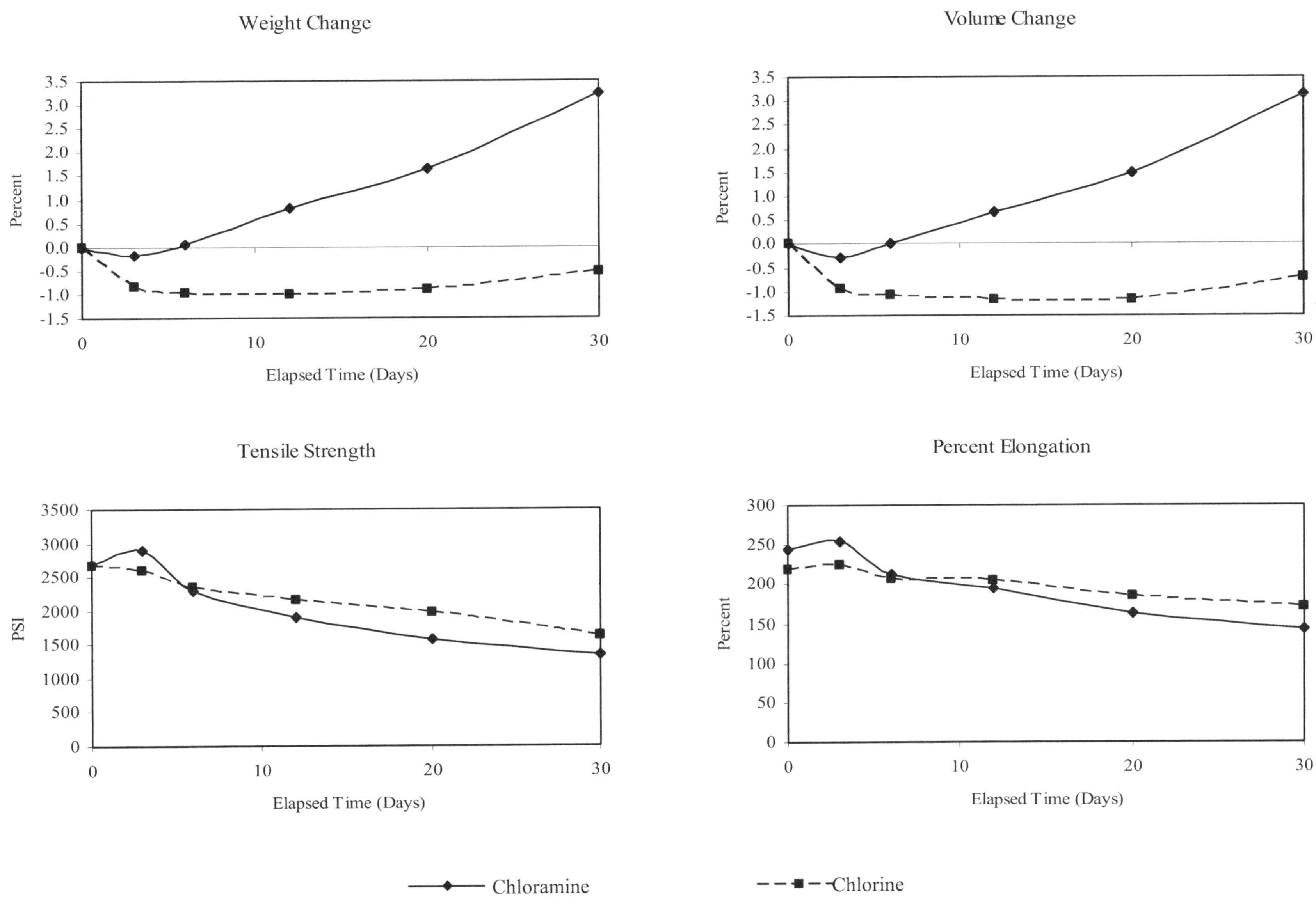

Figure A.119 Chloramine vs. chlorine effect on elastomeric properties at 23°C, 60 ppm, and pH 6.0 of EPDM-P

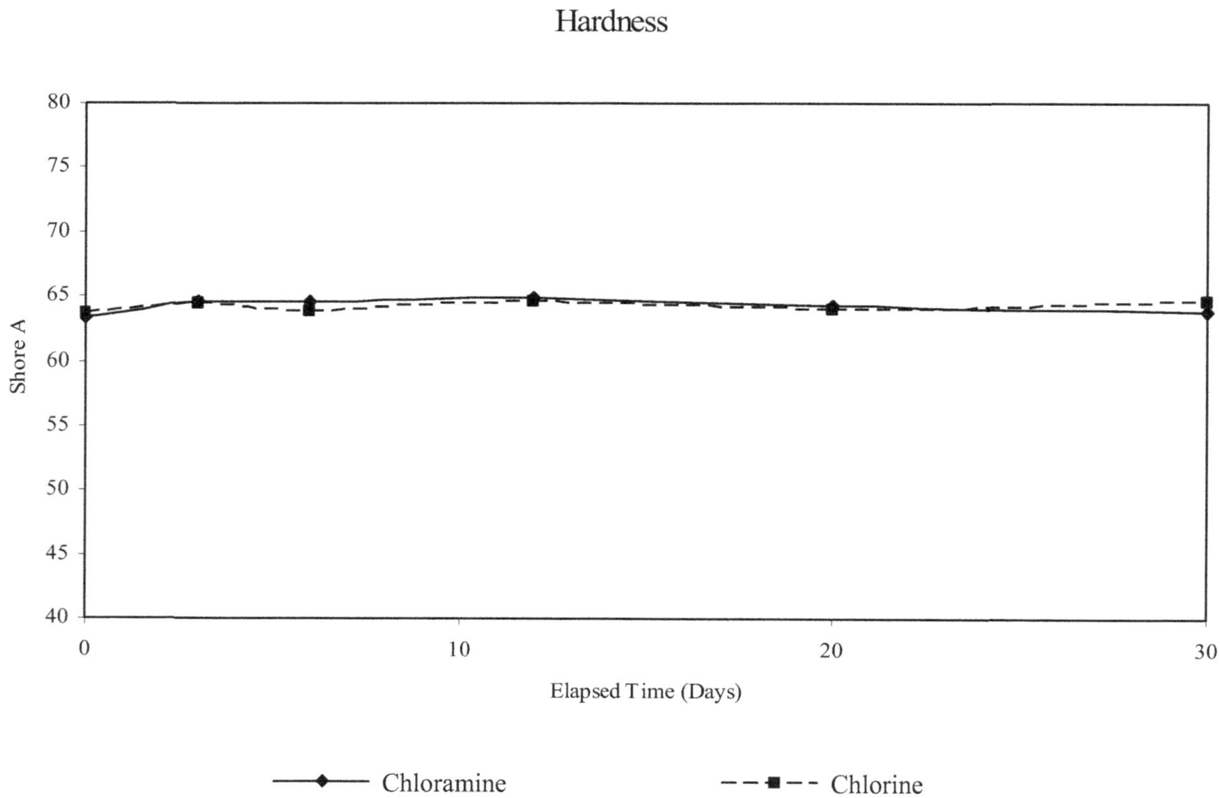

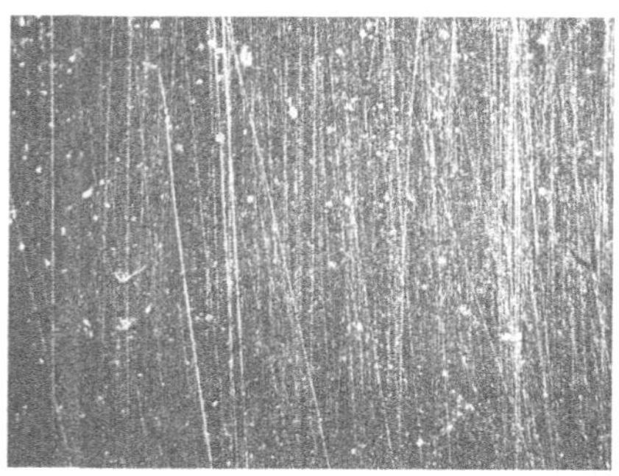

Chloramine at 30 days

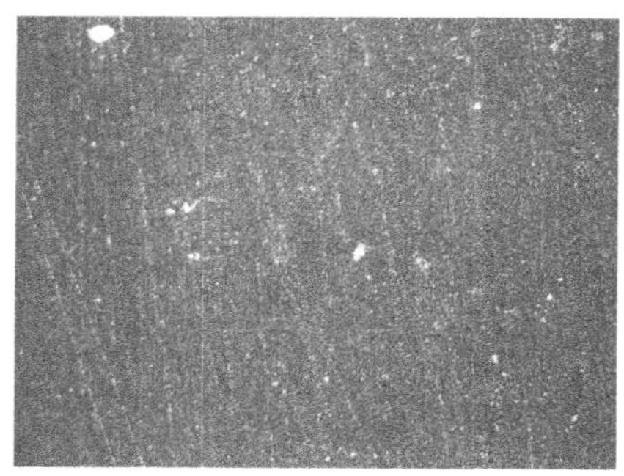

Chlorine at 30 days

Figure A.120 Chloramine vs. chlorine effect on hardness at 23°C, 60 ppm, and pH 6.0 of EPDM-P with 40× magnified pictographs

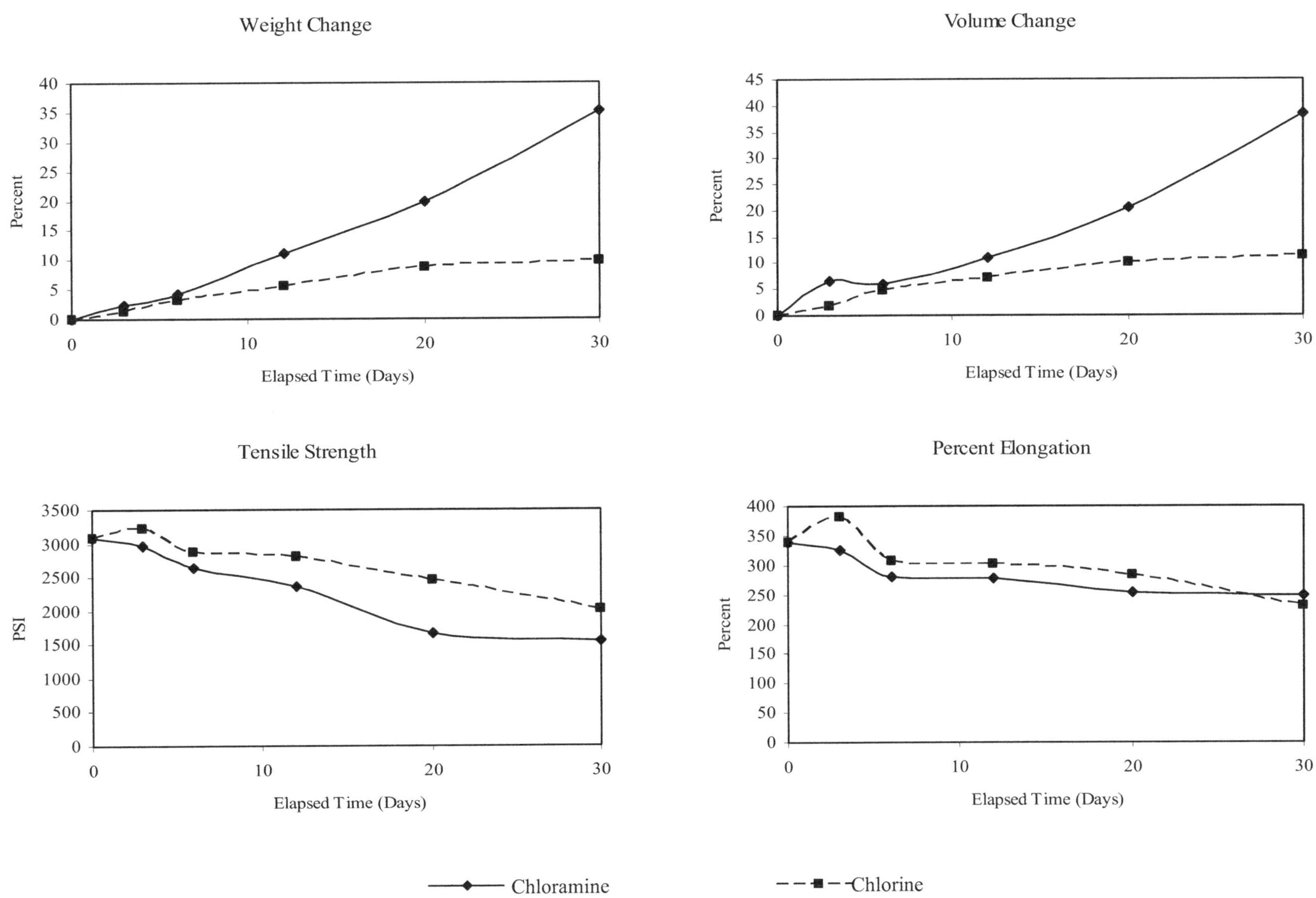

Figure A.121 Loop strain effects on hardness at 45°C, 30 ppm chloramine compared to 30 ppm chlorine at pH 8.3 of natural rubber

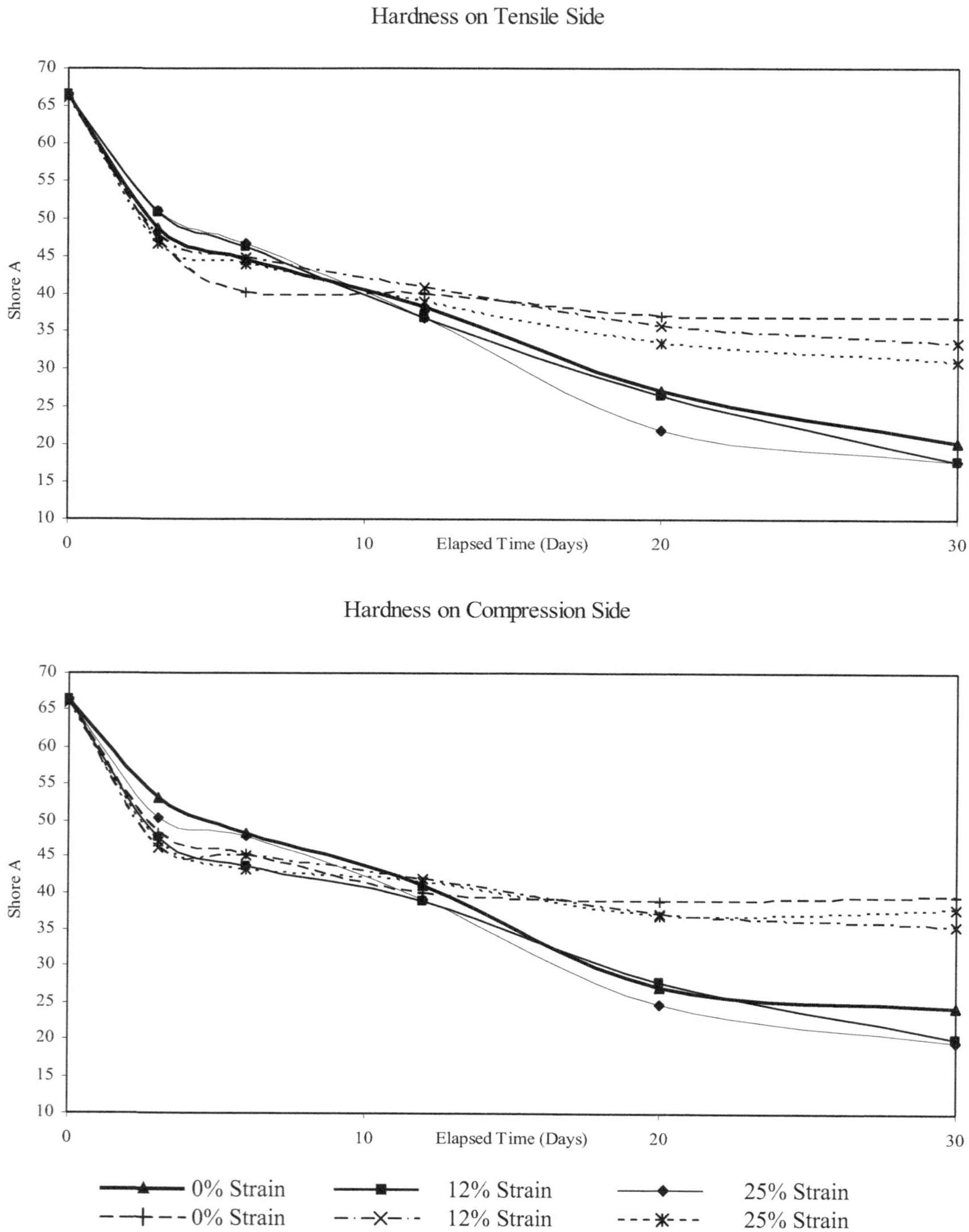

Figure A.122 Loop strain effects on hardness at 45°C, 30 ppm chloramine compared to 30 ppm chlorine at pH 8.3 of natural rubber

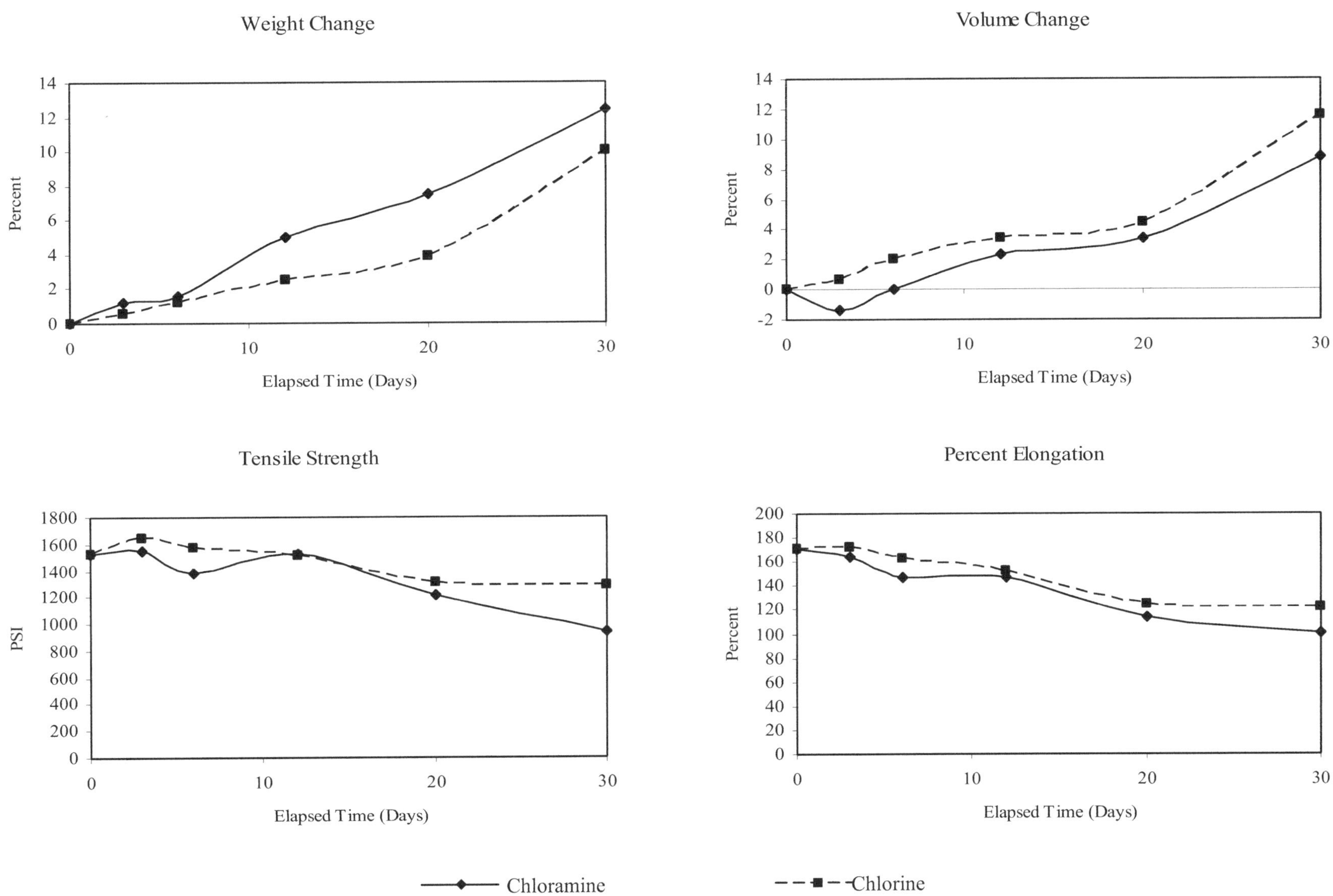

Figure A.123 Loop strain effects on hardness at 45°C, 30 ppm chloramine compared to 30 ppm chlorine at pH 8.3 of EPDM-S

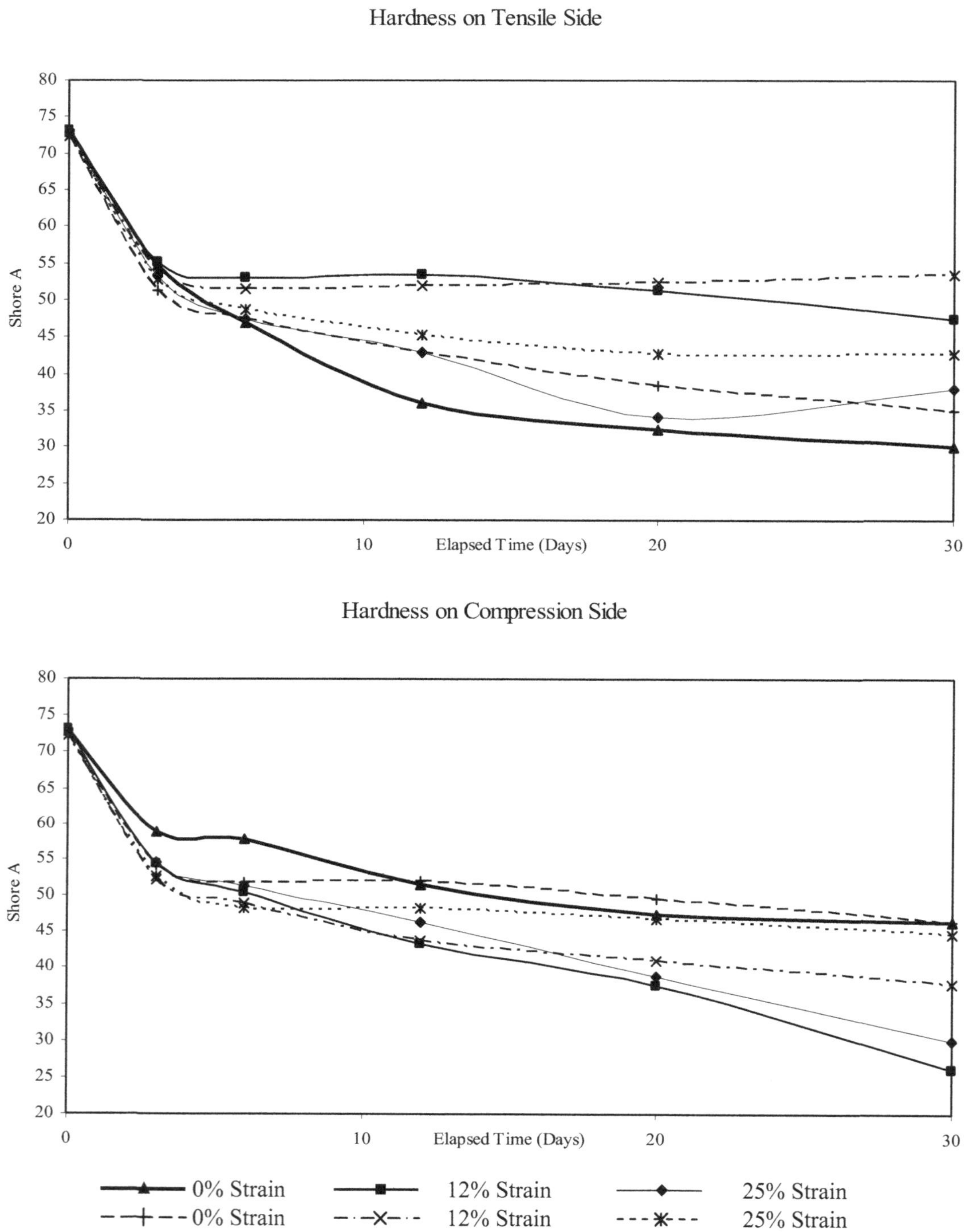

Figure A.124 Loop strain effects on hardness at 45°C, 30 ppm chloramine compared to 30 ppm chlorine at pH 8.3 of EPDM-S

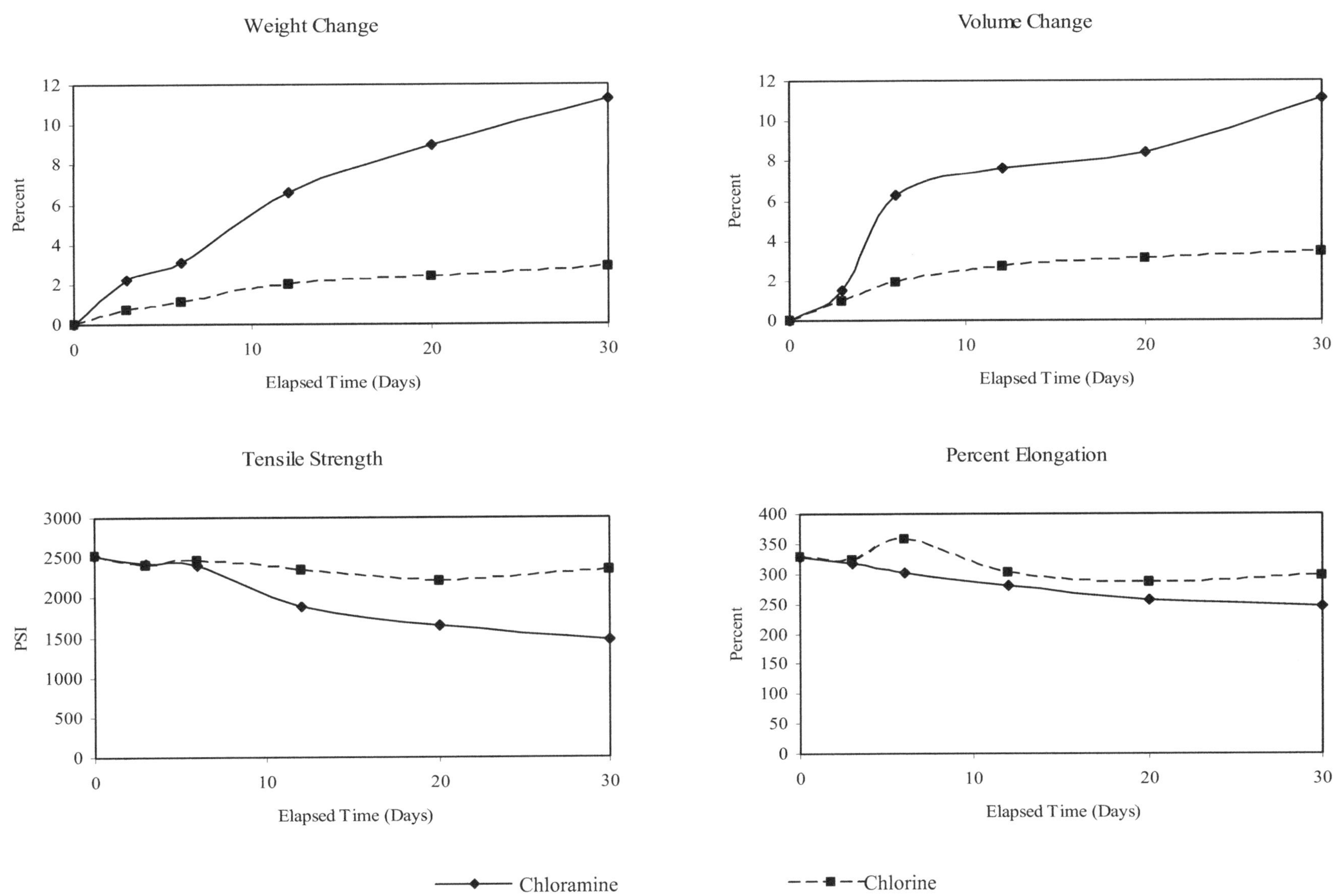

Figure A.125 Loop strain effects on hardness at 45°C, 30 ppm chloramine compared to 30 ppm chlorine at pH 8.3 of nitrile (NBR)

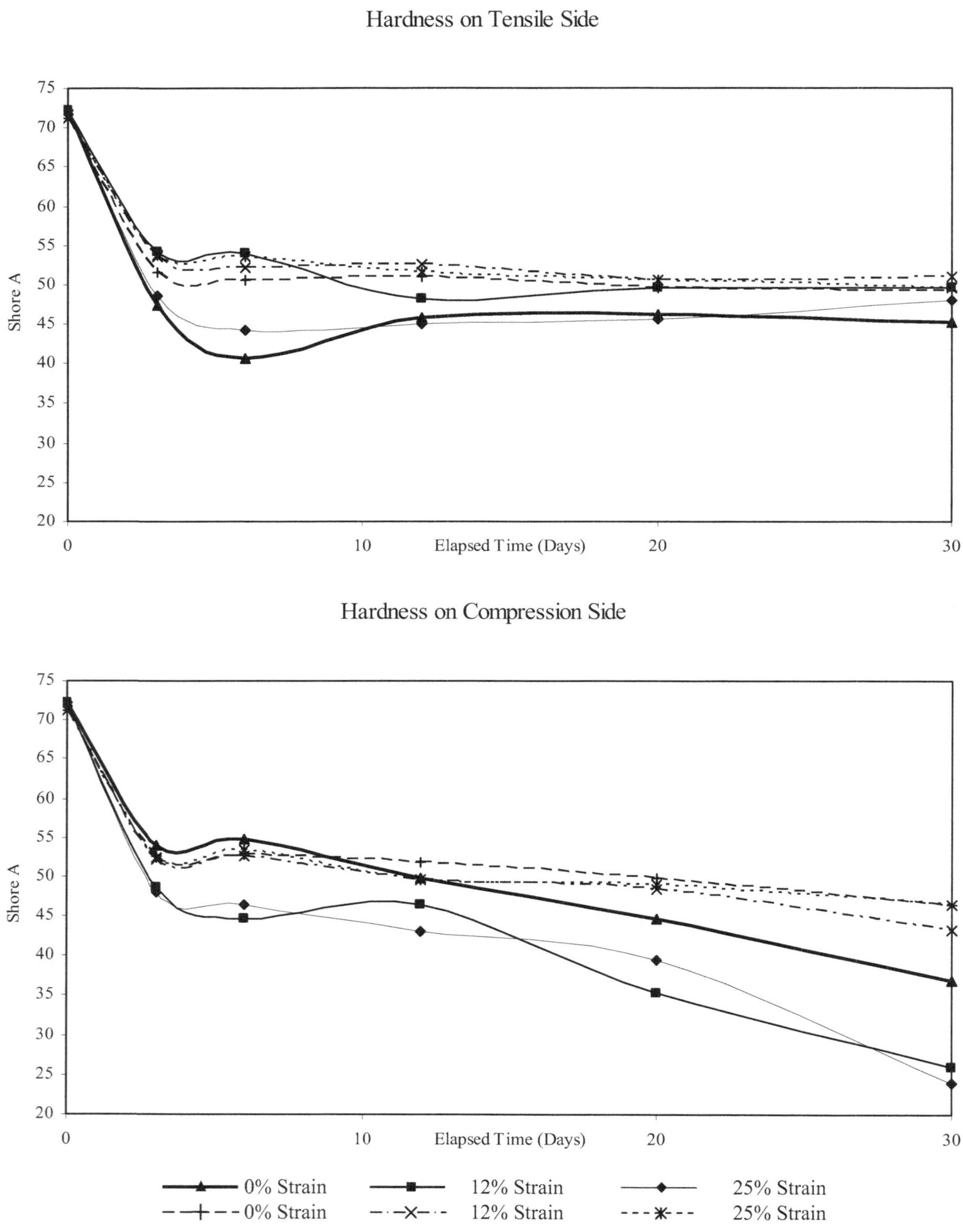

Figure A.126 Loop strain effects on hardness at 45°C, 30 ppm chloramine compared to 30 ppm chlorine at pH 8.3 of nitrile (NBR)

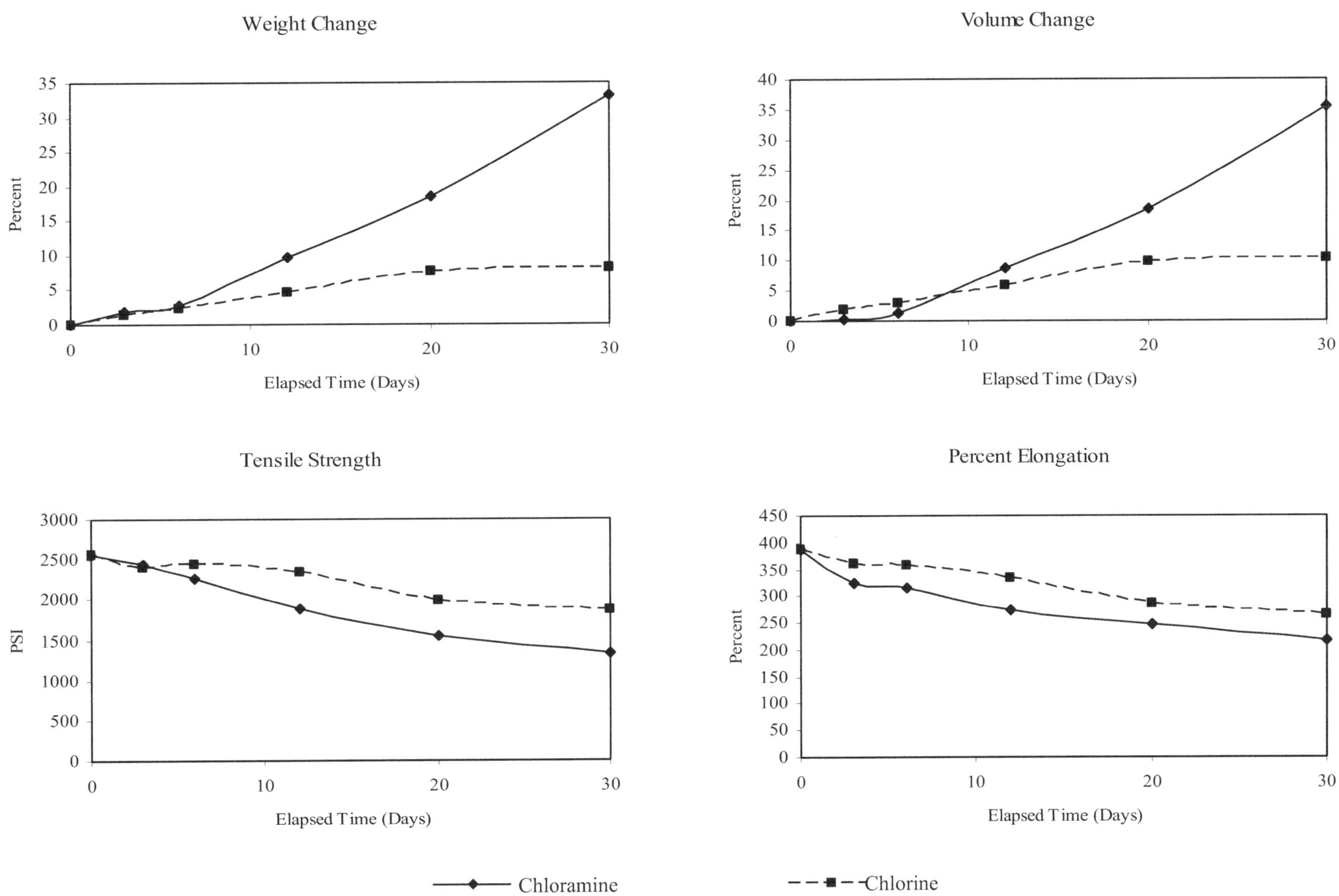

Figure A.127 Loop strain effects on hardness at 45°C, 30 ppm chloramine compared to 30 ppm chlorine at pH 8.3 of SBR

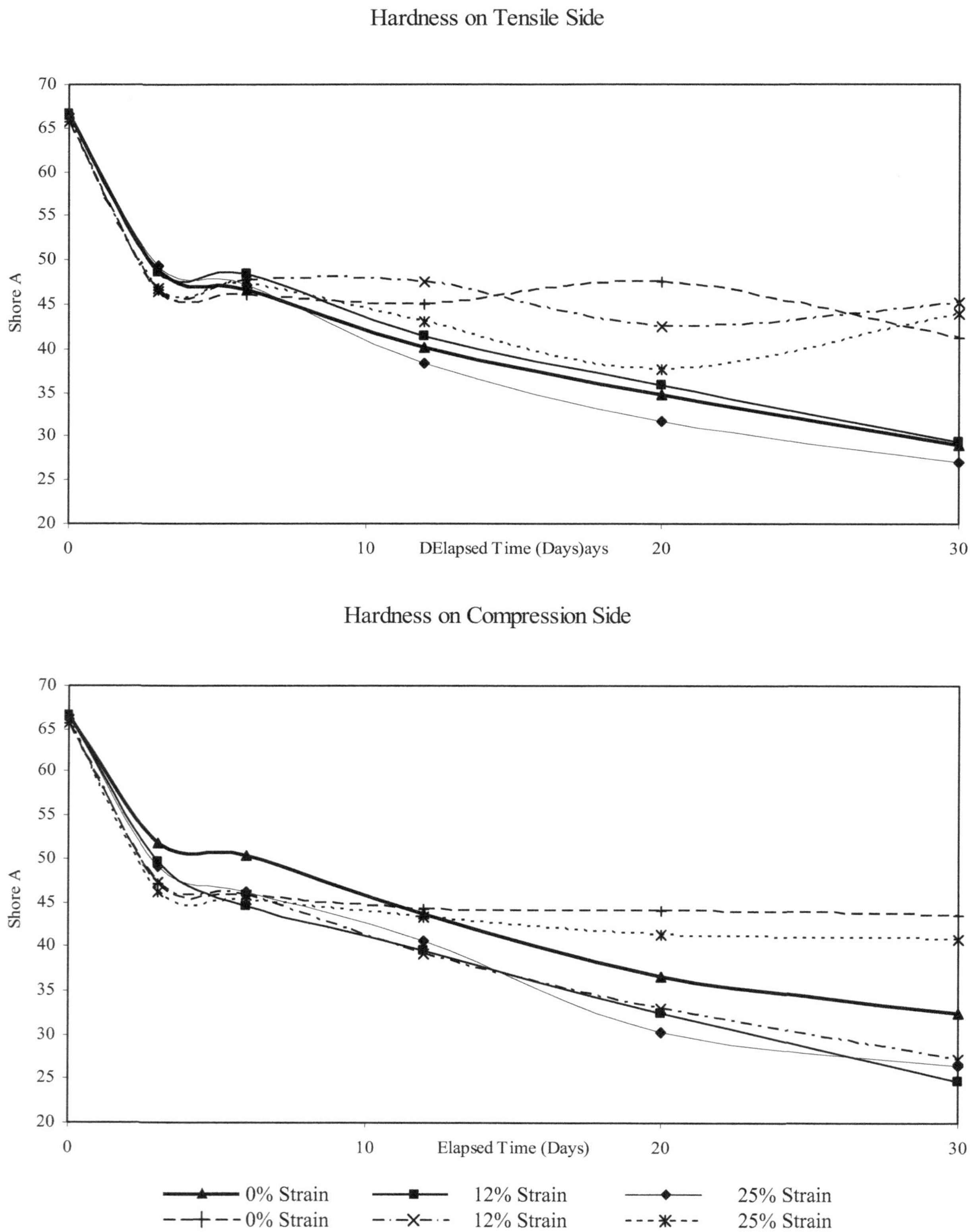

Figure A.128 Loop strain effects on hardness at 45°C, 30 ppm chloramine compared to 30 ppm chlorine at pH 8.3 of SBR

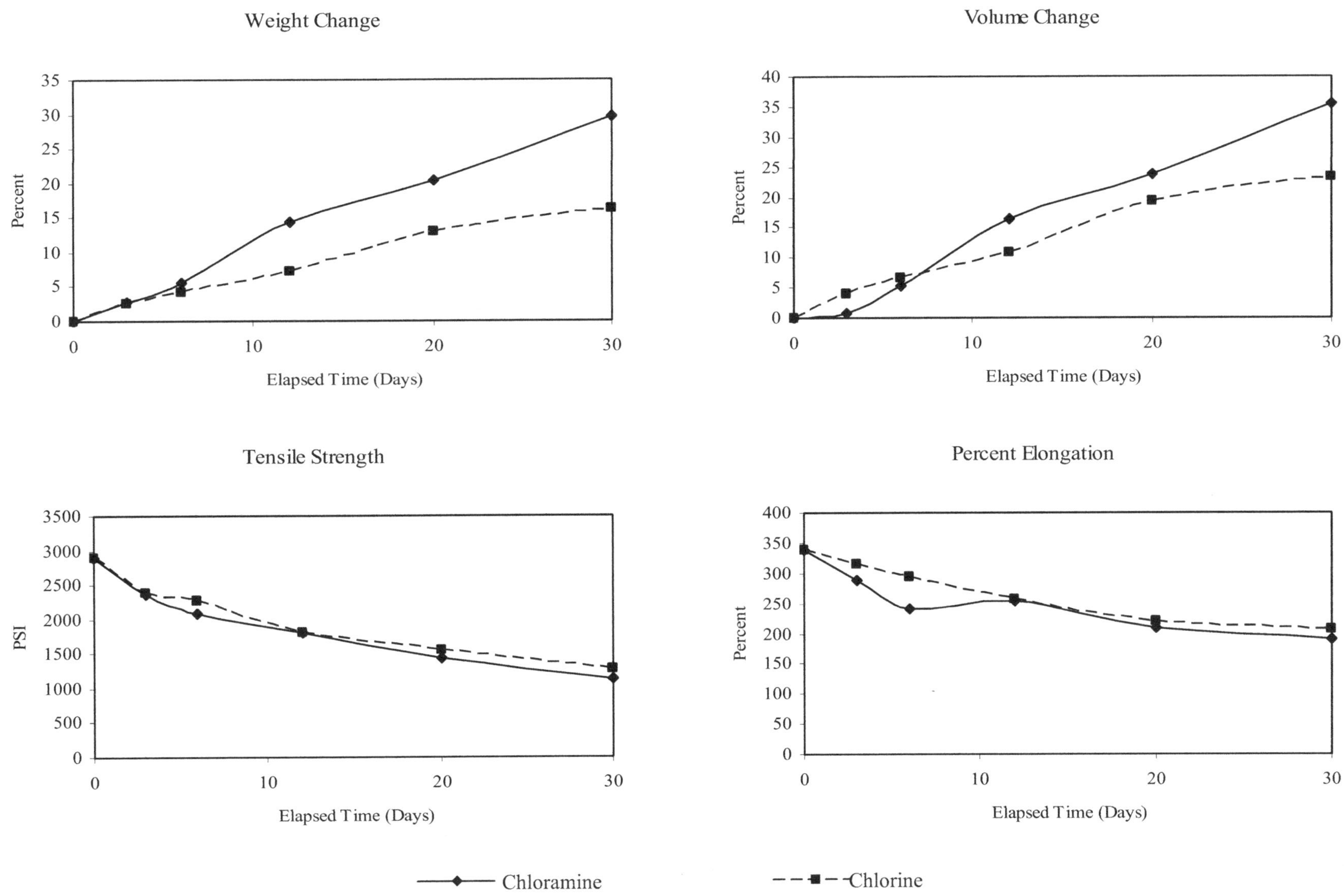

Figure A.129 Loop strain effects on hardness at 45°C, 30 ppm chloramine compared to 30 ppm chlorine at pH 8.3 of neoprene

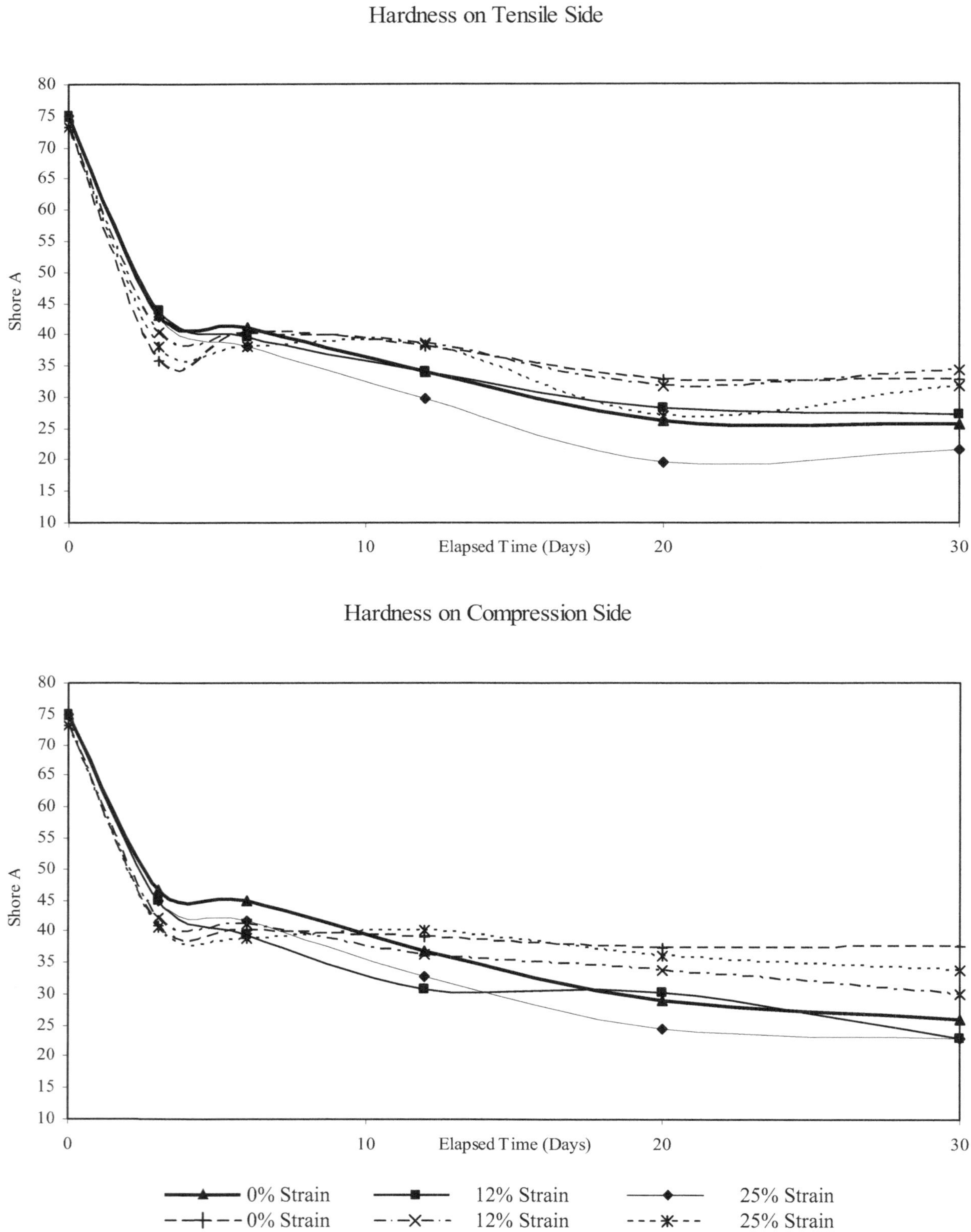

Figure A.130 Loop strain effects on hardness at 45°C, 30 ppm chloramine compared to 30 ppm chlorine at pH 8.3 of neoprene

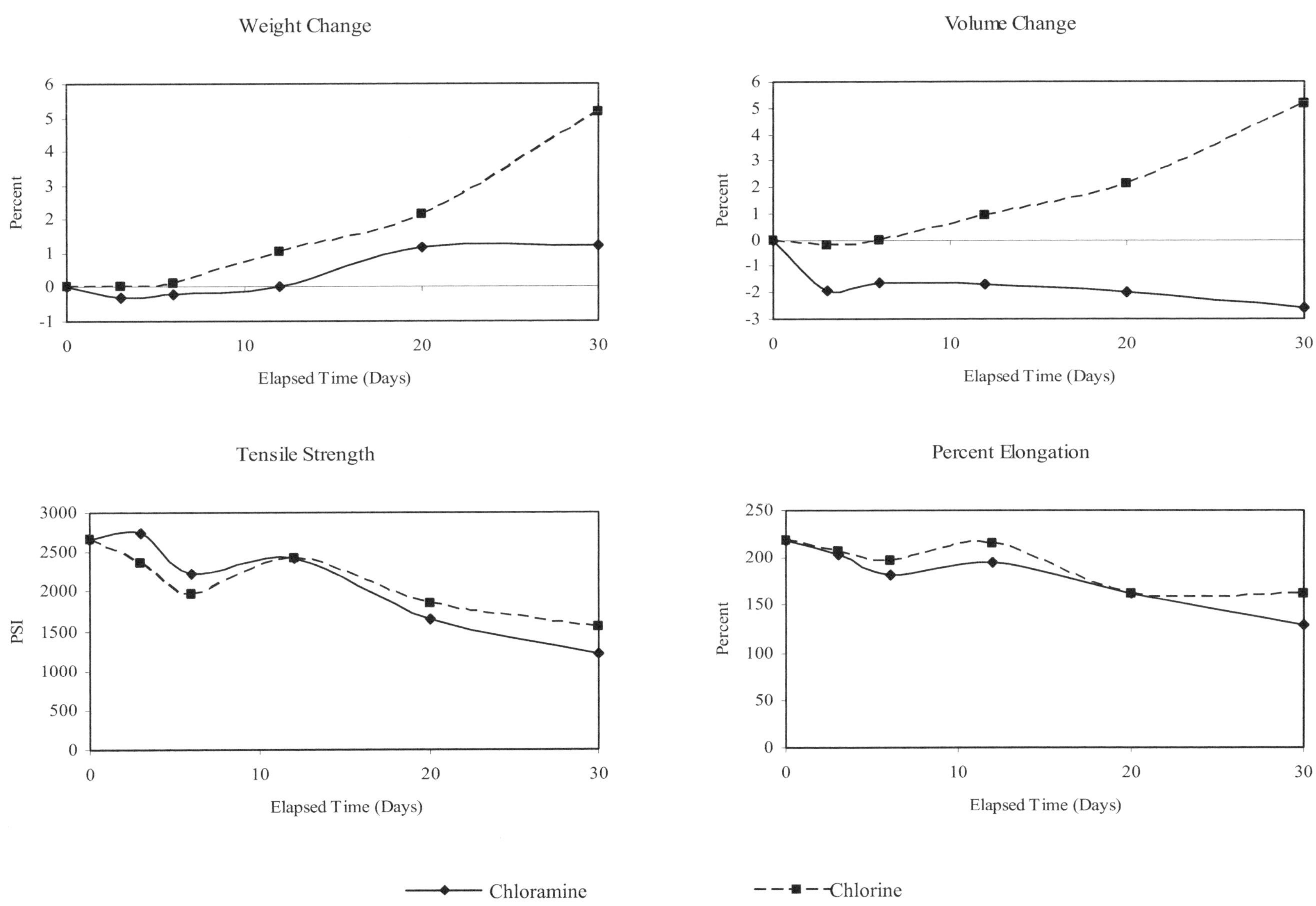

Figure A.131 Loop strain effects on hardness at 45°C, 30 ppm chloramine compared to 30 ppm chlorine at pH 8.3 of EPDM-P

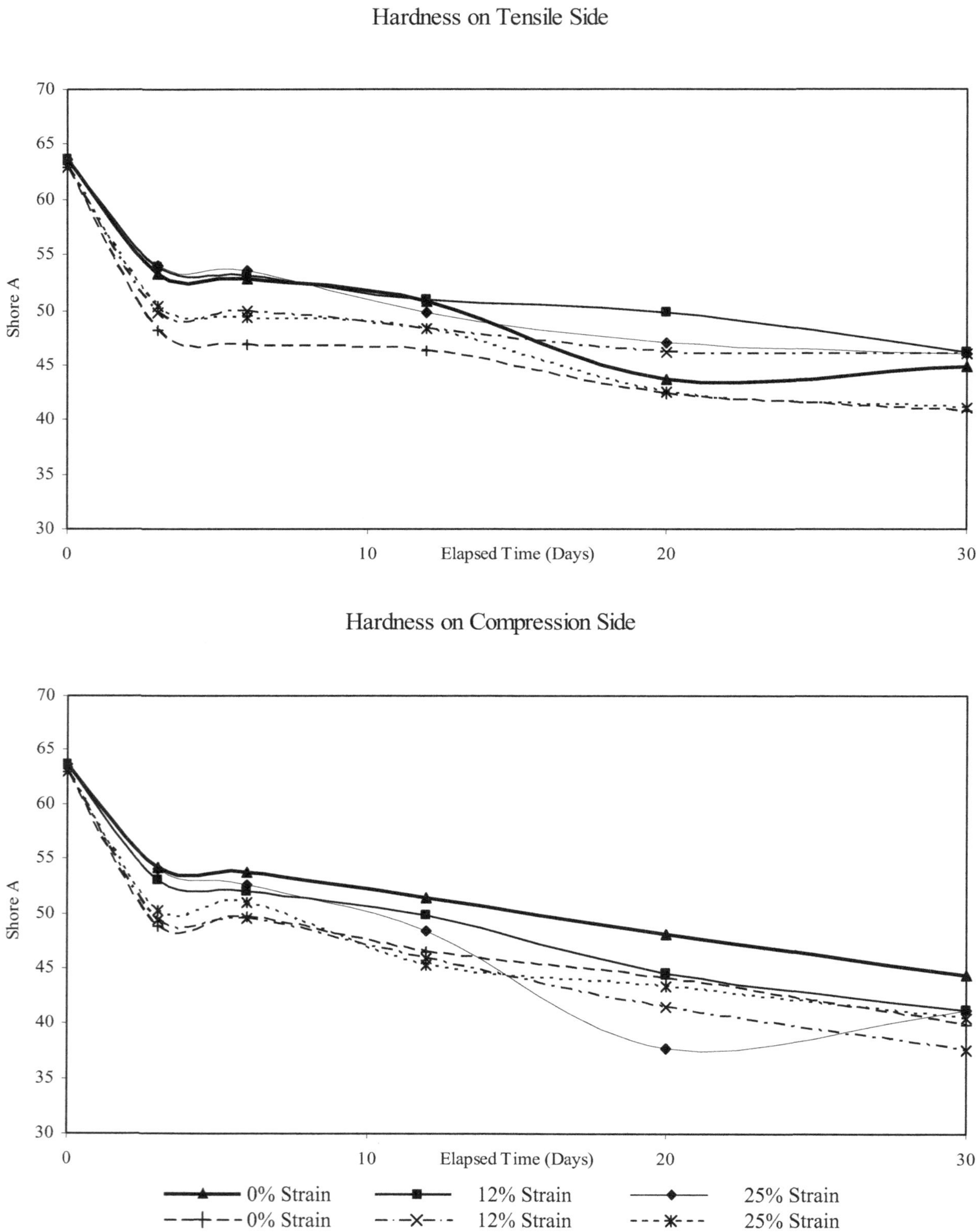

Figure A.132 Loop strain effects on hardness at 45°C, 30 ppm chloramine compared to 30 ppm chlorine at pH 8.3 of EPDM-P

APPENDIX B
FIELD SAMPLES

Component Description	Base Polymer	Date Removed	Service Length (months)	Sample Identifier	Estimated Disinfect. Concentration at Sample Site (ppm)
City of Akron Public Utilities Bureau					
4" compound meter rubber seal	Neoprene	11/9/04	25	-	0.8
Dresser coupling gasket (22" diam)	SBR	10/20/04	602	-	1.7
Prestressed concrete pipe o-ring gasket (22" diam)	Natural Rubber	10/20/04	602	-	1.7
City Utilities, Springfield, MO					
Meter seal	Nitrile	1/1/04	48	-	-
Greater Cincinnati Water Works					
12" American RS Gate Valve	SBR	12/13/04	47	Sheet # 335c Item # 1102	0.65
12" Gasket	SBR	4/19/05	-	-	0.88
2" Curbstop	EPDM-S/Nitrile	12/18/04	36	Br# 284973	0.9
2" Service Saddle	Nitrile	1/21/05	385	WO # 05-006702-002	1.15
3/4" Ferrule	EPDM-S	1/2/05	228	D BRANCH #941982	0.7
6" CJ gasket	Natural Rubber	1/10/05	505	WO O5.003560.001	0.7
6" CJ Gaskets (2)	SBR	2/2/05	974	WO # 05-013603-000	0.83
6" Gasket	SBR	3/4/05	254	-	0.87
6" RS valve top bonnet Gasket	Nitrile	10/4/04	21	-	1.0
8" Water Main	SBR	1/28/05	44	in Ziplock bag, no label	-
9" Water Main	SBR	1/31/05	-	wrapped in blue towel	-
Balata Valve	SBR	2/18/05	902	WO # 05-014863-000	0.79
Balata Valve	SBR	5/4/00	136	WO # 05-018564-001	0.84
Balata Valve	SBR	2/22/05	794	WO # 05-019350-000	0.86
Balata Valve	Unknown	2/17/05	902	WO # 05-014876-000	0.80
Blatta		3/16/05	0	05-027448-000	0.86
Blatta		3/16/05	-	05-027450-000	0.79
Blatta, MJ Rubber, Gaskets		3/8/05	495	04-119468-001	0.89
Bonnet O-Ring		2/25/05	-	04-067994-000	0.78
CJ Gasket	SBR	2/1/05	265	WO # 05-013106-001	0.93
Dual Gasket		3/10/05	278	04-068002-002	0.78

Component Description	Base Polymer	Date Removed	Service Length (months)	Sample Identifier	Estimated Disinfect. Concentration at Sample Site (ppm)
Greater Cincinnati Water Works (continued)					
Dual Purpose Gasket		3/15/05	423	05-024725-003	0.69
Ferrure (Extended)		5/2/05	-	05-043272-001	1.00
Flange Gasket		3/8/05	52	04-084959-014	0.22
Flat Elbow Gasket (K81 Kennedy)	SBR	2/16/05	98	WO # 03-121330-001	1.12
Gasket from Leak Clamp		4/22/05	604	05-040938-00	0.83
Leak Clamp Gasket with Boss		3/8/05	1143	05-020011-001	0.85
Leak Clamp Gasket with Boss		3/7/05	555	05-020103-001	0.89
Meter Seal	EPDM-S	12/22/04	204	88909980	0.84
Meter Seal	EPDM-S	2/25/05	206	88911402	-
Meter Seal	EPDM-S	2/15/05	206	88911488	-
Meter Seal	EPDM-S	2/18/05	206	88911556	-
Meter Seal	EPDM-S	12/22/04	204	88911594	0.83
Meter Seal	EPDM-S	2/18/05	206	88912210	-
Meter Seal	EPDM-S	2/20/05	206	88912318	-
Meter Seal	EPDM-S	2/20/05	206	88912322	-
Meter Seal	EPDM-S	2/18/05	206	88912546	-
Meter Seal	EPDM-S	2/20/05	206	88912506	-
Meter Seal	EPDM-S	3/11/05	206	88912850	-
Meter Seal	EPDM-S	2/15/05	206	88913072	-
Meter Seal	EPDM-S	3/11/05	206	89913107	-
Meter Seal	EPDM-S	?	-	(cream colored tag)	-
Meter Seal	EPDM-S	8/27/04	260	83000735 4C-13	0.87
Meter Seal	EPDM-S	1/5/04	240	84003242 3C-17	0.81
Meter Seal	EPDM-S	4/8/04	219	86791942 4C-3	0.84
Meter Seal	EPDM-S	3/30/04	195	87900857 4C-5	0.86
Meter seal	EPDM-S	11/24/04	203	88911316 3C-53	0.92
Meter Seal	EPDM-S	1/4/04	192	88912212 3C-15	0.94
Meter Seal	EPDM-S	4/6/04	195	88913075 4C-1	0.87
Meter Seal	Nitrile	2/17/05	206	88913089	-
Meter Seal		2/20/05	206	88912856	-

Component Description	Base Polymer	Date Removed	Service Length (months)	Sample Identifier	Estimated Disinfect. Concentration at Sample Site (ppm)
Greater Cincinnati Water Works (continued)					
MJ Gaskets	SBR	1/19/05	59	-	0.81
MJ Rubber		3/10/05	158	04-105398-001	0.78
MJ Rubber		9/3/04	-	04-082817-001	0.78
MJ Rubber		9/3/04	-	04-082817-001	0.78
MJ Rubber		3/11/05	903	04-118683-009	0.89
MJ Rubber		3/8/05	495	04-119468-001	0.89
MJ Rubber		3/8/05	495	04-119468-001	0.89
Schlumberger Water Meter	Nitrile	1/22/05	85	Old Meter # 98004662	0.7
Greater Vancouver Regional District					
12" Robar Coupling	SBR	12/1/04 (est.)	191	-	0.3 to 0.7
Diaphragm Type Globe Valve	SBR	12/1/04 (es.t)	323	-	0.3 to 0.8
Roll Seal Valve	SBR	12/1/04 (est.)	251	-	0.01 to 0.1
Louisville Water Company					
Check Valve	EPDM-S	3/31/05	183	2628482	1.8
Check Valve	Natural Rubber	3/22/05	219	027594	2.0
Check Valve	Nitrile	4/13/05	195	274122	2.0
Check Valve	SBR	4/5/05	215	037534	2.0
Check Valve	SBR	3/21/05	225	26255104	2.0
Flapper Valve		7/1/05	72	B.E. Payne	-
Gate Seat/Glove (4")	EPDM-S	2/25/05	220	273848	2.0
Meter seal	EPDM-S	11/10/04	-	30002174	-
Meter seal	EPDM-S	9/24/04	229	30005796	-
Meter seal	EPDM-S	3/30/00	-	70018779	-
Meter Seal	EPDM-S	3/9/05	-	70028817	-
Meter Seal	EPDM-S	4/1/05	-	70029443	-
Meter Seal	EPDM-S	12/13/04	47	70032173	-
Meter seal	EPDM-S	10/27/04	82	70033851	-
Meter Seal	EPDM-S	2/3/05	-	70035767	-
Meter Seal	EPDM-S	3/16/05	79	70044018	-

Component Description	Base Polymer	Date Removed	Service Length (months)	Sample Identifier	Estimated Disinfect. Concentration at Sample Site (ppm)
Louisville Water Company (continued)					
Meter Seal	EPDM-S	1/24/05	85	70044024	-
Meter seal	EPDM-S	1/14/05	204	88908782	-
Meter seal	EPDM-S	1/1/05	204	88909120	-
Meter seal	EPDM-S	2/1/05	205	88909985	-
Meter seal	EPDM-S	11/1/04	202	88912246	-
Meter seal	EPDM-S	1/1/05	204	88912582	-
Meter seal	EPDM-S	1/28/05	205	88912605	-
Meter seal	EPDM-S	2/1/05	205	88912895	-
Meter seal	EPDM-S	2/1/05	205	88912916	-
Meter seal	EPDM-S	1/1/05	204	88913086	-
Meter seal	EPDM-S	2/1/05	205	88913106	-
Meter seal	EPDM-S	1/1/05	204	88913121	-
Meter seal	EPDM-S	2/1/05	205	88913282	-
Meter seal	Unknown	11/1/04	202	88912247	-
Meter seals in clear zip-lock bag	EPDM-S	10/1/04	84	70034081	-
Meter seals in clear zip-lock bag	EPDM-S	9/29/04	121	30787340	-
Meter seals in clear zip-lock bag	Unknown	9/18/04	55	30782510	-
Meter seals in clear zip-lock bag	Unknown	9/1/04	79	70034080	-
Meter seals in clear zip-lock bag	Unknown	10/3/04	58	70066651	-
Meter seals in clear zip-lock bag		10/3/04	180	29003272	-
Neoprene Flushometer Parts	Nitrile/Neoprene	1/1/05	12	-	-
O-Rings - Revolving Nut, Short Stem Corroded	Nitrile	10/30/03	64	605803	-
O-Rings - Revolving Nut, Short Stem Corroded	Nitrile	11/3/03	47	607073	-
O-Rings - Revolving Nut, Short Stem Corroded	Nitrile	11/7/03	48	607164	-
O-Rings - Revolving Nut, Short Stem Corroded	Nitrile	11/7/03	48	607165	-
O-Rings - Revolving Nut, Short Stem Corroded	Nitrile	10/20/03	42	607615	-

Component Description	Base Polymer	Date Removed	Service Length (months)	Sample Identifier	Estimated Disinfect. Concentration at Sample Site (ppm)
Louisville Water Company (continued)					
O-Rings - Revolving Nut, Short Stem Corroded	Nitrile	11/13/03	18	609643	-
Parker Seals - New Bonnet Revolving Nut O-Rings	Nitrile	n/a	0	n/a	n/a
Parker Seals - New Hold Down Nut O-Rings	Nitrile	n/a	0	n/a	n/a
Parker Seals - New Revolving Nut O-Rings	Nitrile	n/a	0	n/a	n/a
Reduced Pressure Zone/Detector Assemblies	Nitrile	1/24/05	36	Baptist East Hospital	-
RP2s for buliding supply, Crawford Gym (Bldg #26)	Nitrile	12/7/04	126	A008658	-
RP2s for buliding supply, Crawford Gym (Bldg #26)	Nitrile	12/7/04	126	A008664	-
RP2s for buliding supply, Ernst Hall (Bldg #33)	Nitrile	12/20/04	127	BL 0499	-
RP2s for buliding supply, Ernst Hall (Bldg #33)	Nitrile	1/6/05	127	BL 0468	-
RP2s for buliding supply, Life Sciences (Bldg #18)	Nitrile	1/21/05	128	BE 3052	-
RP2s for buliding supply, Life Sciences (Bldg #18)	Nitrile	1/14/05	128	BE 3084	-
Philadelphia Water Department					
3" Meter Gasket	Nitrile	1/19/05	109	PHI 5	1.5
3" Meter Gasket	Unknown	1/18/05	38	PHI 3 Meter Number 8586042	1.0
3" Meter Gasket	Unknown	1/18/05	28	PHI 4 Meter Number 8587021	1.0
Hydrant rubber seat valve	SBR	1/13/05	385	PHI 1 Hydrant Number 37230	1.5
Hydrant rubber seat valve	SBR	1/13/05	32	PHI 2 Hydrant Number 17216	1.5
Pietometer gasket	Nitrile	1/1/05	504	PHI 6	1.0

Component Description	Base Polymer	Date Removed	Service Length (months)	Sample Identifier	Estimated Disinfect. Concentration at Sample Site (ppm)
Portland Water District					
Hydrant Valve Rubber and O-ring	SBR	12/9/04	322	Hydrant 205	1.0
Hydrant Valve Rubber and O-ring	SBR	11/24/04	310	Hydrant 1119	1.0
Meter Based Liner	EPDM-S	11/23/04	166	Serial #R43867894	1.25
Meter Based Liner	EPDM-S	11/23/04	166	Serial #44504779	1.25
Meter Based Liner	EPDM-S	12/1/04	175	Serial #R39169585	-
Meter Based Liner	EPDM-S	11/29/04	176	Serial #R43163643	-
Meter Based Liner	EPDM-S	12/21/04	44	Serial #44821104	-
Meter Based Liner	EPDM-S-P	11/8/04	167	Serial #43866537	0.7
Meter Based Liner and Rubber Washers	EPDM-S	2/23/05	168	SV24F915	0.6
Meter Based Liner and Rubber Washers	EPDM-S	1/24/05	152	SV275502	0.6
Meter Based Liner and Rubber Washers	EPDM-S	2/24/05	71	SV2931871	0.6
Meter Based Liner and Rubber Washers	EPDM-S	2/16/05	192	SV27D2163	0.8
Meter Based Liner and Rubber Washers	EPDM-S	1/7/05	160	SV3211322	1.2
Meter Based Liner and Rubber Washers	EPDM-S	2/25/05	84	SV27D268	0.8
Meter Based Liner and Rubber Washers	EPDM-S	2/16/05	204	SV33NW1550	0.4
Meter Based Liner and Rubber Washers	EPDM-S	3/15/05	193	SV2721320	0.8
Meter Based Liner and Rubber Washers	EPDM-S	2/28/05	126	SV2641010	0.8
Meter Based Liner and Rubber Washers	EPDM-S	1/5/05	99	SV27D1917	1.0
Meter Based Liner and Rubber Washers	EPDM-S	3/25/05	214	SV28S1264	0.5
Meter Based Liner and Rubber Washers	EPDM-S	3/25/05	191	SV29SP2073	1.1
Meter Based Liner and Rubber Washers	EPDM-S	4/14/05	194	-	1.6
Meter Based Liner and Rubber Washers	EPDM-S	3/28/05	127	SV27D3359	1.0
Meter Based Liner and Rubber Washers	EPDM-S	3/2/05	189	SV29SP2586	0.7
Meter Based Liner and Rubber Washers	EPDM-S	4/4/05	144	SV28S2067	0.5
Meter Based Liner and Rubber Washers	EPDM-S	3/1/05	186	SV2713989	1.1
Meter Based Liner and Rubber Washers	EPDM-S	3/30/05	188	SV23C651	0.8
Meter Based Liner and Rubber Washers	EPDM-S	5/2/05	209	SV2816396	0.4
Meter Based Liner and Rubber Washers	EPDM-S	5/12/05	167	SV279983	1.1
Meter Based Liner and Rubber Washers	EPDM-S	5/10/05	181	SV28S1403	0.7
Meter Based Liner and Rubber Washers	EPDM-S	5/9/05	182	SV28S548	0.5

Component Description	Base Polymer	Date Removed	Service Length (months)	Sample Identifier	Estimated Disinfect. Concentration at Sample Site (ppm)
Portland Water District (continued)					
Meter Based Liner and Rubber Washers	EPDM-S	4/12/05	180	SV32W1603	1.1
Meter Based Liner and Rubber Washers	EPDM-S	6/1/05	184	SV23C1111	0.2
Meter Based Liner and Rubber Washers	EPDM-S	6/9/05	169	SV276234	1.5
Meter Based Liner and Rubber Washers	EPDM-S	6/3/05	161	SV21CE1920	0.6
Meter Based Liner and Rubber Washers		6/17/05	193	SV27D3903	0.6
Meter Based Liner and Rubber Washers		6/24/05	169	SV2721196	0.9
Meter Based Liner and Rubber Washers		6/13/05	199	SV271707	1.0
Meter Based Liner and Rubber Washers		4/7/05	175	SV2931338	0.5
Meter Based Liner and Rubber Washers		4/29/05	172	SV21CE320	0.4
Meter Based Liner and Rubber Washers		6/30/05	178	SV2525098	1.4
Meter Based Liner and Rubber Washers		7/13/05	143	SV2445482	0.2
Meter Based Liner and Rubber Washers		7/8/05	173	SV274875	1.0
Meter Based Liner and Rubber Washers		5/18/05	168	SV32W1765	0.5
Meter Based Liner and Rubber Washers		7/6/05	195	SV2721400	0.6
Meter Based Liner and Rubber Washers		6/15/05	195	SV3218367	1.2
Meter Based Liner and Rubber Washers		7/14/05	187	SV27137	1.1
Meter Based Liner and Rubber Washers		5/26/05	178	SV274497	1.1
Meter Based Liner and Rubber Washers		6/14/05	168	SV28S1467	0.4
Meter Based Liner and Rubber Washers		8/4/05	175	SV2919676	0.9
Meter Based Liner and Rubber Washers		7/29/05	192	SV2140477	0.8
Meter Based Liner and Rubber Washers		8/3/05	180	SV2722036	1.1
Meter Based Liner and Rubber Washers		8/1/05	180	SV272095	1.2
Meter Based Liner and Rubber Washers		8/1/05	184	SV32W232	1.2
Meter Based Liner and Rubber Washers		8/10/05	181	SV271097	1.0
Meter Based Liner and Rubber Washers		8/4/05	166	SV2140562	0.9
Meter Based Liner and Rubber Washers		8/2/05	173	SV2721469	1.1
Meter Based Liner and Rubber Washers		8/16/05	199	SV2445220	0.7
Meter Based Liner and Rubber Washers		8/10/05	192	SV2722537	1.1
Meter Based Liner and Rubber Washers		8/24/05	179	SV2933744	0.8
Meter Based Liner and Rubber Washers		4/22/05	176	SV2525291	1.1

Component Description	Base Polymer	Date Removed	Service Length (months)	Sample Identifier	Estimated Disinfect. Concentration at Sample Site (ppm)
Portland Water District (continued)					
Meter Based Liner and Rubber Washers		8/3/05	193	SV25GV222A	1.0
Meter Based Liner and Rubber Washers		8/8/05	174	SV277622	1.1
Meter Based Liner and Rubber Washers		8/17/05	172	SV273859	1.0
Meter Based Liner and Rubber Washers		8/16/05	188	SV275366	1.1
Meter Based Liner and Rubber Washers		8/8/05	197	SV271687	0.9
Meter Based Liner and Rubber Washers		8/1/05	179	SV27920	1.1
Meter Based Liner and Rubber Washers		4/6/05	195	SV2933769	0.9
Meter Based Liner and Rubber Washers		8/24/05	189	SV25GV324	1.0
Meter Based Liner and Rubber Washers		8/9/05	183	SV21CE308	0.9
Meter Based Liner and Rubber Washers		8/2/05	160	SV2715892	1.1
Meter Based Liner and Rubber Washers		9/8/05	166	SV2417796	0.3
Meter Based Liner and Rubber Washers		8/31/05	185	SV274033	0.8
Meter Based Liner and Rubber Washers		9/20/05	189	SV277063	0.9
Meter Based Liner and Rubber Washers		9/12/05	177	SV33NW817	0.4
Meter Based Liner and Rubber Washers		9/8/05	177	SV2914703	0.5
Meter Based Liner and Rubber Washers		9/5/05	177	SV3210241	1.6
Meter Based Liner and Rubber Washers		8/31/05	143	SV27D3332	1
Meter Based Liner and Rubber Washers		8/31/05	200	SV27D3811	0.9
Meter Based Liner and Rubber Washers		8/29/05	207	SV24F1130	0.2
Meter Based Liner and Rubber Washers		8/10/05	173	SV2715805	1.1
Meter Based Liner and Rubber Washers		9/1/05	161	SV2933544	0.7
South Central Connecticut Regional Water Authority					
Bottom Gasket & Outlet Gasket	EPDM-S	7/11/05	43	0058473600	±1
Bottom Gasket & Outlet Gasket	SBR	7/8/05	117	0040494084	±1
Bottom Gasket & Outlet Gasket		9/1/05	0.5	65286834	±1
		9/1/05	148	51825858	±1

Component Description	Base Polymer	Date Removed	Service Length (months)	Sample Identifier	Estimated Disinfect. Concentration at Sample Site (ppm)
St. Paul Regional Water Services					
1-1/4" - 1-1/2" Relief Valve Repair Kit	Nitrile/EPDM-S/Unknown	?	-	RK 009M2 RT11/411/2	-
1-1/4" - 1-1/2" Relief Valve Repair Kit	Nitrile/EPDM-S/Unknown	?	60	1-1/4" - 1-1/2" RK 009 M2 RT	-
1-1/4" - 1-1/2" Relief Valve Repair Kit	Nitrile/EPDM-S/Unknown	?	-	1-1/4" - 1-1/2" RK 009 M2 RT	-
1-1/4" - 2" Relief Valve Repair Kit	Nitrile/EPDM-S/Unknown	?	60	1-1/4" - 2" RK 009 RT	-
2" Relief Valve Repair Kit	Nitrile/EPDM-S/Unknown	?	-	RK 009MS RT 2	-
2-1/2" Relief Valve Repair Kit	Nitrile	?	-	KIT # 29 BFP-RK	-
3/4" - 1" Relief Valve Repair Kit	Nitrile/EPDM-S	?	60	RK 909 RT 3/4-1 (4/20/01)	-
3/4" - 1" Relief Valve Repair Kit	Nitrile/EPDM-S/Unknown	?	60	RK 009 RT 3/4-1 (8/14/03)	-
3/4" - 1" Relief Valve Repair Kit	Nitrile/EPDM-S/Unknown	?	60	RK 009 RT 3/4-1 (10/31/03)	-
3/4" Regulator	Nitrile/EPDM-S-S/Unknown	?	60	009M2	-
3/4" Relief Valve Repair Kit	Nitrile/EPDM-S/Unknown	?	24	RK 009M2 RT 3/4	-
4" Gasket (RP2 Parts)	EPDM-S	4/19/05	177	1990	-
4" Gasket (RP2 Parts)	EPDM-S	4/26/05	85	1998	-
Conbraco 1/2" Boiler	Nitrile/Unknown	?	60	40-003-01	-
RP2 Parts	Nitrile/EPDM-S/Unknown	?	60	RP2 Parts	-
Unknown					
Backflow Preventer	Natural Rubber	?	-	-	-

REFERENCES

Accurate Products. 2007. The Elastomers [Online]. Available: <http://www.accurate-prod.com/sp/elastomers.html>. [cited March 9, 2006]

Allstate Gasket. 2007. Gasket Specifications Guide [Online]. Available: <http://www.allstategasket.com/info_specs_material-properties.asp>. [cited January 2007]

ASTM International. 2002. *Standard Test Method for Rubber Property—Effect of Aqueous Solutions With Available Chlorine and Chloramine.* ASTM D 6284. West Conshohocken, Pa.: ASTM International.

ASTM International. 2003. *Standard Test Method for Rubber Deterioration—Surface Cracking.* ASTM D 518-99. West Conshohocken, Pa.: ASTM International.

AWWA (American Water Works Association). 1953. *American Standard for a Mechanical Joint for Cast Iron Pressure Pipe and Fittings.* ASA/AWWA A21/11/C111. New York: AWWA.

AWWA (American Water Works Association). 1964. *American Standard for a Mechanical Joint for Cast Iron Pressure Pipe and Fittings.* ASA/AWWA A21/11/C111. New York: AWWA.

AWWA (American Water Works Association). 2000. *American National Standard for Rubber-Gasket Joints for Ductile-Iron Pressure Pipe and Fittings.* ANSI/AWWA C111/A21.11-00. Denver, Colo.: AWWA.

AWWA (American Water Works Association). 2006. Water Chlorination/Chloramination Practices and Principles. In *Manual of Water Supply Practices*, Manual M20. 2nd ed. Denver, Colo.: AWWA.

AWWA Committee on Gasket Materials for Water Works Services. 1998. Committee Report: Gasket Materials; Report: Gasket Materials. *Journal AWWA*, 90(11).

Binnig, G., C.F. Quate, and C. Gerber. 1986. Atomic Force Microscope. *Physics Review Letter*, 56:930–933.

Bowen, T. 2006. Personal communication. February 17.

Chernoff, D.A. 1995. High Resolution Chemical Mapping Using Tapping Mode AFM With Phase Contrast. In *Proc. Microscopy and Microanalysis.* Edited by G.W. Bailey, M.H. Ellisman, R.A. Hennigar, and N.J. Zaluzec. New York: Johnes and Begell Publishing.

CIPRA (Cast Iron Pipe Research Association). 1954. *Cast Iron Pipe Research Association Specification for Roll-on Gaskets.* CIPRA 7-54. Chicago: CIPRA.

Dittel, C. 2006. Terramix Rubber Solutions, Costa Rica. Personal communication. 19 April.

Donnet, J.-B., and A. Voet. 1976. *Carbon Black Physics Chemistry and Elastomer Reninforcement.* New York: Mercel Dekker.

Dunn, L. 2007. Personal communication. March 1.

ExxonMobile Chemical. 2004. Resistance to Aqueous Chloramine. Technical correspondence [Online]. Available: <http://www.parker.com/sg/literature/pdf/ord5717.pdf>. [cited January 2006]

ExxonMobil Chemical. 2007. Description and Background, Butyl Polymers [Online]. Available: <http://www.exxonmobilchemical.com/Public_Products/Butyl/Butyl_Polymers/Worldwide/Description_and_Background/Btl_Desc_HistoryInfo_1981_1990.asp>. [cited January 2007]

International Institute of Synthetic Rubber Producers, Inc. 2006a. [Online]. What Is Synthetic Rubber? Available: <http://www.iisrp.com/synthetic-rubber.html>. [cited December 5, 2006]

International Institute of Synthetic Rubber Producers, Inc. 2006b. Brief History and Introduction of Rubber [Online]. Available: <http://www.iisrp.com/WebPolymers/00Rubber_Intro.pdf>. [cited January 2006

International Institute of Synthetic Rubber Producers, Inc. 2006c. Ethylene-Propylene Rubbers and Elastomers [Online]. Available: <http://www.iisrp.com/webpolymers/10EPDMSep11.pdf>. [cited December 5, 2006]

International Rubber Research and Development Board. 2007. History of Natural Rubber [Online]. Available: <http://www.irrdb.com/IRRDB/NaturalRubber/History/History1.htm>. [cited February 7, 2007]

Kirmeyer, G.J., G.W. Foust, G.L. Pierson, J.J. Simmler, and M.W. LeChevallier. 2003. *Optimizing Chloramine Treatment*. 2nd ed. Denver, Colo.: AwwaRF.

Leeflang, K.W.H. 1968. Biologic Degradation of Rubber Gaskets Used for Sealing Pipe Joints. *Journal AWWA*, 60:1070–1076.

Leyden, J. 2003. Akron Rubber Development Laboratory, Inc. Personal communication. December 3.

Loadman, J. 2005. *Tears of the Tree, The Story of Rubber—A Modern Marvel*. New York: Oxford University Press.

Lutz Sales Co. 2007. NBR/Buna-N. [Online]. Available: <http://www.lutzsales.com/buna.asp>. [cited January 2007]

Martin, Y., C.C. Williams, and H.K. Wickramasinghe. 1987. Atomic Force Microscope-Force Ampping and Profiling on a Sub 100-A Scale. *Journal of Applied Physics*, 61(10):4723–4729.

McGuire, M.J., J.L. McLain, and A. Obolensky. 2002. *Information Collection Rule Data Analysis*. Denver, Colo.: AWWA.

Nagdi, K. 1993. *Rubber as an Engineering Material: Guideline for Users*. New York: Hanser Publishers.

Noel, A. 2006. Personal communication. May.

Osswald, T.A., and G. Menges. 2003. *Materials Science of Polymers for Engineers*. 2nd ed. Cincinnati, Ohio: Hanser Gardner Publications.

Parker Hannifin Corporation. 2006. Chloramine Resistant E1257-70, New Chloramine Resistant EPDM Seal Material Is Ideal for Drinking Water Applications [Online]. Available: <http://www.parker.com/sg/literature/pdf/ord5717.pdf>. [cited January 2006]

Plastics Historical Society. 2006. Plastiquarian.com [Online]. Available: <http://www.plastiquarian.com/ind1.htm>. [cited January 2006]

Reiber, S. 1993a. *Chlorine Effects on Distribution System Materials*. Denver, Colo.: AwwaRF and AWWA.

Reiber, S. 1993b. Investigating the Effects of Chloramine on Elastomer Degradation. *Journal AWWA*, 85(8):101.

San Francisco Public Utilities Commission. 2006. [Online]. Available: <http://sfwater.org/detail.cfm/MC_ID/13/MSC_ID/166/MTO_ID/399/C_ID/2215>. [cited April 29, 2006]

Sarai, D.S. 2005. *Basic Chemistry for Water and Wastewater Operators*. Denver, Colo.: AWWA.

Sartomer Company, Inc. 2002. Fundamental Considerations in the Use of Coagents in Peroxide Curable Elastomers. Application Bulletin 4117 [Online]. Available: <http://www.sartomer.com/TechLit/4117.pdf>. [cited February 21, 2007]

Schaffer, J.P., A. Saxena, S.D. Antolovich, T.H. Sanders Jr., and S.B. Warner. 1999. *The Science and Design of Engineering Materials*. 2nd ed. New York: McGraw-Hill.

Simmons, C.L., and P.P. Evanson. 1988. Effects of Domestic Water Systems Additives on Rubber Vulcanizates. *Rubber World*, 32(1):16.

Snoeyink, V.L., and D. Jenkins. 1980. *Water Chemistry*. New York: John Wiley & Sons.

Timco Rubber. 2007. Natural Rubber [Online]. Available: <http://www.timcorubber.com/materials/natural.htm>. [cited January 2007]

Trelleborg, A.B. 2007. Rubber Types [Online]. Available: <http://www.trelleborg.com/template/T005.asp?id=775&lang=2>. [cited February 21, 2007]

USEPA (U.S. Environmental Protection Agency). 2006. Drinking Water Contaminants [Online]. Available: <http://www.epa.gov/safewater/contaminants/index.html#dbps>. [cited November 28, 2006]

ABBREVIATIONS

ALT	alternative
ANSI	American National Standards Institute
a_t	shift factor
AWWA	American Water Works Association
AwwaRF	Awwa Research Foundation
BJLT	common grade of nitrile rubber containing approximately 35% acrylonitrile
°C	degrees Celsius
C	constant (material dependent)
cc	cubic centimeters
C-C	carbon–carbon bonds
CIPRA	Cast Iron Pipe Research Association
Cl_2	chlorine
cm	centimeter
cm^2	units of area–centimeters squared
C-S-C	carbon–sulfur–carbon bonds
C-S-S-C	carbon–sulfur–sulfur–carbon bonds
$C\text{-}S_X\text{-}C$	carbon–sulfur–miscellaneous–carbon bonds
DBP	disinfectant by-products
Dicup 4OKE	dicumyl peroxide
DOTG/Monex	N,N'-di-O-tolylguanidine
DSM	global manufacturer of polymer, chemical, pharmaceutical, and nutrition products based in the Netherlands
dynes	unit of force specified in the centimeter-gram-second system
	change
E	activation energy
EDAX-EDS	energy dispersive X-ray spectroscopy
EPDM	ethylene propylene diene monomer
EPDM-P	ethylene propylene diene monomer—peroxide cured
EPDM-S	ethylene propylene diene monomer—sulfur cured
ET	elapsed time
FKM	fluorocarbon—fluorinated hydrocarbon
FTIR	Fourier transformed infrared spectrometer
HOCl	hypochlorous acid
hrs	hours
i.e.	that is
IIR	butyl rubber—isobutylene-isoprene

K	degradation rate
KJ $mole^{-1}$	kilojoules per mole
L	liter
M	mole
mg	milligrams
M Tuads	tetramethylithiuram disulfide
N	nitrogen
NBR	nitrile
NH_3	ammonia
NH_2Cl	monochloramine
$NHCl_2$	dichloramine
$NHCl_3$	trichloramine
NTU	nephelometric turbidity unit
%	percent
pH	pH
PHR	parts per hundred rubber
ppm	parts per million
pts.	points
PVC	polyvinyl chloride
R	universal gas constant
r_a	degradation parameter
RT	room temperature
SBR	styrene butadiene rubber
SC	triethylene glycol
SDWA	Safe Drinking Water Act
SI	sialstic–polysiloxan
SMR CV 60	standard Malaysian rubber with a constant (Mooney) viscosity of 60
SRF	semi-reinforcing furnace
SRF N762	pelleted furnace carbon black distributed by Cabot Corp. and used in tire applications
Sun 4240	high aromatic content, non-staining oil used in rubber processing, distributed under the Sunthene brand by Sun Oil Company
T	absolute temperature in °K
T	temperature
THMs	trihalomethanes
T_{ref}	reference temperature
USEPA	U.S. Environmental Protection Agency
WLF	Williams-Landel-Ferry

IWA Publishing's authorised EU representative for General Product Safety Regulations is Diane D'Arras, 15 rue Duret, 75116 Paris, France, e-mail: safety@iwap.co.uk.

Printed and bound by CPI Group (UK) Ltd, Croydon, CR0 4YY
17/09/2026
02218799-0008